NOISE

ITS MEASUREMENT, ANALYSIS, RATING AND CONTROL

Noise
its Measurement, Analysis, Rating and Control

J.S. Anderson
and
M. Bratos-Anderson

Published by
Avebury Technical
Ashgate Publishing Limited
Gower House
Croft Road
Aldershot
Hants GU11 3HR
England

Ashgate Publishing Company
Old Post Road
Brookfield
Vermont 05036
USA

British Library Cataloguing in Publication Data

Anderson, J. S.
    Noise; Its Measurement, Analysis, Rating and Control.
    I. Title II. Bratos-Anderson, M.
    620.23

ISBN 0 291 39794 8

Printed in Great Britain at the University Press, Cambridge

REF

Contents

Preface

Noise; its Measurement, Analysis, Rating and Control is intended as a text for students studying noise and the engineering aspects of acoustics, either in Universities and Colleges, or by private study. Many workers in industry have to know about noise, possibly as only a part of their jobs; it is hoped that such readers will find this book helpful, not only for the solution of immediate problems, but also for the purpose of gaining a deeper understanding of the subject.

There are many books available on the subject of noise or acoustics. On the one hand there are many books that provide a practical guide to noise control and give a useful guidance to the solution of real problems, without going into much of the theoretical background. On the other hand the physics of acoustics is dealt with by many excellent texts which consider the theory in detail. It is hoped that this book will provide a bridge between the practical and the theoretical approaches. In addition to practical details we have attempted to include background theory for the reader who really wishes to understand.

Initially we intended the book to be an accompanying text for a short course which has taken place at the City University for many years. The course, with the same title as the present book, is for workers in industry who need to know about noise, or enhance their existing knowledge. The course lasted five days and the participants' initial knowledge was that of a University Graduate or Technician Engineer. Although the book grew from those early intentions to something much more extensive, it remains of value as a text for the course and for other courses of a similar level and duration.

Noise control is also the subject of a 40 hour lecture course in the final year of the honours degree in Mechanical Engineering at the City University. The book is intended for students on that course and should be suitable for similar studies in other universities and related institutions.

The Institute of Acoustics (of Great Britain) has for many years organised a Diploma in Acoustics at various institutions in Britain. The diploma consists of core subjects and several options. The content of the present book covers a good deal of the syllabus of the core subjects and some of the options.

One of the most important developments related to noise control in recent years has been the ratification of the directive of the European Commission for the control of noise in the workplace. The directive applies to all member states of the EC, although individual countries are able to introduce their own regulations to account for varying practices in different countries. In the case of the United Kingdom it is stated in the Noise at Work Regulations 1989 that a 'competent person' should carry out the duties which are required to satisfy the regulations. The Health and Safety Executive, who have overall responsibility for ensuring compliance with the regulations, have proposed in their Guidance Notes a course of study for those wishing to aquire competency. The syllabus required for the training of competent persons is covered in the present book; indeed the book contains additional material, particularly of the theoretical kind, for students who require not just competency but also an understanding in depth.

Where appropriate in the text the concept of sound intensity has been emphasised. Until recently sound intensity has mostly been regarded as a scalar with a direct relation to the sound pressure. However, new equipment has increased the awareness of regarding the time–averaged sound intensity as a vector in space.

We have made use of a number of diagrams from other sources. For permission to reproduce these diagrams we wish to thank the following: Brüel and Kjaer for Figure 3.6, Grunzweig and Hartmann Montage GmbH for Figures 4.15, 8.39 and 8.40, AAF Ltd for Figure 6.30, the Soundcoat Company Inc for Figure 8.30, the editor of the Journal of Sound and Vibration for Figure 6.16 and the editor of the Journal of Noise Control Engineering for Figure 8.23.

The typesetting of this book has been undertaken by Eli Napchan whose skill in deciphering out initial text is very much appreciated. Finally, we would like to thank our publisher, John Hindley, for his patience and encouragement during the preparation of this book.

To the memory of Wacław Bratos

1 The Principles of Sound Propagation

Soun is noght but air y-broken,
And every speche that is spoken,
Loud or privee, foul or fair,
In his substaunce is but air;
For as flaumbe is but lighted smoke,
Right so soun is air y-broke.
But this may be in many wyse,
Of which I wil thee two devyse,
As soun that comth of pype or harpe.
For whan a pype is blowen sharpe,
The air is twist with violence,
And rent; lo, this is my sentence;
Eek, when men harpe-stringes smyte,
Whether hit be moche or lyte,
Lo, with the strook the air to-breaketh;
Right so hit breketh whan men speketh.
Thou wost thou wel what thing is speche.
Geoffrey Chaucer, *The Hous*
of Fame

1.1 Introduction

The first chapter takes the reader through some of the fundamentals of sound theory. Sound is defined as a small time–dependent disturbance in an elastic medium. The propagation of sound is controlled by the wave equation which is derived here in a rigorous manner for a one–dimensional sound wave. In the derivation the equations of state (Section 1.4) and the equations of conservation of mass, momentum and energy in a sound wave are required, see Sections 1.5 to 1.8. Particular emphasis is put on the fact that sound wave propagation can be treated as a reversible and adiabatic (i.e., isentropic) process (Section 1.8). Different forms of the solution of the wave equation for plane waves are presented in Section 1.12; firstly, in a general form, secondly as harmonic waves in terms of trigonometric functions and finally in a phasor representation where use

1

is made of exponential functions. Phasor formalism is particularly useful in problems dealing with radiation from sound sources and, in general, complex vector representation provides a simplified approach, especially in the calculation of time–averaged acoustic quantities (Sections 1.12.2, 1.16.3 and Appendices C and D).

Sound intensity plays an important role in the diagnostics of the sound field and in detecting the sound sources, particularly as specialised equipment is available for its measurement. The sound intensity provides more information about the sound field than the sound pressure, since it is a vector quantity. Sound intensity measurements enable us to determine the sound power of any sound source even in the presence of other sound sources. The theoretical background to sound intensity provided here (Sections 1.14, 1.16.3 and Appendix C) gives more information than is generally found in a textbook of this scope.

There is, unavoidably, in Chapter 1 a certain amount of mathematics. A full understanding of this chapter requires a knowledge of partial differential equations, as well as vector and complex number formalism. The reader who does not wish to become too involved in the intricacies of acoustic theory may concentrate on Section 1.1 and its subsections and Sections 1.9 to 1.12. Sections 1.20 and 1.21 on decibel scales and their manipulation are essential to an understanding of subsequent chapters.

1.1.1 The nature of sound

Sound is all around us. We detect sound with our ears and are able to locate the source of sound. We can also deduce what kind of source is emitting the sound; whether it originates from a loudspeaker, machine or human voice.

All oscillatory motions of small amplitude in elastic media are called sound. The source of the sound could be, for example, the vibrating surface of a machine which causes the air in contact with it to have a displacement. The vibrating surface causes in the air a disturbance of the pressure and density, as well as particle velocity. Each element of air transfers momentum and energy to the adjacent air element, and the pressure disturbance is propagated through the air in the form of a wave. The *element of fluid*, called also the *material volume*, is a fluid volume which is fixed in neither space nor time. It is enclosed by an arbitrary surface, every point of which moves with the local fluid velocity. Although the element of fluid changes its shape and volume during motion, it contains always the same fluid particles, as well as — on average — the same number of molecules.

A *fluid particle* is a macroscopically small element of fluid such that the quantities characterising this fluid element, namely, pressure, density and velocity, are constant throughout the element. The above definition indicates that the fluid particle contains a sufficiently large number of molecules, so that it can be treated as possessing the properties of a continuum medium.

The *local fluid velocity* at time t and at position $\vec{r}$ in the flow field is equal to the *particle velocity* of the fluid particle which at time t is at position $\vec{r} = [x, y, z]$ in space. The fluid particle moves along its trajectory or particle path $\vec{r} = \vec{r}(t)$ with particle velocity $\vec{u} = d\vec{r}/dt$.

The rate of change with time of a fluid particle property, e.g., pressure or particle velocity, contains two parts: an unsteady part, called also the local rate of change, and a convective part. For example, the time derivative of the particle velocity is expressed [1] by the equation,

$$\frac{d}{dt}[\vec{u}(t, \vec{r})] = \frac{\partial \vec{u}}{\partial t} + (\vec{u} \operatorname{grad})\vec{u}, \tag{1.1}$$

where $\vec{u}$ denotes the particle velocity vector for the fluid particle which at time t is at position $\vec{r}$ in the flow field. The derivative $d\vec{u}/dt$, describing the rate of change of the particle velocity, is the substantial (material) derivative of the particle velocity (or the acceleration of the fluid particle) and its unsteady and convective parts are specified, respectively, on the right hand side of equation (1.1).

For one–dimensional flow the equation (1.1) is reduced to:

$$\frac{du}{dt} = \frac{\partial u}{\partial t} + u\frac{\partial u}{\partial x}. \tag{1.1a}$$

Sound waves can only propagate in an elastic medium. Thanks to two factors, *elasticity* and *inertia* of the medium, wave motion develops. Elasticity is the ability of a body to respond to deformation. Inertia, characterised by mass in translational motion or by moment of inertia in rotational motion, is the body resistance to rate of change of its velocity. It is a body property, which results from Newton's first and second laws of motion.

In general sound wave motion is three–dimensional; sound from a source spreads out in all directions. However, in many situations the propagation of a sound wave can be described as a one–dimensional motion, as all quantities which define the sound wave depend only on one space co–ordinate and on time. Such a type of sound wave is called a *longitudinal wave*, since the fluid particles oscillate along the direction of propagation of the sound wave.

Sound propagation through air in a long tube is an example of one–dimensional motion. The sound field parameters depend in this case on one cartesian co–ordinate and on time. This type of sound wave is a *plane wave*.

Consider a long tube, as shown in Figure 1.1. The source of sound waves is a rigid reciprocating piston which is driven by an engine mechanism with a small crank radius. If the length of the crank is short compared with the length of the connecting rod, the motion of the piston is very nearly simple harmonic [2]. As the crank radius is small, the amplitude of piston oscillations is also small.

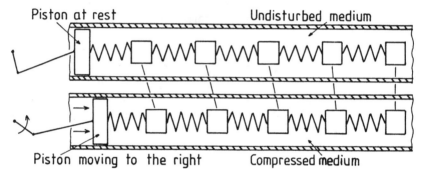

Figure 1.1 Wave motion in a tube.

As already stated, for sound wave propagation the medium must have both inertia and elasticity. The medium in the tube, in the considered case, air, can be treated as a series of fluid elements. Each fluid element can be represented by a spring — which provides the elasticity — and by a body which has mass equivalent to that of the fluid element. The interaction between adjacent fluid elements allows propagation of a disturbance in the medium. During the propagation of the disturbance each fluid element behaves as a mass–spring system. Hence, the whole medium in a tube can be approximated by a series of mass–spring systems.

The mechanism of a successive interaction between fluid elements can be presented in a simplified form as follows. Firstly, as the piston, shown in Figure 1.1, accelerates to the right the fluid element adjacent to the piston is compressed and pushed to the right. Due to its inertia the first fluid element continues to move to the right. Next, the first fluid element compresses the second fluid element which also begins to move to the right, because momentum is transferred from the first to the second. The elasticity of the medium is evinced by restoring forces which tend to pull the displaced fluid elements back to their equilibrium positions and, as a result, cause the oscillation of the elements. The process of momentum and energy transfer from one fluid element to the next one develops in the medium. Hence, the disturbance, which is manifested in changes of fluid pressure, particle velocity and density, propagates to the right throughout the medium. The only motion of the fluid elements is the oscillatory motion. The amplitude of the displacement of the fluid particles in the oscillatory motion depends on the excitation which causes the disturbance. On the other hand, the speed with which the disturbance of small amplitude propagates throughout the medium is a characteristic property of the medium. The disturbance as it propagates is a sound wave.

When the piston is moving continuously forwards and back in simple harmonic motion all fluid elements along the tube are successively compressed and rarefied, and set into oscillatory motion about their mean (or equilibrium) positions. The sound wave, at the frequency of the piston motion, is then propagating down the tube. Every fluid particle of the medium oscillates also with the same frequency about its mean position, i.e., each particle has a particle displacement about its mean position. The velocity with which the fluid particle oscillates about the mean position is the particle velocity. The displacements of the successive fluid particles are not in phase because the disturbance (sound wave) propagates through the fluid with a finite velocity, called the *sound wave velocity* or *sound velocity*.

During the propagation of the sound wave in the medium, the particle velocity is quite small in comparison with the sound wave velocity. It should be noted that the major factors which influence the value of the sound velocity in a medium are its elasticity and density. The sound velocity is large in materials having large elasticity and small density.

A particularly important parameter of sound waves is the *sound pressure*, called also the *acoustic pressure*. The sound pressure at a certain point in an acoustic field is the difference between the instantaneous pressure and the equilibrium pressure at this point. In the considered example of the sound wave propagation through air in a tube the equilibrium pressure is the static pressure. Sound pressure can be measured easily with a microphone and sound level meter, see Section 2.1.

1.1.2 Origins of sound

Any moving body is capable of producing sound. For example, sound is produced by elastic bodies which have flexure, such as car doors or panels of machine tools. It is generally possible to compute the natural frequencies at which a particular plate or panel prefers to vibrate. Sound will normally radiate at these frequencies, although the manner in which the panels radiate sound may be complicated.

Normally, when we think of the origin of sound we have the *vibration* of elastic bodies in mind. However, a rigid body in motion is also capable of producing sound; the oscillating rigid piston, as shown in Figure 1.1, is an example. A sphere set into sudden motion is another example. Two small spheres in collision give a sharp click; the spheres generate sound even though their natural frequencies of vibration, as elastic bodies, can be greater than the upper limit of the audible range of sound. In the case of the spheres the sudden deceleration of one and the sudden acceleration of the other causes an impulsive motion of the air adjacent to the spheres. Consequently, sound of this type is often called *acceleration noise*, and occurs in machinery where impacts take place, as, for example, in presses and forges.

In the case of the oscillating piston which sets up a sound wave in a duct there is normally an open end to the duct — as with the tail pipe of an internal combustion engine. The air at the end of the duct undergoes an oscillation, similar to that caused by a piston, and emits in the open space a sound wave which is heard by a listener. The sound radiation in this case is similar to that from an organ pipe.

Many sounds that we hear are *aerodynamic* in origin. In the turbulent jet, which propels modern aircraft, the turbulent eddies in the jet are the sources of sound; no vibration of elastic bodies is involved. Many of the sounds in nature are aerodynamic, for example, the singing of telegraph wires or the wind whistling through the branches of trees.

1.1.3 Types of sound waves

It is possible to make a geometrical classification of sound waves from the point of view of their symmetry and to specify the following categories:

- (a) plane,
- (b) cylindrical,
- (c) spherical.

In the tube shown in Figure 1.1 plane waves are propagated. The sound wave in the tube has a wave front which is always parallel with the plane of the piston. On the plane wave front values of parameters such as the sound pressure, the particle velocity and the particle displacement are constant. In general plane waves tend to be produced, particularly at high frequencies, by any large, almost plane surface — such as the side of a diesel engine.

Cylindrical waves have a wave front which is the surface of a cylinder. Good approximations of cylindrical waves are sound waves originating from pipes or from a steady flow of traffic on a motorway. Similarly, spherical waves have a wave front in the form of a sphere. They are the most common type of waves and they will be discussed in more detail later in Section 1.16.

A sound source can generate sound waves whose geometrical properties change with distance from the source. For example, close to the side of a diesel engine, the sound waves may be almost plane, but further away the sound waves are spherical. See Section 8.6.3 for further discussion of the spatial distribution of sound in relation to its source.

1.2 General physical model of sound wave motion in a fluid

Sound propagation in a fluid, such as air, is a special case of compressible fluid flow. Fluid is treated as a continuum medium. Every fluid particle, small in macroscopic scale but microscopically large, is characterised by macroscopic quantities, such as density, velocity or pressure, which can be regarded as constant within the whole volume of the fluid particle. The continuum model is applicable when the characteristic size of the fluid region of interest is much greater than a characteristic molecular dimension.

For gases the characteristic molecular dimension is the molecular *mean free path*, which is the distance that, on average, molecule travels between two successive collisions with other molecules. In the case of sound wave propagation in air at standard conditions, the wavelength for audible acoustic waves is much greater than the mean free path; hence the validity of the continuum model is justified.

Fluid is regarded as *ideal*, in other words it is assumed that dissipation caused by fluid viscosity and heat conduction are negligible effects. The acceptability of this assumption is justified in most acoustic problems, since spatial gradients of velocity and temperature in sound waves are small.

It is also assumed that every fluid particle is instantaneously in local thermodynamic equilibrium, and that the relations between the thermodynamic parameters appearing in the motion equations are such as for quasi–equilibrium processes, see Section 1.8.1. In other words fluid particles during the motion are regarded as passing through a continuum of the thermodynamic equilibrium states.

In general the condition for entropy creation should also be investigated to check whether the mathematical model is consistent with the physical reality. However, sound propagates in air approximately as in an ideal fluid and can be treated as an isentropic process, see Section 1.8.

Additionally, we shall consider propagation of sound in air at standard conditions. Hence, the assumption that air can be regarded as a perfect gas, see equations (1.4) and (1.5), is valid.

Finally, the system of equations describing the flow in ideal and perfect gases contains mass, momentum and energy conservation equations, as presented in Sections 1.6, 1.7 and 1.8, and the equations of state for a perfect gas. To solve a physical problem this system of equations must be completed by the appropriate initial and boundary conditions.

Sound propagation in a fluid is by definition a propagation of small disturbances in a compressible fluid. Due to the assumption of small disturbances the nonlinear equations describing ideal fluid motion can be linearised. As a result of the linearisation the system of so–called acoustic equations is obtained. This system can be reduced to the sound wave equation.

In the next sections we shall specify assumptions which allow us to construct the appropriate model of sound propagation. With the aid of these assumptions the equations of fluid motion can be simplified and finally the sound wave equation can be obtained. The sound wave equation will be developed for a plane wave.

1.3 Model of isentropic sound propagation in a fluid

The simplest physical model of acoustic wave propagation in a still, undisturbed medium is based on the following assumptions:

(1) Flow induced by the sound wave is one–dimensional.
(2) Fluid, in which the sound propagates, is compressible, homogeneous and isotropic; it behaves like a perfect gas.
(3) There is no phase transition during the wave motion.
(4) The mean pressure and mean density of the fluid are constant.
(5) The mean velocity of the fluid is equal to zero.
(6) There is no external heat addition to the flow.
(7) Sound is an oscillatory motion characterised by a small amplitude and small gradients with respect to space and time.

Assumptions 6 and 7 imply that sound propagation can be treated as an isentropic process, see Section 1.8.3. It should be noted that to simplify the procedure leading to the sound wave equation, assumptions 4 and 5 are considered. Also a geometrical simplification is introduced by assumption 1.

1.4 The equations of state

Among the parameters characterising the thermodynamic equilibrium state for *simple fluids*, i.e., fluids whose composition is homogeneous and whose properties are isotropic, there are only two independent thermodynamic quantities. Hence, every thermodynamic quantity, which specifies the fluid state in thermodynamic equilibrium [1,3], such as pressure P, density ρ, temperature T, specific entropy s, specific internal energy ϵ, etc., depends on two others.

The equation which describes the relation between these thermodynamic parameters, of which two are independent, is called the *equation of state* and for a simple fluid can be presented, for example, in the form of the equation,

$$f(P, \rho, \mathrm{T}) = 0 \qquad (1.2)$$

or in the form [4],

$$f_1(\rho, \mathrm{T}, \epsilon) = 0. \qquad (1.3)$$

Equation (1.2), called the *thermal equation of state*, may be specified for a *perfect gas* in a simple form. A gas is described as perfect when there is no interaction between molecules, except during collisions, and when the molecules have negligible volumes. The thermal equation of state for a perfect gas is

$$P = R\rho\mathrm{T}, \qquad (1.4)$$

where gas pressure P, density ρ, and absolute temperature T are basic parameters representing the gas in the thermodynamic equilibrium state. R denotes the specific gas constant. A real gas can be treated as perfect when the mean potential energy of an interaction between molecules is much smaller than the mean kinetic energy of the molecules; this is the case if the gas density is sufficiently small or the gas temperature high.

The thermal equation of state has been obtained from empirical laws; namely, the Boyle–Mariotte and Avogadro laws for real gases. These laws are valid at relatively low pressures and at temperatures which are not too low, i.e., for sufficiently rarefied real gases. Statistical theories give more accurate state equations, as they take into account interaction between molecules and molecule size.

The equation of state in the form (1.3) is called the *caloric equation of state*. In the case of a perfect gas equation (1.3) is reduced to a simple relation between the specific internal energy of the gas ϵ, and gas temperature T, namely,

$$\epsilon = F(\mathrm{T}) = c_v \mathrm{T}, \qquad (1.5)$$

where c_v is the specific heat at constant volume, i.e., the heat input per unit mass and per unit increase in temperature at constant volume. The thermal and caloric equations of state for a perfect gas, equations (1.4) and (1.5), provide full information on all thermodynamic properties of the gas in the thermodynamic equilibrium state. The thermal equation of state provides a relationship only between parameters which are the basic and measurable parameters of state.

1.5 Laws of conservation of mass, momentum and energy

Any fluid motion is described by the equations expressing the laws of conservation of mass, momentum and energy; the system of equations is completed by the thermal and caloric equations. The laws of conservation of mass, momentum and energy will be derived in the next sections for an element of fluid.

Fluid flow induced by a disturbance with a small amplitude possess special properties, see Sections 1.2 and 1.3, which lead to a simplification of the conservation equations. Since sound is an oscillatory motion of small amplitude and with small gradients with respect to space and time, the conservation equations for mass and momentum can be linearised and the energy conservation equation reduced to the condition of constant entropy for a fluid particle.

1.6 Law of conservation of mass (continuity equation)

Let us consider a fluid element — where in most cases the element will be of air — in a tube which has a constant cross–sectional area S and an axis parallel to the x co–ordinate, as shown in Figure 1.2. An element Ξ of undisturbed air in the tube is of length dx_0 and has a density ρ_0. The order of magnitude of the length dx_0 is the same as the size of the fluid particle. The fluid element Ξ is bounded by tube walls and by fluid surfaces 1^0 and 2^0, which are perpendicular to the tube axis, as shown in Figure 1.2. Fluid surfaces 1^0 and 2^0 are marked by fluid molecules which at time $t = 0$, when the fluid in the tube is not yet disturbed, are at distances x_0 and $x_0 + dx_0$ from the reference surface, respectively. The reference surface, $x = 0$, is the piston surface at time $t = 0$.

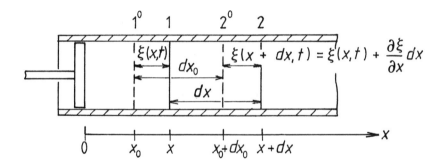

Figure 1.2 Element of fluid in a tube displaced by a disturbance caused by piston movement.

As the piston at time $t = 0$ starts to move to the right along the tube axis, the fluid element Ξ is displaced after a certain time also to the right, due to a disturbance caused by the piston. All fluid molecules situated initially ($t = 0$) at x_0 are at time t at an average position x from the reference surface. Hence, the fluid surface 1^0 was displaced by a distance $\xi(x, t)$. Similarly, all molecules initially at an average position $x_0 + dx_0$ are at time t at an average position $x + dx$ from the reference surface. Thus the fluid surface 2^0 has been displaced by a distance $\xi(x + dx, t)$, see Figure 1.2. Consequently, the fluid element Ξ has at time t a density ρ and is bounded by the tube walls and the fluid surfaces at $x_0 + \xi(x, t)$ and $x_0 + dx_0 + \xi(x + dx, t)$, respectively.

The quantity $\xi(x, t)$ denotes the particle displacement and $u = d\xi/dt$ is the particle velocity. Since $d\xi/dt = \partial\xi/\partial t + u\partial\xi/\partial x$, see also equation (1.1), and the term $u\partial\xi/\partial x$ is small in the sound wave — second order of magnitude — the acoustic approximation, equivalent to neglecting second order terms, results in the relation,

$$u = \frac{d\xi}{dt} \simeq \frac{\partial\xi}{\partial t}. \tag{1.6}$$

As the average number of molecules in the fluid element Ξ, undisturbed by a small disturbance, is the same as in that disturbed by the sound wave, the mass of the fluid element must be the same. Hence,

$$\rho_0 V_0 = \rho V, \tag{1.7}$$

where V_0 is the initial volume of the fluid element Ξ, i.e., the fluid element volume at $t = 0$ when the fluid is undisturbed by a sound wave, ρ_0 is the initial density of the fluid element Ξ and ρ is the density of the fluid element Ξ at time t. V is the volume of the fluid element at time t when the fluid is disturbed by the sound wave. As shown in Figure 1.2, $V = Sdx$ and $V_0 = S[dx - \xi(x + dx, t) + \xi(x, t)]$. Since both the length dx_0 of the fluid particle at the equilibrium position and the length dx of the same fluid particle disturbed by the sound are of small magnitudes, the expression for the fluid element volume V_0 can be approximated by:

$$V_0 = Sdx(1 - \frac{\partial \xi}{\partial x}) \qquad (1.8)$$

or, finally,

$$V_0 = V(1 - \frac{\partial \xi}{\partial x}). \qquad (1.9)$$

Combining equations (1.7) and (1.9) leads to:

$$\rho_0 V(1 - \frac{\partial \xi}{\partial x}) = \rho V. \qquad (1.10)$$

Equation (1.10) expresses the law of mass conservation for the fluid element. Equations (1.7) and (1.10) are two forms of the *mass conservation equation*, called also the *equation of continuity*. The nondimensional variant of formula (1.10), very often used in acoustics [5,6], is

$$\frac{\rho - \rho_0}{\rho_0} = -\frac{\partial \xi}{\partial x}. \qquad (1.11)$$

1.7 Law of conservation of momentum

Let us analyse the forces acting on the fluid element Ξ, see Figure 1.3, as a sound wave propagates through the fluid in the tube. The fluid element Ξ has at time t a density ρ and is bounded by tube walls and surfaces 1 and 2, separated by a distance dx, which is comparable with the size of the fluid particle.

The only forces acting on the fluid element Ξ are due to the change in the sound pressure. Let P_0 denote the equilibrium pressure, i.e., the fluid (air) pressure in the undisturbed fluid, and let p be the sound pressure. The force acting on the surface 1 of the fluid element Ξ due to the presence of the fluid to the left side of the surface 1 is $F_1 = S[P_0 + p(x)]$. Similarly, the force acting on the surface 2 is $F_2 = S[P_0 + p(x) + (\partial p/\partial x)dx]$. The equation of motion for the fluid element can be obtained by equating the

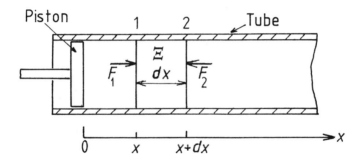

Figure 1.3 Forces acting on a fluid element.

product of the fluid element acceleration $d^2\xi/dt^2$ and the fluid element mass $\rho S dx$ to the net force acting on this fluid element:

$$F_1 - F_2 = \rho S dx \frac{d^2\xi}{dt^2} \tag{1.12}$$

$$-S\frac{\partial p}{\partial x}dx = \rho S dx \frac{d^2\xi}{dt^2}. \tag{1.13}$$

Combining equation (1.13) with (1.10) leads to the equation,

$$-\frac{\partial p}{\partial x} = \rho_0(1 - \frac{\partial \xi}{\partial x})\frac{d^2\xi}{dt^2}. \tag{1.14}$$

The linearisation of equation (1.14), i.e., neglecting the second and higher order of magnitude terms, gives the *momentum conservation equation* for unit volume of fluid in the form,

$$\frac{\partial p}{\partial x} = -\rho_0 \frac{\partial^2\xi}{\partial t^2}. \tag{1.15}$$

1.8 Law of conservation of energy

1.8.1 Adiabaticity and reversibility of the thermodynamic process

In general sound propagation in a fluid is an *adiabatic process*. Adiabatic motion in a fluid occurs when there is no external heat addition to any of the fluid particles in the flow and hence there is no heat energy exchange between a fluid particle and its surrounding.

Sound propagation in a fluid is also a *reversible* process. A reversible process for the thermodynamic system is defined as being infinitely slow and additionally it is a process without any energy dissipation within the system. In general the first condition is not sufficient to satisfy the requirement of the reversibility of the process; in some phenomena energy dissipation still takes place, although the rate of the process tends to zero. In other words a reversible process is a *quasi–static* or *quasi–equilibrium process*, not accompanied by any energy dissipation. During the quasi-equilibrium process the state of the thermodynamic system undergoes a continuum of infinitesimal changes which lead the system successively from one thermodynamic equilibrium state to another.

The name 'reversible' expresses the fact that equal and opposite changes through the set of equilibrium states, which the system undergoes during the process, restore the initial conditions.

All processes in nature are irreversible. However, sound propagation in air is a good approximation of a reversible process. A process which is reversible and adiabatic is called *isentropic*. Sound propagation in air is an example of an isentropic process, since the criteria of adiabaticity and reversibility are fulfilled. This property of sound propagation in air justifies the statement that the sound waves propagate in air as in ideal fluids [7,8]. In ideal fluids the transport coefficients, such as the coefficient of viscosity and the coefficient of thermal conductivity, are equal to zero. The model of an ideal fluid is also applicable to a flow in which velocity and temperature gradients with respect to space are small. As a result of sound propagation small gradients exist in the flow, hence heat transfer, as well as energy dissipation due to viscosity (internal friction), are negligible effects.

1.8.2 Thermodynamics of a reversible process

The energy conservation equation for the thermodynamic system (first law of thermodynamics) applied to the process, during which the thermodynamic state of the system undergoes an infinitesimal change, may be expressed in the form, see also [1,8],

$$dE = dQ + dW. \qquad (1.16)$$

Equation (1.16) indicates that dE, the change of internal energy of the thermodynamic system, is equal to the sum of dQ, the heat supplied to the system, and dW, the work done on this system. If, additionally, the thermodynamic system is a simple fluid (at a pressure P and volume V) and the thermodynamic process is reversible,

$$dW = -PdV. \qquad (1.17)$$

It should be noted that the negative sign is in accordance with convention; the work done on the system (compression of the simple fluid) corresponds with a negative value of dV.

Relating all terms in equation (1.16) to unit mass, we obtain the following expression for the *first law of thermodynamics*:

$$de = dq + \frac{P}{\rho^2} d\rho, \qquad (1.18)$$

where ϵ is the specific internal energy,
        q is the heat supplied per unit mass,
        P is the fluid pressure and
        ρ is the fluid density.

For a perfect gas the equation (1.18) may be combined with equation (1.5) to give:

$$dq = c_v dT - \frac{P}{\rho^2} d\rho. \qquad (1.19)$$

The *second law of thermodynamics* introduces another thermodynamic quantity, namely, *entropy*. Entropy S, like the internal energy of the thermodynamic system ϵ, is a function of state. Let dQ denote an infinitesimal heat supply to the system. The second law of thermodynamics can be formulated as a statement that for every thermodynamic reversible cyclic process [1,3,8–10]:

$$\oint \frac{dQ_{\text{rev}}}{T} = 0 \qquad (1.20)$$

and for an irreversible cyclic process:

$$\oint \frac{dQ_{\text{irrev}}}{T} < 0, \qquad (1.21)$$

where the subscripts imply either reversibility (1.20) or irreversibility (1.21). Hence, if a quantity S, the entropy of the thermodynamic system, is defined by the relation,

$$\oint \frac{dQ_{\text{rev}}}{T} = \oint dS = 0 \qquad (1.22)$$

then dS is an exact (perfect) differential, and

$$dS = \frac{dQ_{\text{rev}}}{T}. \qquad (1.23)$$

Note that dS is an infinitesimal entropy change of the system during the reversible process when an infinitesimal amount of heat is supplied to the system from its surroundings.

It should be also noted that dS is defined in equation (1.23) for a reversible, nonadiabatic process, and that in the reversible process the change of entropy is caused only by the heat supplied to the thermodynamic system from its surroundings. The definition concerns the entropy change, not an absolute entropy. When a process is irreversible,

$$dQ_{\text{irrev}} < T dS. \tag{1.24}$$

In a particular case, when the considered system is adiabatically isolated, $dS > 0$ for an irreversible process and $dS = 0$ for a reversible one. This means that the entropy of an adiabatically isolated system cannot decrease.

Taking into account the definition (1.23) of the entropy change during a reversible process, in which the system undergoes an infinitesimal change, we can rewrite equation (1.18), which expresses energy conservation of the system per unit mass, in the following form,

$$T ds = d\epsilon - \frac{P}{\rho^2} d\rho, \tag{1.25}$$

where s is the specific entropy. For an adiabatically isolated system equations (1.25) can be reduced to the relations,

$$ds = 0 = d\epsilon - \frac{P}{\rho^2} d\rho. \tag{1.26}$$

1.8.3 Isentropic process

Let us consider a flow induced in an ideal fluid by sound. Since sound propagation in an ideal fluid is an adiabatic process, the fluid particle can be treated as a thermodynamic system which is adiabatically isolated from its surrounding. The process of sound propagation in air at standard conditions can be regarded as reversible, except for the propagation of high frequency sound waves, since gradients not only in space but also in time are small [4,11,12]. The *equation of energy conservation* (1.26), when applied to a fluid particle moving in a sound wave, is reduced to the equation,

$$\frac{ds}{dt} = 0 \tag{1.27}$$

or,

$$\frac{d\epsilon}{dt} = \frac{P}{\rho^2}\frac{d\rho}{dt}. \tag{1.28}$$

Equation (1.27), which indicates that the entropy of the fluid particle is constant, defines isentropic flow [1,4]. If the entropy is constant throughout the whole volume of fluid at a certain initial instant, it will maintain the same constant value at all times and for all of the space filled by the fluid. This type of flow is referred to as *homentropic* [4,8]. Hence, for homentropic motion:

$$\frac{ds}{dt} = \text{grad } s = 0. \tag{1.29}$$

Equation (1.27) is also known as a condition for adiabatic motion and very often the name 'isentropic' is reserved for homentropic flow [7,13].

What kind of consequences result from the fact that sound propagation is an isentropic process? We know that the thermodynamic equilibrium state of a system, e.g., of a fluid particle, can be described by two independent thermodynamic variables, say, pressure and temperature. Hence, the state functions, the specific entropy s and the specific internal energy ϵ, can be expressed by these two variables, namely, $s = s(P, T)$ and $\epsilon = \epsilon(P, T)$. The specific entropy and the specific internal energy may be chosen as independent thermodynamic quantities, for example, instead of pressure and temperature, to specify a local thermodynamic equilibrium state. Especially, it is convenient to describe an adiabatic and reversible, i.e., isentropic, process in terms of the entropy and a basic thermodynamic variable, say, density. For such a case we can treat gas pressure as a function of entropy and density, namely, $P = P(s, \rho)$. During an isentropic process pressure undergoes changes, hence,

$$dP = \left(\frac{\partial P}{\partial \rho}\right)_s dp + \left(\frac{\partial P}{\partial s}\right)_\rho ds = \left(\frac{\partial P}{\partial \rho}\right)_s d\rho \tag{1.30}$$

or, in other words,

$$P = P(\rho). \tag{1.31}$$

To specify the relation between pressure and density for the isentropic process in a perfect gas, let us combine equation (1.28), which expresses the condition for a reversible, adiabatic process, with the equations of state for a perfect gas. For perfect gases the thermal and caloric equations of state are equations (1.4) and (1.5), respectively.

As a result we have the equation,

$$\frac{c_v}{R}\frac{dP}{P} - \frac{(c_v + R)}{R}\frac{d\rho}{\rho} = 0 \tag{1.32}$$

which after integration becomes

$$\ln\left(\frac{P}{\rho^\gamma}\right) = \text{const.} \tag{1.33}$$

Hence, for an isentropic process:

$$\frac{P}{\rho^\gamma} = \text{const.,} \tag{1.34}$$

where $\gamma = (c_v + R)/c_v = c_p/c_v$ and is the ratio of c_p, the specific heat at constant pressure, to c_v, the specific heat at constant volume. Since $P = \rho RT$, the relation between temperature and density for an isentropic process is

$$\frac{T}{\rho^{\gamma-1}} = \text{const.} \tag{1.35}$$

Considering sound propagation through the undisturbed air in a tube, let us denote the pressure of the undisturbed air as P_0 (the equilibrium pressure) and the sound pressure as p. Let also ρ_0 be the density of the undisturbed air (the equilibrium density) and ρ density of air disturbed by sound. Then $P(x,t) = P_0 + p(x,t)$, $\rho(x,t) = \rho_0 + \Delta\rho(x,t)$, and the relation (1.34) can be presented in the form,

$$\frac{P}{\rho^\gamma} = \frac{P_0}{\rho_0{}^\gamma} = \text{const.} \tag{1.36}$$

Sound propagates in a fluid (air) as a disturbance characterised by a small amplitude, hence pressure or density variations in a sound wave are small in relation to the mean values of pressure or density. In terms of fluid density the condition for a small amplitude of the disturbance is

$$|\rho - \rho_0| = |\Delta\rho| \ll \rho_0 \tag{1.37}$$

and, correspondingly, in terms of fluid pressure is

$$|P - P_0| = |p| \ll P_0. \tag{1.38}$$

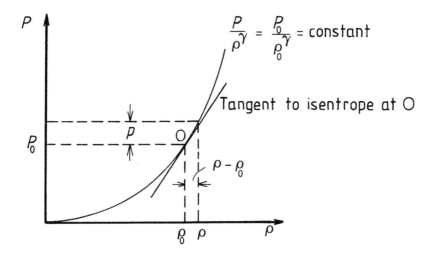

Figure 1.4 Variation of pressure with density in an isentropic process.

See also equations (1.49) and (1.50). The particle velocity in an oscillatory fluid motion caused by sound is also small, see Section 1.10 and particularly the condition (1.61).

The curve which describes the adiabatic, reversible process in a gas such as air is shown in Figure 1.4. The axes of the co–ordinate system are gas density ρ and gas pressure P. Shown on the curve is the point O, with co–ordinates (ρ_0, P_0), which corresponds to the initial state of the gas, undisturbed by sound.

As a sound wave passes through the gas, the pressure P and density ρ of a gas particle in the sound wave vary according to the relation (1.36). The gas particle undergoes thermodynamic changes which lead it from one equilibrium state to another. These thermodynamic equilibrium states, specified by ρ and P, correspond to the points situated on the curve very close to the point O (up or down the curve from O). Sound pressure p and the change in density $\rho - \rho_0$ are small in relation to P_0 and ρ_0, thus the deviations from O on the curve are also small. Because of this, the curve close to O can be approximated by a straight line which is the tangent to the curve at O.

Differentiation of P with respect to ρ leads to:

$$\left(\frac{dP}{d\rho}\right)_s = \frac{P}{\rho^\gamma}\gamma\rho^{\gamma-1} \tag{1.39}$$

$$\left(\frac{dP}{d\rho}\right)_s = \frac{\gamma P}{\rho}. \tag{1.40}$$

At the point O:

$$\left(\frac{dP}{d\rho}\right)_s\bigg|_{\rho=\rho_0} = \frac{\gamma P_0}{\rho_0} = c_0{}^2, \tag{1.41}$$

where $c_0{}^2$ is a constant and is the slope of the isentrope (the adiabatic curve) at point O. The constant $c_0{}^2$ depends only on the properties of the medium, undisturbed by sound, such as pressure P_0, density ρ_0 and also the ratio of specific heats γ.

Thus, as the sound in the medium is a pressure (density) perturbation of small amplitude (condition (1.37)), the gas pressure P, as given by Taylor's expansion,

$$\begin{aligned} P = P_0 &+ \left[\left(\frac{dP}{d\rho}\right)_s\right]_{\rho=\rho_0} (\rho - \rho_0) \\ &+ \frac{1}{2}\left[\left(\frac{d^2 P}{d\rho^2}\right)_s\right]_{\rho=\rho_0} (\rho - \rho_0)^2 + \ldots, \end{aligned} \tag{1.42}$$

can be approximated by the expression,

$$P = P_0 + \left[\left(\frac{dP}{d\rho}\right)_s\right]_{\rho=\rho_0} (\rho - \rho_0) \tag{1.43}$$

or,

$$P = P_0 + c_0{}^2(\rho - \rho_0). \tag{1.44}$$

Hence, with an accuracy to the first order of magnitude in respect to the amplitude of the sound wave, we obtain

$$\frac{p}{\rho - \rho_0} = c_0{}^2, \tag{1.45}$$

where

$$c_0{}^2 = \left[\left(\frac{dP}{d\rho}\right)_s\right]_{\rho=\rho_0} \tag{1.46}$$

and p is the sound pressure. This linear approximation, called the *acoustic approximation*, results from neglecting second and higher order terms in equation (1.42), namely, term $(1/2)c_0{}^2\rho_0(\gamma - 1)[(\rho - \rho_0)/\rho_0]^2$ and higher order terms.

The expression (1.45) can be presented in the form,

$$p = (\Delta\rho)c_0{}^2,$$
(1.47)

where $\Delta\rho = \rho - \rho_0$, hence,

$$p = \frac{\rho - \rho_0}{\rho_0}c_0{}^2\rho_0.$$
(1.48)

It should be noted that, since sound propagation is an isentropic process, the disturbances of pressure and density in a sound wave are not independent of each other but related by equation (1.47).

Since the condition (1.37) denotes that disturbances in the gas are small, the relation (1.48) leads to the equivalent form of (1.37), namely,

$$|p| \ll c_0{}^2\rho_0.$$
(1.49)

Taking into account relation (1.41), which is applicable for a perfect gas, we obtain:

$$|p| \ll \gamma P_0,$$
(1.50)

which denotes that the condition (1.38) is also fulfilled. Combining the relation (1.48) with the equation (1.11) leads to:

$$p = -\rho_0 c_0{}^2 \frac{\partial \xi}{\partial x},$$
(1.51)

where $\rho_0 c_0{}^2$ is constant for every considered medium.

Equation (1.51) is one of the ways of expressing *Hooke's law* in a form applicable for small deformations of a medium [12,15,16]. Hooke's law states that, as long as the deformation is elastic, the strain (relative deformation) is a linear function of the stress. Equation (1.51), valid for an elastic deformation of the medium, can be presented in the form,

$$p = -\mathcal{K}_0 \frac{\partial \xi}{\partial x},$$
(1.52)

where $\mathcal{K}_0 = \rho_0 c_0{}^2 = \gamma P_0 = \rho_0[(\partial P/\partial \rho)_s]_{\rho=\rho_0}$ and is the isentropic (adiabatic) bulk modulus at $\rho = \rho_0$ and where $\mathcal{K} = \rho(\partial P/\partial \rho)_s$ is the isentropic (adiabatic) bulk modulus of elasticity [1,4,17], called also volume elasticity [17] or coefficient of cubic elasticity [18]. Often these names are used in relation to $\mathcal{K}_0$, see [18]; also compare with [10,12,14,19].

1.9 Acoustic plane wave equation

The sound wave equation for a plane wave is obtained finally by combining equations (1.15) and (1.51). Elimination of either p or ξ leads to the equation (1.53) or to equation (1.54):

$$\frac{\partial^2 \xi}{\partial t^2} = c_0^2 \frac{\partial^2 \xi}{\partial x^2} \qquad (1.53)$$

$$\frac{\partial^2 p}{\partial t^2} = c_0^2 \frac{\partial^2 p}{\partial x^2}. \qquad (1.54)$$

Both equations are equivalent forms of the acoustic plane wave equation. The sound wave equation (1.53) is a second order partial differential equation, written in terms of the particle displacement ξ. If we know the solution for ξ, the other quantities associated with sound propagation, such as sound pressure, relative density change $(\rho - \rho_0)/\rho_0$ and particle velocity, can be obtained from the formulae (1.51), (1.11) and (1.6), respectively.

In practice the sound wave equation written in terms of sound pressure is mostly applied, because sound pressure can be obtained directly from measurements.

1.10 Velocity of sound

We consider again a long tube, of cross–section S, filled by a homogenous, elastic medium, e.g., air. A piston can move inside the tube along its axis, as shown in Figure 1.5.

Let us assume that the piston is displaced gradually and slowly during a small range of time Δt by a small distance Δx with an average velocity $u = \Delta x/\Delta t$. The piston pushes and compresses the medium ahead of it. The displacement of the piston results in a small displacement of fluid adjacent to the piston, as well as an increase of fluid pressure (and density), and finally causes the propagation of a small disturbance or sound wave in the air. The sound wave front, moving in the undisturbed medium with an average (in the time range Δt) velocity c_a, reaches after time Δt the position $x = x_s = c_a\Delta t$, i.e., a location which is at a distance $(c_a - u)\Delta t$ from the piston, as shown in Figure 1.5. We consider the fluid column bounded by the moving piston and by the moving wave front. At $t = 0$, as the piston starts to move, the fluid volume and mass within the column are both equal to zero. At $t = \Delta t$ the volume of the fluid column amounts to $S(c_a - u)\Delta t$ and the mass is equal to $\rho S(c_a - u)\Delta t$, where ρ denotes the average density within the considered fluid between the moving wave

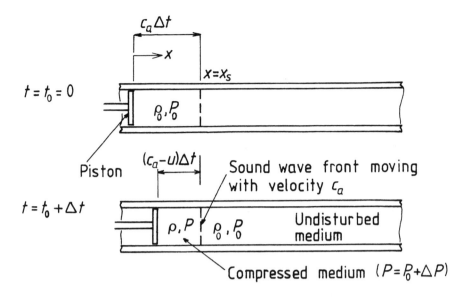

Figure 1.5 Propagation of a sound wave generated by movement of a piston inside a tube.

front and the moving piston, see Figure 1.5. One can also conclude on the basis of the mass conservation law that

$$\rho S(c_a - u)\Delta t = \rho_0 S c_a \Delta t. \tag{1.55}$$

Therefore,

$$\rho(c_a - u) = \rho_0 c_a. \tag{1.56}$$

Hence, the rate of increase of the mass of fluid bounded by the moving piston and the moving sound wave front is $\rho_0 S c_a \Delta t / \Delta t$ or $\rho_0 S c_a$ and the rate of increase of the momentum of the considered fluid is $\rho_0 S c_a u$. The net force acting on the fluid between the piston and the sound wave front is $S\Delta P$, where ΔP denotes the fluid pressure increment within the considered fluid (between the piston and the sound wave front). As a consequence of Newton's second law of motion we obtain:

$$S\Delta P = \rho_0 c_a S u, \tag{1.57}$$

where the expression on the right side of the equation (1.57) is the rate of change of momentum of the fluid bounded by the piston and the sound wave front. Hence,

$$\Delta P = c_a \rho_0 u. \tag{1.58}$$

Taking into account the fact that the density of the fluid between the piston and the sound wave front ρ can be presented as equal to $\rho_0 + \Delta\rho$, we obtain from equation (1.56),

$$(\rho_0 + \Delta\rho)(c_a - u) = \rho_0 c_a \tag{1.59}$$

or,

$$u\rho_0 = \Delta\rho(c_a - u). \tag{1.60}$$

If the piston moves with the velocity u, which is much smaller than the sound velocity c_a, i.e.,

$$u \ll c_a \tag{1.61}$$

then equation (1.60) can be reduced to:

$$\frac{\Delta\rho}{\rho_0} = \frac{u}{c_a}. \tag{1.62}$$

The condition (1.61) (equivalent to condition $\Delta\rho \ll \rho_0$) implies, that the acoustic approximation is valid, see Section 1.8.3. In other words, it is the condition for the acoustic perturbation. The relations (1.58) and (1.62) lead finally to:

$$\Delta P = c_a{}^2 \Delta\rho \tag{1.63}$$

or,

$$c_a{}^2 = \frac{\Delta P}{\Delta\rho}. \tag{1.64}$$

Taking into account the fact that the sound propagation is an isentropic process and passing to the limit when Δt tends to zero, one obtains the velocity of the sound wave front c_s in the form,

$$c_s{}^2 = \lim_{\Delta t \to 0} c_a{}^2 = \left[\left(\frac{dP}{d\rho}\right)_s\right]_{\rho=\rho_0}. \tag{1.65}$$

Comparing equation (1.65) with equation (1.46), we can see that the velocity of propagation of the disturbance in the undisturbed medium c_s is equal to the constant c_0, see also equation (1.41). Hence, the sound propagates in the undisturbed medium with a constant velocity, which may now be denoted by c_0,

$$c_0 = c_s = \sqrt{\left[\left(\frac{dP}{d\rho}\right)_s\right]_{\rho=\rho_0}}. \tag{1.66}$$

Strictly speaking, c_0 is the velocity with which the *sound wave front* propagates in the undisturbed medium. Consequently, the *local velocity of sound* is defined as

$$c = \sqrt{\left(\frac{dP}{d\rho}\right)_s}.$$ (1.67)

Taking into account the definition of the volume elasticity $\mathcal{K}$, see equation (1.52), we can present equation (1.67) as follows:

$$c = \sqrt{\frac{\mathcal{K}}{\rho}}$$ (1.68)

and conclude from equation (1.68) that the sound speed depends upon density and the volume elasticity of the medium in which it propagates.

For a fluid, which can be treated as a perfect gas, we obtain from equations (1.4), (1.40) and (1.67):

$$c^2 = \gamma R \mathrm{T}.$$ (1.69)

For relatively large amplitudes of oscillation, the sound speed depends upon the amplitude [8,14]. Expanding the derivative $(dP/d\rho)_s$ in a Taylor series around ρ_0, we obtain:

$$c^2 = c_0{}^2 \left[1 + \frac{(\gamma - 1)(\rho - \rho_0)}{\rho_0} + \frac{(\gamma - 1)(\gamma - 2)(\rho - \rho_0)^2}{2\rho_0{}^2} + \dots \right].$$ (1.70)

We can conclude from equation (1.70) that for large amplitudes of oscillation in regions of compression ($\rho > \rho_0$) the value of the sound velocity is higher than in regions of rarefaction ($\rho < \rho_0$). However, for sufficiently small amplitudes of disturbance, when $|\rho - \rho_0|/\rho_0 \ll 1$, the approximation,

$$c^2 \simeq c_0{}^2$$ (1.71)

can be applied.

The formula (1.69) presents the velocity of sound in air as a function of temperature. For air $\gamma = 1.404$ and $R = 286.9\,\mathrm{J\,kg^{-1}K^{-1}}$, hence,

$$c = 20.07\sqrt{\mathrm{T}},$$ (1.72)

where T is the absolute temperature of the air. Numerical values of sound velocity in air for different temperatures are shown in Table 1.1.

TABLE 1.1 Velocity of sound c at different temperatures

T [K]		c [m/s]
253	$(-20°\ \text{C})$	319.2
273	$(\ \ \ 0°\ \text{C})$	331.6
293	$(\ \ 20°\ \text{C})$	343.5
313	$(\ \ 40°\ \text{C})$	355.1

A sound velocity value which applies to room conditions, and which is used in many practical applications, is 344 m/s.

1.11 Time-dependent characteristics of sound waves

1.11.1 Harmonic waves

If the piston motion, see Figure 1.1, is simple harmonic the plane sound wave generated is a simple harmonic wave, see also Section 1.12.1. A harmonic sound wave is *deterministic* sound. For a harmonic sound wave the value of the sound pressure, for example, can be determined at a certain position in the sound field for any time, because the sound pressure time history is specified by an explicit mathematical expression. Obviously, the time history in this case is repeatable, if the amplitude, frequency and phase are maintained constant in time.

Frequency of a sound wave   At a fixed point in the tube shown in Figure 1.1 the particle displacement or the sound pressure varies with time in a sinusoidal manner, with period T, as shown in Figure 1.6a. The reciprocal of the period is the frequency f. If the period T is expressed in seconds the unit of frequency is *Hertz*. 1 Hertz denotes 1 cycle per second. The frequency f is related to the angular or circular frequency ω, expressed in radians per second (rad/s), by:

$$\omega = 2\pi f,\tag{1.73}$$

where

$$f = \frac{1}{T}.\tag{1.74}$$

In general *audible sound* has frequencies within 20 Hz to 20 kHz. The lower and upper limits of the frequencies of audible sound are different for each person; the upper limit in particular becomes lower as people age, see Section 7.2.2. Sound with frequencies greater than the upper limit of the audible range is referred to as *ultra–sound*, whereas sound with frequencies smaller than the lower limit is called *infra–sound*.

Amplitude of sound pressure wave   The amplitude of an undamped harmonic wave does not change with time; the amplitude of a sound wave is indicated in Figure 1.6a. It is important to be aware that for sound waves the amplitude of the sound pressure is small compared with atmospheric pressure. Some people are able to hear very faint sounds with an amplitude for the sound pressure of approximately 10^{-5} Pa at certain frequencies.

On the other hand, a sound with an amplitude of 100 Pa is beyond or close to the threshold of pain, but it may be just possible to withstand such a sound for a very short time without damage to the ears. But even a sound with an amplitude of 100 Pa is small compared with the standard atmospheric pressure which is 1.013×10^5 Pa.

Note that the unit of pressure in the International System (SI) of units is the Pascal (Pa) or Newton/m^2 (N/m^2). One Pascal is the same as 10 μbar or 10 dynes/cm^2. The British unit of lbf/in^2 is equal to 6894 Pa.

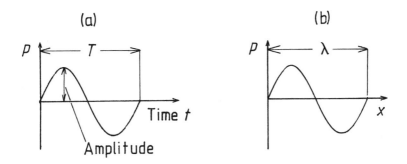

Figure 1.6 Sound pressure as harmonic function of (a) time and (b) distance.

Wavelength   It is shown in Figure 1.6a how the sound pressure of a plane harmonic sound wave varies with time at a given point in the tube. Another approach is to consider how the sound pressure varies with distance along the tube at a given instant of time, as in Figure 1.6b. The sound pressure changes with distance sinusoidally and the values of sound pressure are repeated every *wavelength* λ. The wavelength and the period are related through the sound wave velocity c_0 by the relation,

$$c_0 = \lambda/T \qquad (1.75)$$

or,

$$c_0 = \lambda f. \qquad (1.76)$$

The wave velocity of sound in air is close to 344 m/s, see Table 1.1. Thus a harmonic sound wave of frequency 1000 Hz has a wavelength of 344 mm and a sound wave of frequency 100 Hz has a wavelength of 3.44 m. It is sometimes more useful to analyse acoustic phenomena with reference to wavelength rather than frequency, particularly when the size of a source or the distance away from a measurement point is considered.

1.11.2 Deterministic sound

In Section 1.11.1 the properties of a harmonic wave generated by an oscillating piston were described. The simple harmonic wave is called also *monochromatic* since sound pressure, particle velocity and particle displacement in the sound wave depend upon time only through functions of the single, circular frequency ω. The monochromatic sound wave is often described as a *pure tone*. In general, however, every arbitrary sound wave can be presented as the sum of its monochromatic components. Sound as a function of time (at a certain point in a sound field) is often referred to as a *signal*. Strictly speaking, the term 'sound signal' implies that the time–dependent parameters describing the sound are directly related to the time–varying electrical quantities.

To know how the sound energy is distributed among the monochromatic components of the sound signal, it is essential to obtain the frequency spectrum of the signal.

The specification of different types of sound signals is based on the characteristic features of their frequency spectra. Thus, sinusoidal, complex periodic and almost periodic signals are deterministic and have discrete spectra. A *complex periodic signal* comprises a series of pure tones, the frequency of which are integer multiples of a fundamental — for example, the components of the series could have frequencies of 50, 100, 150 and 200 Hz, etc., where 50 Hz is the fundamental frequency. When there is no integer relationship between the frequencies the sound is *almost periodic* — for example, pure tones at 50, 90 and 135 Hz. Transient sound can also be deterministic; it has always a spectrum which is continuous.

The deterministic *transient* sound wave is a type of signal in which the sound pressure reaches a peak value and then decays to zero; for example, sound pressure dependent on time in the form of an exponential function, damped sinusoid or rectangular pulse.

1.11.3 Random sound wave

Measurements in a factory are most likely to deal with random sound waves. Random sound has an unpredictable and hence also unrepeatable character. We cannot predict values of the signal parameters at any instant in time. As regards spectral representation, all frequencies are present in random sound. Random sound of short duration is often referred to as transient, if it has a clearly defined beginning and end. During the operation of a machine, such as a punch press, there are many transients separated by short intervals of time, but each transient is never exactly repeated.

A random signal is often called *noise*. More generally noise is *unwanted sound*. Although the sound from a fan may be harmonic, there may be situations in which its sound is unwelcome and can be regarded as noise.

1.12 Solution of the wave equation for plane waves

Let us consider equation (1.54), the one-dimensional sound wave equation for plane waves expressed in terms of sound pressure p. This wave equation, which is a second order partial differential equation, has a general solution in the form [13,20,21]:

$$p = F_1(t - x/c_0) + F_2(t + x/c_0), \qquad (1.77)$$

where F_1, F_2 are arbitrary functions of the arguments $t - x/c_0$ and $t + x/c_0$, respectively, and where c_0 is the sound velocity. It is assumed that the functions F_1, F_2 have continuous derivatives of the first and second order. The fact that equation (1.77) is a general solution can be checked by substitution of (1.77) into the wave equation (1.54).

The first term in the expression for p, namely $F_1(t - x/c_0)$, describes a disturbance (sound wave) travelling with velocity c_0 in the direction of increasing values of x. The second term $F_2(t + x/c_0)$ describes a disturbance moving in the direction of the decreasing values of x. When $F_2 = 0$ the sound pressure is expressed by the formula,

$$p = p_+(t - x/c_0) = F_1(t - x/c_0). \qquad (1.78)$$

This type of disturbance progressing only in one direction can be generated, e.g., by the slow movement of a piston in an infinite tube. Such a wave is called *a freely progressing, progressive* or *simple wave* [22].

Function F_1 is constant, if the argument of this function is constant, namely if $t - x/c_0 = \text{const}$. Hence, along the straight line which is defined by the equation $x = c_0 t + C$, where C is a constant, the function F_1 maintains a constant value. A family of lines $x = c_0 t + C$, where the constant C plays the role of a parameter, is called a family of *characteristics*. In other words the characteristics are lines in the x, t plane along which small disturbances propagate. Figure 1.7 illustrates the fact that the shape of small disturbances (sound waves) does not change in time. These disturbances move with a constant velocity c_0.

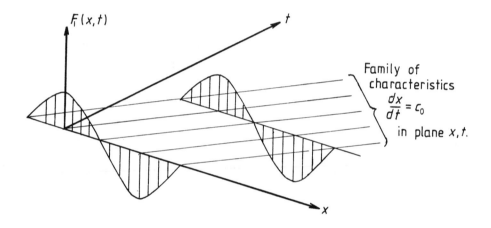

Figure 1.7 Propagation of small disturbances (plane waves).

Functions F_1 and F_2 can be specified from the initial conditions, for example from the initial, small disturbances of pressure and particle velocity. In the case when the sound field is reduced to a sound wave propagating only in one direction, $F_2 = 0$, knowledge of the initial disturbance of only one parameter — for example, of fluid pressure — is sufficient to specify F_1, see equation (1.78).

If we introduce the new variables, $\theta = t - x/c_0$ and $\tau = t + x/c_0$, the general solution of the sound wave equation in terms of sound pressure can be written as

$$p = p_+(t - x/c_0) + p_-(t + x/c_0) = p_+(\theta) + p_-(\tau). \qquad (1.79)$$

Similarly, the general solution of the sound wave equations in terms of particle displacement can be presented in the form,

$$\xi = \xi_+(t - x/c_0) + \xi_-(t + x/c_0) = \xi_+(\theta) + \xi_-(\tau). \qquad (1.80)$$

Terms with $+$ or $-$ indices describe sound waves progressing in the direction of the positive values of x, and in the direction of the negative values of x, respectively. Since

$$\frac{\partial \xi_+}{\partial t} = \frac{\partial \xi_+}{\partial \theta} \frac{\partial \theta}{\partial t} = \frac{\partial \xi_+}{\partial \theta}$$

and

$$\frac{\partial \xi_+}{\partial x} = \frac{\partial \xi_+}{\partial \theta} \frac{\partial \theta}{\partial t} = -\frac{1}{c_0} \frac{\partial \xi_+}{\partial \theta}$$

then

$$\frac{\partial \xi_+}{\partial x} = -\frac{1}{c_0} \frac{\partial \xi_+}{\partial t}. \qquad (1.81)$$

Hence, taking into account the relation (1.51) and remembering that $\partial \xi_+/\partial t = u_+$, we obtain:

$$p_+(t - x/c_0) = \rho_0 c_0 u_+(t - x/c_0). \qquad (1.82)$$

A similar procedure with respect to the term describing the wave travelling in the direction of decreasing values of x, leads to the formula,

$$p_-(t + x/c_0) = -\rho_0 c_0 u_-(t + x/c_0). \qquad (1.83)$$

Finally, taking into account equations (1.82) and (1.83), we obtain:

$$u = u_+(t - x/c_0) + u_-(t + x/c_0)$$
$$= \frac{1}{\rho_0 c_0}[p_+(t - x/c_0) - p_-(t + x/c_0)] \qquad (1.84)$$

or

$$u = \frac{1}{\rho_0 c_0}[F_1(t - x/c_0) - F_2(t + x/c_0)]. \qquad (1.85)$$

A sound wave, which results from a certain initial disturbance of flow at time $t = 0$, is called a *free sound wave*. A *forced sound wave* occurs when, for example, a piston in a tube moves periodically with time. In many physical problems the term $F_1(t - x/c_0)$ in equation (1.77) represents an *incident sound wave* which is moving away from a certain sound source, and the term $F_2(t + x/c_0)$ can be identified as a *reflected wave* which is returning towards the source after reflection from some obstacle downstream.

1.12.1 Harmonic solution of the wave equation for plane waves

The expression (1.77) is a general solution of the wave equation (1.54) which describes the propagation of a plane sound wave. Let us assume that the sound wave generated in a tube is harmonic, that is, sinusoidal. This type of sound wave is produced by a reciprocating piston, as described in Section 1.1.1 and shown in Figure 1.1. Additionally, we assume that the tube is finite and terminated by a wall. When the piston executes harmonic movement with small amplitude A_p, the motion of the piston is described by the expression $x_p(t) = A_p \sin \omega t$, where x_p denotes the piston displacement and $A_p \ll \lambda/(2\pi)$. This means that the velocity of the piston u_p is also a harmonic function of time, namely $u_p = dx_p/dt = A_p \omega \cos \omega t$, and is very small compared to the sound velocity, i.e., $A_p \omega \ll c_0$. Disturbances caused by the piston propagate as a sound wave in the direction of the positive values of x. In the (x, t) plane the sound wave paths are straight line characteristics C_+, as shown in Figure 1.8.

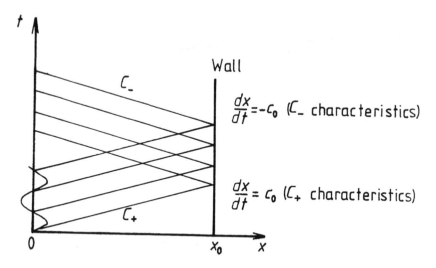

Figure 1.8 Propagation of small disturbances in a tube terminated by a wall.

The sound wave generated in this case by harmonic vibrations of a piston and propagating in the direction of the positive values of x is described in terms of particle displacement in the form of a harmonic (sinusoidal) function of time and position in space, as follows:

$$\xi_+ = A_1 \sin(\omega t - \frac{\omega x}{c_0}) + A_2 \cos(\omega t - \frac{\omega x}{c_0}) \qquad (1.86)$$

or,

$$\xi_+ = A \sin(\omega t - \frac{\omega x}{c_0} + \phi_0), \qquad (1.87)$$

where A is the amplitude and ω is the circular frequency of the harmonic wave. Note that $A_1 = A \cos \phi_0$ and $A_2 = A \sin \phi_0$. The quantity, which for plane waves has the form, $\Phi(x,t) = \omega t - \omega x/c_0 + \phi_0$, is called the *phase*. $\Phi_0 = \Phi(x,0) = -\omega x/c_0 + \phi_0$ is the *initial phase*, i.e., the phase at $t = 0$. Finally, ϕ_0 is the *phase angle* which defines the process phase when $t = 0$ and $x = 0$. The phase $\Phi(x, t)$ which appears in the equation (1.87) is the phase for the particle displacement.

In general the equation $\Phi_p(\vec{r}, t) = \text{const.}$, where Φ_p denotes the phase of sound pressure wave, defines the *sound wave fronts* at different instants, see also Appendix B. In the case of a plane, harmonic wave:

$$\Phi_p(x,t) = \omega t - \frac{\omega}{c_0} x + \phi_p, \qquad (1.88)$$

where constant ϕ_p is the phase angle. For an observer moving with velocity

$$c_{p,ph} = \frac{dx}{dt} = -\frac{\partial \Phi_p}{\partial t} / \frac{\partial \Phi_p}{\partial x} \qquad (1.89)$$

the phase Φ_p maintains a constant value. The velocity $c_{p,ph}$ is called the *sound pressure phase velocity* and for plane, harmonic (monochromatic) waves is equal to:

$$\frac{\omega}{k} = \frac{\omega \lambda}{2\pi} = c_0 \qquad (1.90)$$

since ω and k are defined as

$$\omega = \frac{\partial \Phi_p}{\partial t} \qquad (1.91)$$

and

$$k = -\frac{\partial \Phi_p}{\partial x}, \qquad (1.92)$$

where k is called the *wave number* and λ is the wavelength corresponding to the frequency of sound ω. The quantity k is the magnitude of the *wave vector* or the *propagation vector* $\vec{k}$, defined locally (for any type of wave) — see equation (1.92) — as the gradient of Φ_p in space [8,13,20].

Taking into account the equations (1.6) and (1.86), we obtain for the particle velocity:

$$u_+ = A\omega \cos(\omega t - kx + \phi_0). \tag{1.93}$$

Since at time $t = 0$ the position of the piston is at $x = 0$, i.e., $x_p(0) = 0$, the phase angle ϕ_0 defining the process phase when $t = 0$ and $x = 0$ should be assumed as equal to 0. As a result we obtain:

$$\xi_+ = A \sin(\omega t - kx) \tag{1.94}$$

and

$$u_+ = A\omega \cos(\omega t - kx). \tag{1.95}$$

If u_0 denotes the velocity of the piston at $t = 0$,

$$u_p(0) = A_p\omega = A\omega = u_0 \tag{1.96}$$

and hence,

$$A = \frac{u_0}{\omega} = A_p. \tag{1.97}$$

Therefore, equations (1.94) and (1.95) are reduced, respectively, to equations,

$$\xi_+ = A_p \sin(\omega t - kx) \tag{1.98}$$

and

$$u_+ = A_p\omega \cos(\omega t - kx). \tag{1.99}$$

Finally, since $p_+ = \rho_0 c_0 u_+$, see equation (1.82), the sound pressure in the sound wave travelling in the direction of the positive values of x changes with time and position, as follows:

$$p_+ = \rho_0 c_0 \omega A_p \cos(\omega t - kx). \tag{1.100}$$

It is obvious that for a plane wave progressing in the direction of the positive values of x the sound pressure p_+ and the particle velocity u_+ are in phase with each other. However, they lead the particle displacement ξ_+ by $\pi/2$ radians.

Reflected wave   The reflected sound wave, travelling in the direction of the negative values of x, expressed in terms of particle displacement,

$$\xi_- = B \sin(\omega t + kx + \phi), \tag{1.101}$$

is characterised by an amplitude B which is equal to or less than A, the amplitude of the particle displacement of the incident wave. ϕ denotes the phase angle for the reflected wave.

In general the reflected wave is not in phase with the incident wave. The relationship between the reflected and the incident waves depends upon the obstacles or the termination which cause the reflection. If a plane sound wave, travelling along the tube, meets a rigid cap terminating the tube, then the reflected wave has the same amplitude as the incident wave at the cap. For non-rigid terminations the amplitude of the reflected wave is less than that of the incident wave and the waves are out of phase.

The distributions of particle displacements with time for the incident and reflected waves at the point of reflection are shown in Figure 1.9. In Figure 1.9 the incident and reflected waves are shown separately to illustrate differences in amplitudes and phases. In reality they combine to form a standing wave, see Sections 1.15 and 4.4.

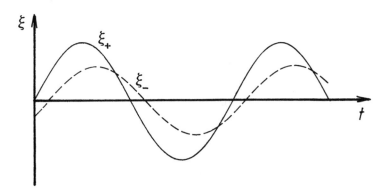

Figure 1.9 The particle displacement distributions for the incident and reflected waves.

In terms of the particle velocity the reflected wave, called also the negative wave, may be expressed by the following formula,

$$u_- = B\omega \cos(\omega t + kx + \phi) \tag{1.102}$$

and in terms of the sound pressure p_-, see equation (1.83), as follows:

$$p_- = -\rho_0 c_0 B\omega \cos(\omega t + kx + \phi). \tag{1.103}$$

In the case of the reflected wave the particle velocity leads the particle displacement by a phase angle $\pi/2$ radians, but the sound pressure lags the particle displacement by the phase angle $\pi/2$ radians.

Characteristic impedance   The fundamental parameters for acoustic waves are sound pressure and particle velocity. Their ratio is called the *specific acoustic impedance*. For plane, freely progressing waves the ratio of the sound pressure to the particle velocity depends only on the properties of the medium in which the sound wave propagates. In this case the specific acoustic impedance is a resistance which is also called the characteristic impedance or resistance of the medium. Hence, for the incident plane wave, progressing in the direction of the positive values of x, we have

$$\frac{p_+}{u_+} = \rho_0 c_0 \qquad (1.104)$$

and for the reflected plane wave,

$$\frac{p_-}{u_-} = -\rho_0 c_0. \qquad (1.105)$$

The expressions (1.104) and (1.105) apply not only to harmonic waves but also to any plane wave, see equations (1.82) and (1.83). The characteristic impedance $\rho_0 c_0$ is a frequently occurring quantity in the theory of acoustics.

1.12.2 Exponential form of the solution of the wave equation

Let us consider the general solution of the wave equation (1.77) for a plane wave. F_1 and F_2 are arbitrary functions of the arguments $t - x/c_0$ and $t + x/c_0$, respectively, except that both functions have continuous derivatives of the first and second order. In the previous section we have considered the harmonic solution of the plane wave equation, in which F_1 and F_2 are sinusoidal functions of their arguments.

The general solution of the plane wave equation in a class of real harmonic functions can be presented as the real part of a complex expression, namely,

$$p = \text{Re}\{\widetilde{\mathbf{p}}\} = \text{Re}\{\mathbf{A}e^{i(\omega t - kx)} + \mathbf{B}e^{i(\omega t + kx)}\}. \qquad (1.106)$$

The tilde or wavy line $\sim$ above a boldface letter denotes that the quantities are complex and additionally harmonic. As we shall describe later, these quantities are called *phasors*. Boldface type indicates a complex quantity (complex vector). The sound pressure p is the real part of the complex quantity $\widetilde{\mathbf{p}}$. Hence, only the real part of the quantity $\widetilde{\mathbf{p}}$ possesses physical meaning.

Formally, the quantity $\widetilde{\mathbf{p}}$ is the solution of the wave equation,

$$\frac{\partial^2 \widetilde{\mathbf{p}}}{\partial t^2} = c_0{}^2 \frac{\partial^2 \widetilde{\mathbf{p}}}{\partial x^2} \qquad (1.107)$$

in a class of complex harmonic functions. If the dependence of the quantity $\tilde{p}$ on time is simple harmonic, the time dependence may be separated from the space dependence, namely,

$$\tilde{p}(x\,,t) = p(x)e^{i\omega t}. \tag{1.108}$$

Putting $\tilde{p}(x,t)$ from equation (1.108) into (1.107), we obtain:

$$\frac{\partial^2 p(x)}{\partial x^2} + k^2 p(x) = 0, \tag{1.109}$$

where $k = \omega/c_0$.

Next, we postulate a solution for $p(x)$ in the form,

$$p(x) = A_0 e^{\sigma x}, \tag{1.110}$$

where σ is a constant [23]. Introducing equation (1.110) into (1.109) leads to the *characteristic equation*,

$$\sigma^2 + k^2 = 0 \tag{1.111}$$

and finally to a solution for σ,

$$\sigma = \pm ik. \tag{1.112}$$

Hence, the general solution of the wave equation in the class of complex harmonic functions is

$$\tilde{p}(x\,,t) = e^{i\omega t}\left(A e^{-ikx} + B e^{ikx}\right), \tag{1.113}$$

where A and B are constants. Taking into account the fact that every complex number may be presented in an exponential form [24], the solution (1.113) may be reduced to:

$$\tilde{p} = A e^{i(\omega t - kx + \phi_A)} + B e^{i(\omega t + kx + \phi_B)}, \tag{1.114}$$

where the amplitudes $A = |A|$ and $B = |B|$ are real quantities. They are moduli (absolute values) of A and B, respectively. The angles ϕ_A and ϕ_B are arguments of the complex quantities A and B, respectively. They define the phase of the process for both waves (positive and negative ones) at $t = 0$ and $x = 0$. Finally, making use of Euler's formula for complex quantities, we obtain from the last relation (1.114):

$$\begin{aligned}\tilde{p} = {} & A\cos(\omega t - kx + \phi_A) + B\cos(\omega t + kx + \phi_B) \\ & + i\{A\sin(\omega t - kx + \phi_A) + B\sin(\omega t + kx + \phi_B)\}. \end{aligned} \tag{1.115}$$

The complex exponential method is widely used in acoustics and vibration problems [25,26]. It possesses several advantages over the trigonometric method; namely, mathematical convenience, simplicity in finding phase relations between different acoustic quantities and graphic representation of the considered complex quantities in the form of the rotating vectors (phasors). It should be emphasised, however, that only the real part of the final solution obtained by this method is physically meaningful.

Complex vectors   Complex quantities may be visualised in the complex plane in the form of *complex vectors*. The plane containing the real and imaginary co–ordinate axes is called the complex plane, see Figure 1.10.

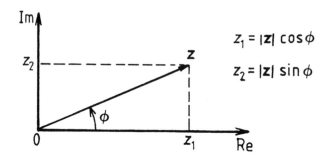

Figure 1.10 Complex vector.

Every complex quantity is represented by a point in the complex plane. This point is determined by a position vector which lies in the complex plane and joins the origin of the co–ordinate system with the point. Hence, every complex magnitude can be represented in the complex plane either by a point or by a complex vector. For example, the complex quantity $\mathbf{z} = z_1 + iz_2$ can be shown, see Figure 1.10, as a vector of length $z = |\mathbf{z}| = \sqrt{z_1^2 + z_2^2}$; the complex vector magnitude $z = |\mathbf{z}|$ is called also the absolute value or the modulus of the complex quantity. The vector $\mathbf{z}$ subtends an angle $\phi = \arctan(z_2/z_1)$ with the positive real axis. The angle ϕ is called the argument of the complex quantity. Every complex quantity (complex vector) can be presented in algebraic ($\mathbf{z} = z_1 + iz_2$), trigonometric ($\mathbf{z} = |\mathbf{z}|(\cos\phi + i\sin\phi)$) or exponential — called also polar — ($\mathbf{z} = |\mathbf{z}|e^{i\phi}$) forms.

Phasors   Let us consider a complex vector whose position in relation to the positive real axis changes with time. Such a complex vector, in an exponential form $\widetilde{\mathbf{A}} = \mathbf{A}e^{i\omega t} = |\mathbf{A}|e^{i(\omega t + \Phi_0)}$ is a rotating vector, named also a phasor. It rotates in the complex plane with constant angular velocity ω, see also Section 1.11.1. The angle Φ_0 is the initial phase, i.e., argument of $\tilde{\mathbf{A}}$ at time $t = 0$. The angle Φ_0 describes the initial position of the phasor $\tilde{\mathbf{A}}$ in relation to the positive real axis. The angle $(\omega t + \Phi_0)$ denotes the position of the phasor $\tilde{\mathbf{A}}$ at a certain time t, see Figure 1.11.

It should be emphasised that $\sim$ above a boldface letter denotes a phasor (rotating vector). This notation is used to distinguish phasors from the other complex vectors (marked by boldface letters) whose positions in the complex plane do not depend on time. The projection of the phasor $\tilde{\mathbf{A}}$ on the real axis, i.e., its real part, is equal to $|\mathbf{A}| \cos(\omega t + \Phi_0)$ and varies harmonically with time. Similarly, its imaginary part $|\mathbf{A}| \sin(\omega t + \Phi_0)$ changes also harmonically with time.

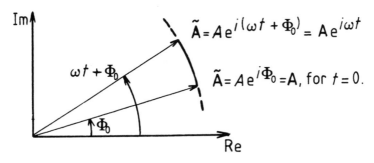

Figure 1.11 Phasor.

Multiplication of the complex vector by the unit vector $e^{i\phi} = \cos\phi + i\sin\phi$ is equivalent to its rotation counter–clockwise by an angle ϕ. Hence, as shown in Figure 1.11, the rotation of the complex vector $\mathbf{A}$ by the angle ωt means its multiplication by the unit complex vector (phasor) $e^{i\omega t}$. The greatest achievement of the complex quantity (complex vector) formalism lies in the simplicity of such operations as differentiation and integration. For example, differentiation of the complex quantity (phasor) $\mathbf{A}e^{i\omega t} = Ae^{i(\omega t + \phi)}$ with respect to time is reduced to the multiplication of the rotating vector $Ae^{i(\omega t + \phi)}$ by a factor $i\omega$. As a result, the complex vector $\omega Ae^{i(\omega t + \phi + \pi/2)}$ of length ωA is obtained. This vector subtends the angle $\omega t + \phi + \pi/2$ with the positive real axis, because multiplying any complex vector by i rotates it $\pi/2$ radians counter–clockwise. The phasor convention may be applied to scalars, e.g., such as sound pressure, as well as to quantities, such as particle displacement $\vec{d}$ or particle velocity $\vec{u}$, which are vectors in the formalism of real vectors.

Real vector formalism Every vector (real vector) in three–dimensional space can be expressed by unit vectors along co–ordinate axes, i.e., by principal or co–ordinate vectors, which are denoted in this book by the symbols $\vec{e}_x = [1, 0, 0]$, $\vec{e}_y = [0, 1, 0]$, $\vec{e}_z = [0, 0, 1]$ and are related to the cartesian rectangular co-ordinate system.

Thus, in three–dimensional space vectors of particle displacement $\vec{d}$ and particle velocity $\vec{u}$ are, respectively,

$$\vec{d} = \xi\vec{e}_x + \eta\vec{e}_y + \zeta\vec{e}_z = [\xi, \eta, \zeta],$$
$$\vec{u} = u_x\vec{e}_x + u_y\vec{e}_y + u_z\vec{e}_z = [u_x, u_y, u_z],$$

where ξ, η, ζ and u_x, u_y, u_z denote components of the vectors of the particle displacement and particle velocity in the cartesian rectangular co-ordinate system. In one–dimensional problems, as in the case of the propagation of plane waves in a tube, $\vec{d} = \xi\vec{e}_x$ and $\vec{u} = u_x\vec{e}_x$. Therefore, it is convenient to operate on the component of the vector rather than the vector itself, see for example, equations (1.6), (1.11) and (1.15). The notation u, ξ indicates in this case x co–ordinate components of vectors $\vec{u}$, $\vec{d}$ or vector magnitudes, since $|\vec{u}| = u_x$, $|\vec{d}| = \xi$. The transition from the formalism of phasors or complex vectors to real vector convention may be illustrated by the formulae,

$$\vec{u} = \text{Re}(\vec{\tilde{u}}). \qquad (1.116)$$

In the case of the one–dimensional problem, see also equation (1.120), formula (1.116) is reduced to:

$$\vec{u} = \{\text{Re}(\tilde{u})\}\vec{e}_x = \text{Re}\{\tilde{u}\vec{e}_x\}. \qquad (1.117)$$

$\text{Re}(\tilde{u})$ denotes the real part of the phasor $\tilde{u}$ and $\text{Re}\{\tilde{u}\vec{e}_x\}$ means the real part of the phasor–vector $\tilde{u}\vec{e}_x$. For a scalar, such as sound pressure, similarly, we have

$$p = \text{Re}(\tilde{p}). \qquad (1.118)$$

Let us assume that the monochromatic plane wave propagates in space in an arbitrarily chosen direction, but not along one of the co–ordinate axes of the cartesian rectangular co–ordinate system, see Figure 1.12.

For a certain time the constant phase surface for a plane wave is defined by the equation $\vec{k} \cdot \vec{r} = \text{constant}$, where $\vec{k}$ (see equation (1.92)) is the propagation vector or wave vector which is normal to the constant

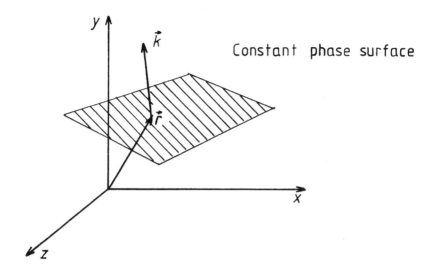

Figure 1.12 Propagation of plane sound wave in an arbitrary direction.

phase surface and $\vec{r} = x\vec{e}_x + y\vec{e}_y + z\vec{e}_z$ is a position vector. In terms of particle velocity the wave equation solution has the form,

$$\widetilde{\mathbf{u}} = \mathbf{A}e^{i(\omega t - \vec{k}\cdot\vec{r})}. \qquad (1.119)$$

The magnitude of the wave vector $k = |\vec{k}| = \sqrt{k_x^2 + k_y^2 + k_z^2}$ is equal to ω/c_0 and k_x/k, k_y/k, k_z/k are the direction cosines of $\vec{k}$ with respect to the x, y, z co–ordinate axes, respectively. In this case, see equation (B.5), the transition from phasor formalism to real vector formalism can be applied as follows:

$$\vec{u} = -\mathrm{Re}(\widetilde{\mathbf{u}})\frac{\nabla\Phi_0}{|\nabla\Phi_0|} = \{\mathrm{Re}(\widetilde{\mathbf{u}})\}\frac{\vec{k}}{|\vec{k}|} = \mathrm{Re}\{\widetilde{\mathbf{u}}\frac{\vec{k}}{|\vec{k}|}\}. \qquad (1.120)$$

It should be noted that the complex vectors convention can be used not only for first order quantities, such as particle displacement, sound pressure, etc., but also for second order quantities, e.g., square of the particle displacement ξ^2 or sound intensity, provided that the second order quantities are time averages [8,26].

1.13 Sound intensity

The energy conservation equation for the sound wave, or the acoustic energy balance equation, see equation (A.20) of Appendix A, has the form,

$$\frac{\partial E_a}{\partial t} = -\mathrm{div}(p\vec{u}), \tag{1.121}$$

where $E_a = \rho_0 u^2/2 + (c_0{}^2/\rho_0)(\Delta\rho)^2/2$ is the acoustic energy density (sound energy density) — the acoustic energy per unit volume — which contains as the first term the kinetic energy and as the second term the potential energy, both per unit volume.

The vector quantity $\vec{I} = p\vec{u}$, the product of the sound pressure p and the corresponding particle velocity $\vec{u}$ (at the same point in the sound field), appears in equation (1.121). This is the *energy flux density vector* for the sound wave. Its magnitude is equal to the sound energy flux through the unit surface perpendicular to the direction of the fluid velocity (particle velocity). In general the component $I_n = \vec{I}\cdot\vec{n}$ of the energy flux density vector in the direction of any unit vector $\vec{n}$ is the sound energy flux through a unit surface which is perpendicular to the direction of the unit vector $\vec{n}$. In acoustics the sound energy flux density vector $\vec{I}$, is called the *instantaneous sound intensity*. However, very often the name sound (acoustic) intensity is used in relation to the instantaneous sound intensity. The instantaneous sound intensity is also known as the acoustic energy flux vector or acoustic energy transport vector [8,11,14,28,29].

Taking into account the definition of the instantaneous sound intensity, we can rewrite the acoustic energy balance equation (1.121) as

$$\frac{\partial E_a}{\partial t} + \mathrm{div}\vec{I} = 0. \tag{1.122}$$

The acoustic energy balance equation is valid in any source free region of space.

Let the sound intensity $\vec{I}$ be presented in the form,

$$\vec{I} = p|\vec{u}|\vec{e}_u \tag{1.123}$$

where $\vec{e}_u = \vec{u}/|\vec{u}|$ is the unit vector whose direction is the same as the particle velocity $\vec{u}$. It should be mentioned that for longitudinal waves, as in the case of plane waves, the unit vector $\vec{e}_u$ has the same direction as the wave vector $\vec{k}$.

The time average of the instantaneous sound intensity or the *time-averaged sound intensity* is defined as follows:

$$\overline{\vec{I}} = \overline{p\vec{u}} = \frac{1}{T}\int_0^T p\vec{u}\,dt, \tag{1.124}$$

where T is the period of the harmonic wave. (Note that the lines above $\vec{I}$ and the product $p\vec{u}$ indicate time averages.) This definition is equivalent to:

$$\overline{\vec{I}} = \lim_{\tau \to \infty} \frac{1}{\tau} \int_0^\tau p\vec{u}\,dt. \tag{1.125}$$

The definition (1.125) has more universal character than the definition of the time-averaged intensity for the harmonic sound wave (1.124). It is valid for any type of sound. Both definitions lead to the same results; namely, in any sound source free region of space, see (1.122),

$$\mathrm{div}\,\overline{\vec{I}} = 0. \tag{1.126}$$

Hence, the time-averaged sound intensity $\overline{\vec{I}}$ 'creates' a solenoidal vector field.

Integrating the equation (1.126) over the fluid volume V, which is fixed in space and sound source free, we have

$$\int_V \mathrm{div}\,\overline{\vec{I}}\,dV = 0 \tag{1.127}$$

and finally taking into account Gauss's (Green's) theorem [24,27], we obtain:

$$\oint_S \overline{\vec{I}} \cdot \vec{n}\,dS = 0, \tag{1.128}$$

where $\vec{n}$ is the unit vector normal to dS, the infinitesimal element of surface S, and where S is the surface enclosing the volume V.

It should be mentioned that in some references [8,14] the name 'sound intensity' is reserved for the time–averaged sound intensity. In this book we shall use the term time–averaged sound intensity. (Note that the terms acoustic intensity and sound intensity are synonymous.)

Sound power   The magnitude of the sound intensity vector has dimensions of power per unit area. Hence, we can define the sound power (acoustic power) W crossing the surface S (closed or open) as follows:

$$W = \int_S \overline{\vec{I}} \cdot \vec{n}\,dS = \int_S \overline{I_n}\,dS, \tag{1.129}$$

where $\overline{I_n}$ is the time–averaged intensity vector component in the direction normal to the infinitesimal surface dS. It is obvious that equation (1.128) leads to the conclusion that the acoustic power crossing the surface which encloses completely the source free space is equal to zero; $W = 0$. On the other hand the value of the acoustic power crossing any surface which completely encloses a space with acoustic sources is the same for any such surfaces. This property of the acoustic power enables us to perform sound power measurement in the far field.

1.14 Sound energy density and sound intensity for plane waves

The sound energy density for plane, freely progressing waves is given (see Appendix A and particularly equations (A.21) and (A.22)) by the formula,

$$E_a = \rho_0 u^2 = \frac{c_0^2}{\rho_0}(\Delta\rho)^2. \tag{1.130}$$

Let us consider plane, harmonic, freely progressing waves, defined in terms of particle velocity u of amplitude U_0, where

$$u = U_0 \cos(\omega t - kx). \tag{1.131}$$

The time average of the sound energy density for such a harmonic wave is

$$\overline{E_a} = \frac{1}{T}\int_0^T E_a dt = \overline{\rho_0 u^2} = \rho_0 U_0^2 \frac{1}{T}\int_0^T \cos^2(\omega t - kx)dt, \tag{1.132}$$

where T is the period of the oscillation. Hence,

$$\overline{E_a} = \rho_0 U_0^2 \frac{1}{T}\int_0^T \frac{1}{2}\{1 + \cos[2(\omega t - kx)]\}dt = \frac{\rho_0 U_0^2}{2}. \tag{1.133}$$

The instantaneous sound intensity vector $\vec{I} = p\vec{u}$ for a plane, freely progressing wave can with the aid of equations (B.8), (1.130) and (1.47) be written as:

$$\vec{I} = [p^2/(\rho_0 c_0)](-\nabla\Phi_0)/|\nabla\Phi_0| = [p^2/(\rho_0 c_0)]\vec{a} = \rho_0 c_0 u^2 \vec{a}, \tag{1.134}$$

where $\vec{a}$ is the unit vector along the direction of plane wave propagation. The formulae (1.130) and (1.134) lead to the relation,

$$\vec{I} = c_0 E_a \vec{a}. \tag{1.135}$$

Hence, the magnitude of the sound energy flux density vector (sound intensity vector) is equal to the product of the sound energy density and the sound velocity c_0. The plane, progressing sound wave transports energy with velocity c_0. Averaging both sides of the equation (1.135) with respect to time, we obtain:

$$\vec{\overline{I}} = c_0 \overline{E_a} \vec{a}. \tag{1.136}$$

Both formulae (1.135) and (1.136) hold for plane, freely progressing sound waves.

The instantaneous sound intensity can be presented also in another form, see equation (1.134),

$$\vec{I} = \frac{p^2}{\rho_0 c_0} \vec{a}, \tag{1.137}$$

where p denotes the sound pressure. Averaging both sides of equation (1.137) with respect to time, leads to

$$\vec{\overline{I}} = \frac{\vec{a}}{\rho_0 c_0} \frac{1}{T} \int_0^T p^2 \, dt = \frac{\vec{a}}{\rho_0 c_0} \overline{p^2}. \tag{1.138}$$

Finally, considering the plane, harmonic, freely progressing wave in terms of sound pressure $p = p_0 \cos(\omega t - kx)$, we obtain the expression for $\vec{\overline{I}}$ in the form:

$$\vec{\overline{I}} = \frac{p_0^2 \vec{a}}{\rho_0 c_0} \frac{1}{T} \int_0^T \cos^2(\omega t - kx) dt = \frac{1}{2} \frac{p_0^2}{\rho_0 c_0} \vec{a}, \tag{1.139}$$

where p_0 is the amplitude of the harmonic disturbance of pressure.

1.14.1 Root mean-squared values

Since we are considering harmonic waves the time averages over period T of the acoustic quantities, such as sound pressure and particle velocity, are zero. Therefore, instead of the usual average over time τ, defined for a time–varying quantity $f(t)$ as

$$\overline{f(t)} = \frac{1}{\tau} \int_0^\tau f(t) dt,$$

the root mean–squared (rms) average should be applied to acoustical quantities. The root mean–squared average over time τ of the time–dependent quantity $f(t)$ called also the root mean–squared value of $f(t)$ over time τ or simply the root mean square of $f(t)$ over time τ is specified by the equation,

$$f_{\rm rms} = \sqrt{\overline{f^2}} = \sqrt{\frac{1}{\tau} \int_0^\tau f^2(t)dt}. \tag{1.140}$$

Hence, the root mean square of the sound pressure $p_{\rm rms}$ over time τ (sometimes called the effective pressure) is given by:

$$p_{\rm rms} = \sqrt{\overline{p^2}} = \sqrt{\frac{1}{\tau} \int_0^\tau p^2(t)dt}. \tag{1.141}$$

Since for the plane, harmonic, freely progressing sound wave, we have

$$p = p_0 \cos(\omega t - kx) \tag{1.142}$$

and τ is assumed to be equal to T, the period of the harmonic wave, then

$$p_{\rm rms} = \frac{p_0}{\sqrt{2}}. \tag{1.143}$$

Figure 1.13 shows the sound pressure change in time as in equation (1.142) and the rms of the sound pressure.

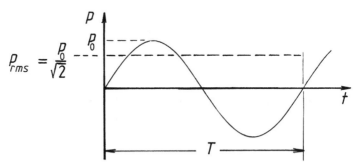

Figure 1.13 Root mean square of a harmonic function.

In most theoretical considerations of sound signals, as well as in many problems concerning, for example, random sound, a more universal definition of the root mean–squared average $f_{\rm rms}$ is applicable, namely,

$$f_{\text{rms}} = \sqrt{\overline{f^2}} = \sqrt{\lim_{\tau \to \infty} \frac{1}{\tau} \int_0^\tau f^2(t)dt}. \qquad (1.144)$$

When the time over which averaging takes place is not mentioned the root mean–squared value of the signal $f(t)$ should be understood either as defined by the formula (1.144) or, in the case of the harmonic sound wave, by the formula (1.140) with the additional assumption that $\tau = T$.

Since $\overline{p^2} = p_{\text{rms}}^2$ the time–averaged sound intensity for plane waves can be expressed in the form,

$$\vec{\overline{I}} = \frac{p_{\text{rms}}^2}{\rho_0 c_0}\vec{a}. \qquad (1.145)$$

1.15 Standing plane waves

In previous sections we considered freely progressing plane sound waves. Such waves are generated by a piston oscillating in an infinitely long tube and, as was shown in Figure 1.1, sound waves are travelling from the piston to infinity. Backward travelling waves do not occur in this case. If we attempt to measure the pressure in the tube and view the result on an oscilloscope as a trace of pressure against time, we would always see a similar trace with the same amplitude, whatever position we chose. The wave is referred to as plane, because the wave front is always perpendicular to the axis of the tube, i.e., to the direction of sound propagation.

On the other hand termination of the tube by a flat wall, perpendicular to the direction of propagation of the incident plane sound wave, results in a reflected wave. In general the amplitude and the phase of the reflected wave differ from those of the incident wave. These differences depend upon the wall properties. If the reflecting surface is 'rigid' or 'hard', the incident wave will be reflected with unchanged amplitude and with no phase change in terms of sound pressure. Both sound waves, namely the incident wave and the reflected one, interfere with each other and form a standing wave.

Let us assume a co–ordinate system, as shown in Figure 1.1, and consider the incident wave moving from the piston in the direction of increasing values of x. In terms of sound pressure the incident sound wave can be expressed, see Section 1.12.1., in the form

$$p_+ = p_0 \cos(\omega t - kx)$$

and the sound wave reflected from the rigid wall as

$$p_- = p_0 \cos(\omega t + kx).$$

Due to the interference of both waves the resulting sound pressure is

$$p = p_+ + p_- = p_0[\cos(\omega t - kx) + \cos(\omega t + kx)].$$

Finally,

$$p = [2p_0 \cos kx] \cos \omega t. \tag{1.146}$$

At any point x the sound pressure fluctuates harmonically in time except at the points at which the sound pressure is equal to zero for all time.

In terms of particle velocity the standing wave can be expressed, see Section 1.12, by:

$$u = u_+ + u_- = \frac{p_0}{\rho_0 c_0}[\cos(\omega t - kx) - \cos(\omega t + kx)]$$

$$= [2\frac{p_0}{\rho_0 c_0} \sin kx] \sin \omega t. \tag{1.147}$$

Using the expressions for p and u from equations (1.146) and (1.147) and putting them into equation (1.124), allows us to deduce that the time–averaged sound intensity is equal to zero for any value of x, because

$$\overline{\vec{I}} = \vec{a}\frac{p_0{}^2}{\rho_0 c_0} \sin(2kx)\frac{1}{T}\int\limits_0^T \sin(2\omega t)dt = 0.$$

The sound pressure, see equation (1.146), may be expressed also as follows:

$$p = [2p_0 \cos kx] \cos \omega t = \{\text{sgn}(\cos kx)2p_0|\cos kx|\} \cos \omega t \tag{1.148}$$

or

$$p = \text{sgn}(\cos kx)A \cos \omega t, \tag{1.149}$$

where $A = 2p_0|\cos kx|$ and can be treated as the amplitude of the harmonic oscillation in time. The dependence of amplitude A upon the distance x is shown in Figure 1.14.

Thus the amplitude of the resulting standing wave depends upon the distance x. For certain locations the amplitude of the sound pressure is

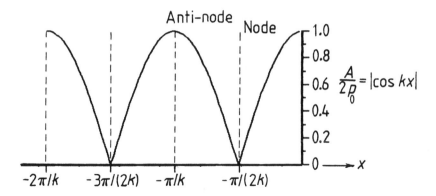

Figure 1.14 Standing wave; the dependence of the amplitude of the standing wave upon spatial co–ordinate x.

zero; these points are sound pressure *nodes*. Thus a sound pressure measuring instrument would detect at the nodes no evidence of a sound wave. At other locations there are sound pressure *anti–nodes* where the sound pressure amplitude is twice that of the incident sound wave. Standing waves of this kind can be formed in rooms where reflections occur from the walls, see also Chapter 4.

1.16 Three-dimensional sound wave equation

In the previous sections plane sound waves were considered. Propagation of plane sound waves is a one–dimensional problem, since the flow parameters at every point in the sound field depend in this case on time and only one cartesian co–ordinate; this type of flow is described by the one–dimensional wave equation (1.54). In general, however, sound can propagate in all directions in space and its propagation is governed by the three–dimensional wave equation. The procedure leading from the mass, momentum and energy conservation equations for the sound wave to the wave equation does not depend on the geometry of the problem; for three–dimensional sound waves the procedure is the same as for one–dimensional waves. However, in the case of a plane wave the particle displacement vector $\vec{d}$ has only one component, namely $\vec{d} = \xi \vec{e}_x$, where $\vec{e}_x$ is the unit vector for this case in the direction of the wave propagation. For a three–dimensional problem the particle displacement $\vec{d}$ has three components ξ, η, ζ in the cartesian co–ordinate system and can be written in the form: $\vec{d} = \xi \vec{e}_x + \eta \vec{e}_y + \zeta \vec{e}_z$, where $\vec{e}_x, \vec{e}_y, \vec{e}_z$ are unit vectors along the cartesian rectangular co–ordinate axes. Hence, for the three–dimensional problem, instead of equation (1.11), we have:

$$\frac{\rho - \rho_0}{\rho_0} = -\operatorname{div} \vec{d} \qquad (1.150)$$

or, using nabla notation (Hamilton operator) [24],

$$\frac{\rho - \rho_0}{\rho_0} = -\nabla \cdot \vec{d} \qquad (1.150a)$$

or,

$$\frac{\rho - \rho_0}{\rho_0} = -\left(\frac{\partial \xi}{\partial x} + \frac{\partial \eta}{\partial y} + \frac{\partial \zeta}{\partial z}\right), \qquad (1.150b)$$

which is the acoustic approximation of the mass conservation equation for three–dimensional sound wave motion.

Equation (1.51), which resulted from the continuity equation and the assumption that sound propagation is an isentropic process, in the case of a three–dimensional sound wave is

$$p = -\rho_0 c_o{}^2 \operatorname{div} \vec{d} \qquad (1.151)$$

or,

$$p = -\rho_0 c_o{}^2 \nabla \cdot \vec{d}. \qquad (1.151a)$$

Finally, the momentum conservation equation for the three–dimensional sound wave has the forms,

$$\rho_0 \frac{\partial^2 \vec{d}}{\partial t^2} = -\operatorname{grad} p, \qquad (1.152)$$

$$\rho_0 \frac{\partial^2 \vec{d}}{\partial t^2} = -\nabla p. \qquad (1.152a)$$

Combining equations (1.151a) and (1.152a) provides us with the three–dimensional wave equation,

$$\frac{\partial^2 \vec{d}}{\partial t^2} = c_0{}^2 \nabla^2 \vec{d}, \qquad (1.153)$$

where

$$\nabla^2 \vec{d} = \nabla^2 \xi \vec{e}_x + \nabla^2 \eta \vec{e}_y + \nabla^2 \zeta \vec{e}_z = \left(\frac{\partial^2 \xi}{\partial x^2} + \frac{\partial^2 \xi}{\partial y^2} + \frac{\partial^2 \xi}{\partial z^2}\right)\vec{e}_x$$

$$+(\frac{\partial^2\eta}{\partial x^2} + \frac{\partial^2\eta}{\partial y^2} + \frac{\partial^2\eta}{\partial z^2})\vec{e}_y + (\frac{\partial^2\zeta}{\partial x^2} + \frac{\partial^2\zeta}{\partial y^2} + \frac{\partial^2\zeta}{\partial z^2})\vec{e}_z. \tag{1.154}$$

The vector equation (1.153) is the sound wave equation in terms of the particle displacement. In terms of sound pressure the three–dimensional sound wave equation has the form,

$$\frac{\partial^2 p}{\partial t^2} = c_0{}^2\nabla^2 p, \tag{1.155}$$

where

$$\nabla^2 p = \frac{\partial^2 p}{\partial x^2} + \frac{\partial^2 p}{\partial y^2} + \frac{\partial^2 p}{\partial z^2}.$$

In terms of particle velocity $\vec{u}$ the three–dimensional sound wave equation is

$$\frac{\partial^2 \vec{u}}{\partial t^2} = c_0{}^2\nabla^2 \vec{u}. \tag{1.156}$$

In certain acoustic problems it is convenient to express the sound wave equation in spherical co–ordinates (r, θ, ψ) or cylindrical co–ordinates (σ, ψ, z), see Figure 1.15.

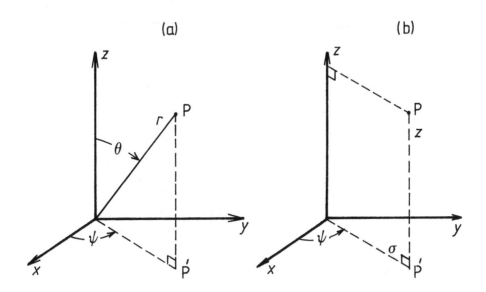

Figure 1.15 Relation between (a) spherical co–ordinates (r, θ, ψ) (b) cylindrical co–ordinates (σ, ψ, z) and rectangular cartesian co-ordinates.

In spherical co–ordinates the sound wave equation (1.155) is written as [6,24]

$$\frac{\partial^2 p}{\partial t^2} = c_0{}^2 \left[\frac{1}{r^2} \frac{\partial}{\partial r} (r^2 \frac{\partial p}{\partial r}) + \frac{1}{r^2} \frac{1}{\sin \theta} \frac{\partial}{\partial \theta} (\sin \theta \frac{\partial p}{\partial \theta}) + \frac{1}{r^2 \sin^2 \theta} \frac{\partial^2 p}{\partial \psi^2} \right].$$
(1.157)

The same sound wave equation expressed in cylindrical co–ordinates is [24]

$$\frac{\partial^2 p}{\partial t^2} = c_0{}^2 \left[\frac{1}{\sigma} \frac{\partial}{\partial \sigma} (\sigma \frac{\partial p}{\partial \sigma}) + \frac{1}{\sigma^2} \frac{\partial^2 p}{\partial \psi^2} + \frac{\partial^2 p}{\partial z^2} \right].$$
(1.158)

1.16.1 Spherical sound waves from a point source

The classification of three–dimensional waves as spherical or cylindrical depends upon the symmetry of the wave fronts or, in other words, upon the symmetry of their surfaces of constant phase. The symmetry, or lack of symmetry, of sound waves is a result of the characteristics of the sound source and the properties of the obstacles with which the sound waves interact. The surfaces of constant phase (the wave fronts) of spherical waves are spheres, whatever the directional distribution of the sound. In general for spherical waves the amplitudes of sound pressure, particle velocity and particle displacement depend upon the distance from the sound source, as well as upon the direction; hence the surfaces of equal amplitude of, say, sound pressure are not usually spheres. Only in the case of *spherical waves of zero order*, see for details [17], are the surfaces of constant phase and the surfaces of constant amplitude of sound pressure concentric spheres. This type of spherical wave originates from a *point sound source* or *monopole*; it can also be generated by a pulsating sphere (see Section 1.17 and Appendix C). The point sound source radiates sound equally in all directions and the amplitude of the sound pressure in such a spherical wave changes in inverse proportion to the distance from a certain central point, which is the location of the point source. In the sound field from the point sound source parameters such as sound pressure, particle velocity and particle displacement, each of which fulfils the one–dimensional wave equation, depend only upon time and the distance r from the point source, see Figure 1.16.

For spherical waves generated by the point source the sound wave equation (1.157) is reduced to:

$$\frac{\partial^2 p}{\partial t^2} = c_0{}^2 \frac{1}{r^2} \frac{\partial}{\partial r} \left(r^2 \frac{\partial p}{\partial r} \right)$$
(1.159)

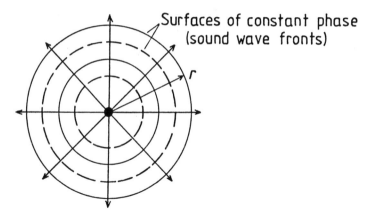

Figure 1.16 Propagation of spherical waves from a point source.

$$\frac{\partial^2 p}{\partial t^2} = c_0{}^2 \left(\frac{2}{r} \frac{\partial p}{\partial r} + \frac{\partial^2 p}{\partial r^2} \right) \tag{1.160}$$

and finally

$$\frac{\partial^2 (pr)}{\partial t^2} = c_0{}^2 \frac{\partial^2 (pr)}{\partial r^2}. \tag{1.161}$$

As with equation (1.54) for plane waves, the sound wave equation (1.161) has a general solution in the form,

$$pr = f_+(r - c_0 t) + f_-(r + c_0 t)$$
$$p = \frac{f_+(r - c_0 t)}{r} + \frac{f_-(r + c_0 t)}{r}, \tag{1.162}$$

where the first term represents an outgoing wave of arbitrary form (a wave diverging with velocity c_0 from the point $r = 0$) and the second term represents an incoming wave, or a wave converging on the point $r = 0$. The general solution (1.162) is finite everywhere except at the point $r = 0$. In physical problems the finite size of the sound source should be taken into account, hence in equation (1.162) only the solution around the central point $r = 0$ is considered.

If the sound is generated in a free field the term representing the incoming wave should be eliminated; the solution becomes

$$p = \frac{f_+(r - c_0 t)}{r}. \tag{1.163}$$

Spherical pressure waves from the point source do not change their shape as they move outwards, but their amplitude is inversely proportional to distance r; in contrast, plane waves in an ideal fluid maintain a constant amplitude.

For spherical waves from the point source the equation (1.156) is reduced to:

$$\frac{\partial^2 u_r}{\partial t^2} = c_0^2 (\nabla^2 \vec{u})_r \qquad (1.164)$$

where $\vec{u} = u_r \vec{e}_r$ and $(\nabla^2 \vec{u})_r$ denotes the radial component of the vector $\nabla^2 \vec{u}$. Equation (1.164) can also be presented as

$$\frac{\partial^2 u_r}{\partial t^2} = c_0^2 \left(\nabla^2 u_r - 2\frac{u_r}{r^2} \right) \qquad (1.165)$$

$$\frac{\partial^2 u_r}{\partial t^2} = c_0^2 \left(\frac{\partial^2 u_r}{\partial r^2} + \frac{2}{r}\frac{\partial u_r}{\partial r} - 2\frac{u_r}{r^2} \right) \qquad (1.166)$$

or finally

$$\frac{\partial^2 (u_r r)}{\partial t^2} = c_0^2 \left\{ \frac{\partial^2 (u_r r)}{\partial r^2} - 2\frac{u_r r}{r^2} \right\}. \qquad (1.167)$$

It can be deduced from equation (1.165) that the particle velocity components in spherical co-ordinates do not satisfy the sound wave equation in the way that sound pressure does; compare also (1.161) with (1.167).

1.16.2 Harmonic spherical sound waves from a point source

In the class of real harmonic functions the solution (1.163) is

$$p = \frac{A}{r} \cos(\omega t - \vec{k}\cdot\vec{r} + \beta), \qquad (1.168)$$

where A is a real constant and β is a phase angle.

Using complex (phasor) notation the solution (1.168) can be presented as

$$p = \mathrm{Re}\,\{\tilde{\mathbf{p}}\} = \mathrm{Re}\left\{ \frac{A}{r} e^{i(\omega t - \vec{k}\cdot\vec{r})} \right\}, \qquad (1.169)$$

where $A = |\mathbf{A}|$, $\vec{k}$ is the wave vector and $\vec{k} = k\vec{a} = (\omega/c_0)\vec{a}$; $\vec{a}$ is a unit vector along the direction of wave propagation, and for spherical waves, emitted by the point source, $\vec{a} = \vec{r}/|\vec{r}|$. The magnitude of the wave vector $k = |\vec{k}| = \omega/c_0 = 2\pi/\lambda$ is the wave number. The vector $\vec{r}$ is a position vector. The complex harmonic solution of the wave equation (1.107) for spherical waves generated by the point source is in terms of the sound pressure phasor,

$$\widetilde{\mathbf{p}}(r,\,t) = \frac{\mathbf{A}}{r}e^{i(\omega t - \vec{k}\cdot\vec{r})} = \frac{A}{r}e^{i(\omega t - \vec{k}\cdot\vec{r} + \beta)}, \tag{1.170}$$

where $\mathbf{A}$ is a complex constant (complex vector).

Let us find the relationship between phasors $\widetilde{\mathbf{u}}_r$ and $\widetilde{\mathbf{p}}$ which enables us to determine the difference in phase between the particle velocity and the sound pressure. Applying the momentum conservation equation (1.15) in the complex notation, namely,

$$\rho_0 \frac{\partial \widetilde{\mathbf{u}}_r}{\partial t} = -\frac{\partial \widetilde{\mathbf{p}}}{\partial r} \tag{1.171}$$

and putting the expression for $\widetilde{\mathbf{p}}$ from equation (1.170) into equation (1.171), we obtain

$$\frac{\partial \widetilde{\mathbf{u}}_r}{\partial t} = \frac{\widetilde{\mathbf{p}}}{\rho_0}\left(\frac{1}{r} + ik\right). \tag{1.172}$$

Finally, after integration of equation (1.172) with respect to time we have

$$\widetilde{\mathbf{u}}_r = \frac{\widetilde{\mathbf{p}}}{\rho_0 \omega r}\left(\frac{1}{i} + kr\right) \tag{1.173}$$

$$\widetilde{\mathbf{u}}_r = \frac{\widetilde{\mathbf{p}}}{\rho_0 c_0 kr}\,(kr - i) \tag{1.174}$$

$$\widetilde{\mathbf{u}}_r = \frac{\widetilde{\mathbf{p}}}{\rho_0 c_0} - \frac{i\widetilde{\mathbf{p}}}{\rho_0 c_0 kr}. \tag{1.175}$$

The complex vector (complex quantity) $kr - i$ may be presented in exponential form as follows:

$$kr - i = \left(1 + k^2 r^2\right)^{1/2} e^{i\phi}, \tag{1.176}$$

where

$$\phi = \arccos\left\{\frac{kr}{(1 + k^2 r^2)^{1/2}}\right\} = \arctan\left\{\frac{-1}{(kr)}\right\}. \tag{1.177}$$

Hence, the particle velocity phasor for spherical waves from a point source can be also presented in the forms,

$$\widetilde{\mathbf{u}}_r = \widetilde{\mathbf{p}}\frac{(1 + k^2 r^2)^{1/2}}{\rho_0 c_0 kr}e^{i\phi} \tag{1.178}$$

$$\widetilde{\mathbf{u}}_r = \frac{\widetilde{\mathbf{p}}}{\rho_0 c_0 \cos\phi}e^{i\phi}, \tag{1.179}$$

where ϕ is the angle by which the particle velocity phasor leads in phase the sound pressure phasor. Since $\tilde{\mathbf{u}}_r = \partial\tilde{\mathbf{d}}/\partial t$ then

$$\tilde{\mathbf{d}} = -\frac{i\tilde{\mathbf{p}}}{\rho_0 c_0 \omega \cos\phi} e^{i\phi}. \qquad (1.180)$$

Finally,

$$\tilde{\mathbf{d}} = \frac{\tilde{\mathbf{p}} e^{i(\phi - \pi/2)}}{\rho_0 c_0^2 k \cos\phi}. \qquad (1.181)$$

On the other hand taking into account the equation (1.173), we can present the expression for the particle displacement phasor $\tilde{\mathbf{d}}$ as

$$\tilde{\mathbf{d}} = -\frac{\tilde{\mathbf{p}}}{\rho_0 r \omega^2}(1 + ikr) \qquad (1.182)$$

$$\tilde{\mathbf{d}} = \frac{\tilde{\mathbf{p}}(1 + k^2 r^2)^{1/2}}{\rho_0 c_0^2 k^2 r} e^{i\vartheta}, \qquad (1.183)$$

where

$$\vartheta = \arctan(kr) = \arccos\left\{\frac{-1}{(1 + k^2 r^2)^{1/2}}\right\}. \qquad (1.184)$$

The angle ϑ, the difference in phase between the particle displacement and sound pressure phasors is equal to $\phi - \pi/2$. The results indicated by formulae (1.179) and (1.183) are illustrated in Figure 1.17.

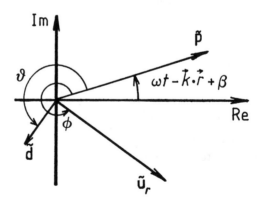

Figure 1.17 The relative locations of phasors $\tilde{\mathbf{p}}$, $\tilde{\mathbf{u}}_r$ and $\tilde{\mathbf{d}}$ for a spherical wave generated by a point sound source.

For plane waves freely progressing in the direction of the positive values of x both the sound pressure and the particle velocity are in phase. They lead the particle displacement by $\pi/2$ radians.

Figure 1.17 illustrates the fact that, in general, for spherical waves from a point sound source the sound pressure, particle velocity and particle displacement are all out of phase. However, the angles ϕ and ϑ are functions of $\vec{k} \cdot \vec{r} = kr$, where $k = \omega/c_0$ and r is the distance of the considered fluid particle from the centre of the sound source. Hence, differences between the phases of the sound pressure and the particle velocity and between the phases of the sound pressure and the particle displacement depend upon values of kr. It should be noted that the difference between the phases of the particle velocity and the particle displacement maintains the same value of $\pi/2$ radians.

The particle velocity, see equation (1.175), consists of two terms, namely, the *active* component of the particle velocity, which is always in phase with the sound pressure, and the *reactive* component of the particle velocity, which lags in phase the sound pressure by $\pi/2$ radians.

The sound pressure amplitude of a spherical wave from the point source, see equation (1.168), varies proportionally to the inverse of r. However, the particle velocity amplitude u_0 changes with r in a more complicated manner; namely from equation (1.178) we have

$$u_0(r) = \frac{A}{\rho_0 c_0} \frac{(1 + k^2 r^2)^{1/2}}{kr^2}. \tag{1.185}$$

1.16.3 Intensity, sound power and energy density of spherical waves

The vector of the instantaneous sound intensity, defined in Section 1.13, can be presented for spherical waves from a point source as

$$\vec{I} = p\vec{u} = [\mathrm{Re}\,(\widetilde{\mathbf{p}})\,\mathrm{Re}\,(\widetilde{\mathbf{u}})]\,\vec{a}, \tag{1.186}$$

where $\vec{a} = -\nabla \Phi_0 / |\nabla \Phi_0|$, see equation (B.5). For the spherical wave from a point source $\vec{a} = \vec{k}/|\vec{k}| = \vec{e}_r = \vec{r}/|\vec{r}|$ and $\widetilde{\mathbf{u}} = \widetilde{\mathbf{u}}_r$, where $\vec{r}$ is the position vector (radius vector). The time–averaged acoustic intensity is

$$\overline{\vec{I}} = \overline{[\mathrm{Re}(\widetilde{\mathbf{p}})\mathrm{Re}(\widetilde{\mathbf{u}})]}\vec{a}. \tag{1.187}$$

The magnitude of the time–averaged acoustic intensity $\overline{I}$ may also be determined in a much simpler manner than by (1.187). Namely, using complex vector formalism [8,25,26], we can present $\overline{I}$, see equation (1.187) in the form,

$$\overline{I} = \text{Re}(\widetilde{\mathbf{p}})\text{Re}(\widetilde{\mathbf{u}}) = \frac{1}{2}\text{Re}\left(\widetilde{\mathbf{p}}^*\widetilde{\mathbf{u}}\right) = \frac{1}{2}\text{Re}\left(\widetilde{\mathbf{p}}\widetilde{\mathbf{u}}^*\right),\qquad(1.188)$$

where $\widetilde{\mathbf{p}}^*$ and $\widetilde{\mathbf{u}}^*$ are the complex conjugates of $\widetilde{\mathbf{p}}$ and $\widetilde{\mathbf{u}}$, respectively.

Taking into account equations (1.170) and (1.179), we obtain for the product of the two phasors the expressions,

$$\widetilde{\mathbf{p}}^*\widetilde{\mathbf{u}} = \frac{A^2 e^{i\phi}}{\rho_0 c_0 r^2 \cos\phi}\qquad(1.189)$$

$$\widetilde{\mathbf{p}}^*\widetilde{\mathbf{u}} = \frac{A^2}{\rho_0 c_0 r^2 \cos\phi}(\cos\phi + i\sin\phi),\qquad(1.190)$$

where A is a real constant. Hence, putting the expression for $\widetilde{\mathbf{p}}^*\widetilde{\mathbf{u}}$ into equation (1.188), one obtains for the harmonic spherical wave which originated from the point source:

$$\overline{I} = \frac{A^2}{2\rho_0 c_0 r^2},\qquad(1.191)$$

or, finally,

$$\overline{\overline{I}} = \frac{p_0{}^2(r)}{2\rho_0 c_0}\vec{a},\qquad(1.192)$$

where $p_0(r) = A/r$, denotes the sound pressure amplitude of the spherical wave from the point sound source.

As was already mentioned, the sound pressure amplitude of the considered undamped spherical wave at a certain point in space is inversely proportional to the distance of this point from the point source. In an undamped plane wave the amplitudes of sound pressure and particle velocity are constant. However, the relation between the time–averaged intensity and the sound pressure amplitude is the same for a spherical wave from a point source, as well as for a plane wave, see equation (1.139) and compare with equation (1.192). Since also the relation (1.143) holds, the formula (1.145) is valid for both spherical (from a point source), as well as plane, sound waves.

Finally, considering the definition of the *sound power* crossing a certain surface, equation (1.129), we can conclude that the time-averaged sound energy flux through a closed spherical surface of radius r surrounding a point sound source (sound power W) is independent of the radius of the surface, as

$$W = 4\pi r^2 \vec{\overline{I}} \cdot \frac{\vec{r}}{|\vec{r}|} = 4\pi r^2 \overline{I} \qquad (1.193)$$

$$W = 2\pi r^2 \frac{p_0{}^2}{\rho_0 c_0} = \frac{2\pi A^2}{\rho_0 c_0}. \qquad (1.194)$$

The *energy density of the sound wave*, see equation (1.45), Appendix A and Section 1.14, can be presented in the form,

$$E_a = \frac{1}{2}\rho_0 u^2 + \frac{1}{2}\frac{p^2}{\rho_0 c_0{}^2}. \qquad (1.195)$$

Since $u = \mathrm{Re}(\tilde{u})$ and $p = \mathrm{Re}(\tilde{p})$ then the time–averaged quantities $\overline{u^2}$ and $\overline{p^2}$ are

$$\overline{u^2} = \overline{\mathrm{Re}(\tilde{u})\mathrm{Re}(\tilde{u})} = \frac{1}{2}\tilde{u}^*\tilde{u} \qquad (1.196)$$

and

$$\overline{p^2} = \overline{\mathrm{Re}(\tilde{p})\mathrm{Re}(\tilde{p})} = \frac{1}{2}\tilde{p}^*\tilde{p}, \qquad (1.197)$$

respectively. Finally, since for spherical sound waves generated by the point source, see equations (1.170) and (1.179),

$$\overline{u^2} = \frac{A^2}{2\rho_0{}^2 c_0{}^2 r^2 \cos^2 \phi} \qquad (1.198)$$

and

$$\overline{p^2} = \frac{1}{2}\frac{A}{r^2}$$

the *time–averaged sound energy density* $\overline{E_a}$ is given by:

$$\overline{E_a} = \frac{A^2}{4\rho_0 c_0{}^2 r^2}\left(\frac{1}{\cos^2 \phi} + 1\right) \qquad (1.199)$$

and, since ϕ is obtained from equation (1.177), equation (1.199) may be written as

$$\overline{E_a} = \frac{p_0{}^2}{2\rho_0 c_0{}^2}\left(1 + \frac{1}{2k^2 r^2}\right) \qquad (1.200)$$

or, see equation (1.192), as

$$\overline{E_a} = \frac{\overline{I}}{c_0}\left(1 + \frac{1}{2k^2r^2}\right).$$

(1.201)

Finally, from equation (1.201), we obtain:

$$\overline{\vec{I}} = \frac{c_0\overline{E_a}}{1 + 1/(2k^2r^2)}\vec{a}.$$

(1.202)

The relation between the time–averaged sound intensity and the time–averaged sound energy density (1.202) is different from the analogical relation (1.136), fulfilled by the time-averaged sound intensity vector and the time–averaged sound energy density for a plane sound wave. However, for the case when $kr \gg 1$ both relations are the same. More details about sound intensity are given in Appendix B.

1.17 Simple and point sound sources

The sound radiation from a point source has no directional characteristics, see Section 1.16.1. Also, the sound field generated by a sound source of a finite size can possess in a certain situation all the attributes of a sound field which originated from a point source, despite the size or irregular shape of the sound source. Such a sound field is the simplest sound field generated by a physical object. All the sound field parameters fulfil the one–dimensional wave equations and they are functions, as in the case of the sound field from a point source, only of time t and r, the distance from the central point of the sound source. This type of sound source is called a *simple source*.

    Sound waves generated by a simple source are spherical, because their surfaces of constant phase are concentric spheres. The surfaces of constant amplitude are also in this case concentric spheres. A simple sound source can be realised by a periodically and radially *pulsating sphere*, the radius of which changes equally in all directions during the pulsation. In other words all points on the sphere's surface vibrate in phase, see also Appendix C. Additionally, a, the radius of the pulsating sphere, must be much smaller than λ, the wavelength of sound generated by the sphere. The last condition denotes that the sound source is acoustically *compact*, i.e., $a \ll c_0/\omega = \lambda/(2\pi)$ or $ka \ll 1$.

    Simple sound sources are good approximations for a certain class of pulsating sources regardless of the different shapes and sizes of these pulsating radiators. Pulsating sound sources of different shapes and sizes but of equal strength, which additionally fulfil the compactness condition,

generate identical sound fields at distances which are much greater than
the dimensions of the sources.

The *sound source strength* $\mathbf{Q}$ of any sound source with a harmonically
time-varying volume, whose surface points do not necessarily vibrate in
phase, is in general a complex quantity and is defined by the equation
given below:

$$\mathbf{Q}e^{i\omega t} = \int_S \vec{\tilde{u}} \cdot d\vec{S} \tag{1.203}$$

or,

$$\mathbf{Q}e^{i\omega t} = \int_S e^{i[\omega t + \Phi_0(\vec{r})]} \vec{U}_0(\vec{r}) \cdot d\vec{S}, \tag{1.204}$$

where $\vec{\tilde{u}}$ is the velocity vector–phasor of the vibrating surface element $d\vec{S}$
and $\int_S \vec{\tilde{u}} \cdot d\vec{S}$ denotes the phasor of the *volume velocity* through the sound
source surface S. Note that $\vec{U}_0$ is the amplitude of the velocity (real)
vector of the vibrating surface element $d\vec{S}$ and $\vec{U}_0 \cdot d\vec{S}$ is the amplitude of
the volume velocity through the infinitesimal surface $d\vec{S}$.

For the sound sources whose surface points vibrate harmonically in
phase the equation (1.204) can be simplified to

$$\mathbf{Q}e^{-i\Phi_{os}} = \int_S \vec{U}_0 \cdot d\vec{S}, \tag{1.205}$$

where Φ_{os} denotes the value of the initial phase, see Section 1.12.1, which
is constant for all points located at the vibrating surface S and $\int_S \vec{U}_0 \cdot d\vec{S}$
is the maximum rate of volume flow at the surface S, i.e., the amplitude
of the volume velocity across the source surface. Since

$$Q = |\mathbf{Q}| = \mathbf{Q}e^{-i\Phi_{os}} = \int_S \vec{U}_0 \cdot d\vec{S}$$

the name 'sound source strength' can be associated with the magnitude
of the complex sound source strength which in this case is equal to the
amplitude of the volume velocity across the source surface.

The sound source strength of the pulsating sphere (whose surface points vibrate radially in phase) is

$$Q = 4\pi a^2 U_0, \qquad (1.206)$$

where a is the radius of the sphere and U_0 is the normal component of the amplitude of the velocity vector on the surface of the sphere.

A *point sound source* is a mathematical idealisation of a simple sound source. It is a sound source concentrated at a point. Its complex strength $\mathbf{Q}_p$ is defined as the maximum rate of volume flow through an infinitesimal sphere surrounding the point source:

$$\mathbf{Q}_p = \lim_{r \to 0} \left(4\pi r^2 \mathbf{u}_{r0} \right), \qquad (1.207)$$

where $\tilde{\mathbf{u}}_r = \mathbf{u}_{r0} e^{i\omega t}$, see equation (1.179) for specification of $\tilde{\mathbf{u}}_r$.

Taking into account the formulae (1.175) and (1.170), describing a sound field from a point source in terms of the particle velocity and sound pressure, respectively, we have

$$\mathbf{Q}_p = \lim_{r \to 0} \left\{ 4\pi r^2 \frac{\mathbf{A}}{r} e^{-i\vec{k}\cdot\vec{r}} \left(\frac{1}{\rho_0 c_0} - \frac{i}{\rho_0 c_0 k r} \right) \right\} = \frac{4\pi \mathbf{A}}{i\omega\rho_0} \qquad (1.208)$$

and from equation (1.208) we obtain:

$$\mathbf{A} = \frac{i\omega\rho_0 \mathbf{Q}_p}{4\pi}. \qquad (1.209)$$

Hence the sound field from a point source of the strength $\mathbf{Q}_p$ may be described in terms of sound pressure as follows:

$$\tilde{\mathbf{p}} = \frac{i\omega\rho_0 \mathbf{Q}_p}{4\pi r} e^{i(\omega t - \vec{k}\cdot\vec{r})}. \qquad (1.210)$$

It should be mentioned that the sound field from a simple source is specified also by the formula (1.210), but in this case $\mathbf{Q}_p$ is real and equal to Q_s, the strength of the simple source: $\mathbf{Q}_p = Q_p = Q_s$, see Appendix C.

Taking into account equations (1.210) and (1.175), we obtain the time–averaged sound intensity at a distance r from the point sound source:

$$\bar{I} = \frac{1}{2} \mathrm{Re} \left(\tilde{\mathbf{p}}\tilde{\mathbf{u}}^* \right) = \frac{\omega^2}{32\pi^2 r^2 c_0} \frac{\rho_0 Q_p^2}{}, \qquad (1.211)$$

where $Q_p = |\mathbf{Q}_p|$.

Hence, equation (1.211) leads to the expression for the sound power W_p of a point source:

$$W_p = 4\pi r^2 \overline{I} = \frac{\rho_0 \omega^2 Q_p{}^2}{8\pi c_0}. \tag{1.212}$$

In most published works simple and point sources have interchangeable meaning, see [8,11], compare also with [14,25,31]. In this book a simple source is treated as a physical object while a point source has only a mathematical meaning.

1.17.1 Radiation from a hemispherical sound source

A simple sound source mounted in an infinite baffle and radiating sound into space on one side of the infinite plane only forms a *simple hemispherical sound source*, see Figure 1.18.

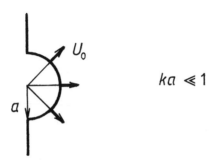

Figure 1.18 Simple hemispherical sound source.

The strength of a pulsating sound source of hemispherical shape of radius a is

$$Q_H = 2\pi a^2 U_0, \tag{1.213}$$

where U_0 is the magnitude of the amplitude of the velocity on the hemispherical surface of the sound source. Formula (1.213) results from the definition of sound source strength as the amplitude of the volume velocity across the source surface. It implies that the strength of a hemispherical source is only half as great as that of a similar spherical source characterised by the same radius and the same amplitude of surface velocity. For a *point hemispherical sound source*, see also (1.207), we have

$$\mathbf{Q}_{pH} = \lim_{r \to 0} \left(2\pi r^2 \mathbf{u}_{r0} \right) \tag{1.214}$$

and combining formulae (1.170) and (1.175) with (1.214), we obtain for the sound pressure in the field of the point hemispherical sound source:

$$\widetilde{\mathbf{p}} = \frac{i\omega\rho_0\mathbf{Q}_{pH}}{2\pi r}e^{i(\omega t - \vec{k}\cdot\vec{r})}. \tag{1.215}$$

The amplitude of sound pressure, see equation (1.215), is twice as great as that from a point sound source of the same strength $\mathbf{Q}_p = \mathbf{Q}_{pH}$. The time–averaged sound intensity at a certain distance r from a point hemispherical sound source of strength $\mathbf{Q}_{pH}$ is

$$\overline{I} = \frac{\omega^2\rho_0 Q_{pH}^2}{8\pi^2 r^2 c_0} \tag{1.216}$$

and the sound power W_{pH} radiated by this sound source is

$$W_{pH} = \frac{\omega^2\rho_0 Q_{pH}^2}{4\pi c_0}. \tag{1.217}$$

1.18 Impedance

The concept of an impedance is widely used in acoustics and mechanics, see also Appendix C. In general impedance is a complex quantity, defined as a ratio of two phasors, namely the ratio of the pressure or force phasor to the velocity or volume velocity phasor.

There are several kinds of impedance applicable in acoustics. The *acoustic impedance* at a surface S (of a medium acting on surface S) is the ratio of a sound pressure phasor averaged over the surface to the volume velocity phasor through the surface S, namely,

$$\mathbf{Z} = \frac{1}{S}\int_S \widetilde{\mathbf{p}}\,dS \bigg/ \int_S \vec{\widetilde{\mathbf{u}}}\cdot d\vec{S}. \tag{1.218}$$

The surface S at which the acoustic impedance is considered does not need necessarily to be a physical surface, but can be a hypothetical surface in a medium, where sound is propagated.

The *specific acoustic impedance* $\mathbf{z}$ is the ratio of the sound pressure phasor at a certain point of a sound field to the particle velocity phasor at that point. Hence,

$$\mathbf{z} = \frac{\widetilde{\mathbf{p}}}{\widetilde{\mathbf{u}}}. \tag{1.219}$$

When

$$\tilde{p} = \frac{1}{S} \int_S \tilde{p} \, dS$$

and

$$\tilde{u} = \frac{1}{S} \int_S \vec{\tilde{u}} \cdot d\vec{S}$$

the specific acoustic impedance is related to the acoustic impedance, see equation (1.218), by the expression

$$z = SZ. \tag{1.220}$$

The specific acoustic impedance depends on the medium in which sound propagates, as well as on the type of sound field. In the case of progressive plane waves the specific acoustic impedance is a real quantity, see equations (1.104) and (1.105).

For spherical waves from a point sound source, see equation (1.179), we have that

$$z = \frac{\tilde{p}}{\tilde{u}} = \rho_0 c_0 \cos \phi e^{-i\phi} \tag{1.221}$$

$$= \rho_0 c_0 \left(\cos^2 \phi - i \sin \phi \cos \phi \right)$$

$$= \rho_0 c_0 \left(\frac{k^2 r^2}{1 + k^2 r^2} + i \frac{kr}{1 + k^2 r^2} \right) \tag{1.222}$$

$$= r_z + i x_z.$$

For very small values of kr, $kr \ll 1$ (i.e., for low frequencies or small distances from the sound source) both terms of the specific acoustic impedance are approaching zero. When $kr \gg 1$ the specific acoustic resistance r_z approaches a value of $\rho_0 c_0$, but the specific acoustic reactance is close to zero, see Figure 1.19. When $kr = 1$ both terms are equal to $(1/2)\rho_0 c_0$; for the specific acoustic reactance this value is the maximum.

The *mechanical impedance* Z_m is a ratio of the phasor of the driving force acting on a mechanical or mechanical–acoustic system to the phasor of the resulting velocity of the system:

$$Z_m = \frac{\tilde{F}}{\tilde{u}}. \tag{1.223}$$

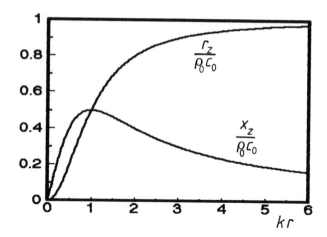

Figure 1.19 Resistive r_z and reactive x_z components of the specific acoustic impedance for a spherical wave from a point source, shown as a function of kr.

For a vibrating mechanical–acoustic system the resultant mechanical impedance $\mathbf{Z}_m$ is the sum of the mechanical impedance of the source $\mathbf{Z}_{ms}$ and the radiation impedance of the medium, $\mathbf{Z}_r$. The radiation impedance appears as a consequence of the reaction of the medium on sound radiation, see Appendix C.

1.19 Far and near acoustic fields

The concept of a far (distant) and a near acoustic field can be introduced even for spherical waves generated by a point source, see equations (1.170) and (1.175). The region of the sound field in which $kr \gg 1$ is called the *far* acoustic field and the region in which $kr \ll 1$ is the *near* acoustic field. If $kr \gg 1$, the formula (1.175) leads to the relation,

$$\tilde{\mathbf{u}}_r = \frac{\tilde{\mathbf{p}}}{\rho_0 c_0},\qquad(1.224)$$

which indicates that the particle velocity and the sound pressure are in phase, $\phi = 0$, see equation (1.177). When $kr \gg 1$, ϑ is approaching $3\pi/2$ radians, see equation (1.184) and Figure 1.20, hence the sound pressure leads the particle displacement by $\pi/2$ radians.

If $kr \ll 1$ the value of ϕ is close to $3\pi/2$ radians, i.e., the sound pressure leads the particle velocity almost by $\pi/2$ radians. The phase

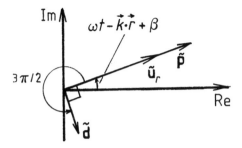

Figure 1.20 The relative locations of phasors $\tilde{p}$, $\tilde{u}_r$ and $\tilde{d}$ in the complex plane for the case when kr approaches ∞ ($kr \gg 1$).

angle ϑ approaches π radians, hence the particle displacement lags the sound pressure by π radians, see Figure 1.21.

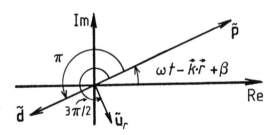

Figure 1.21 The relative locations of phasors $\tilde{p}$, $\tilde{u}_r$ and $\tilde{d}$ in the complex plane for the near field when kr approaches 0 ($kr \ll 1$).

In the far acoustic field ($kr \gg 1$), the *active* component of the particle velocity dominates the reactive. In the near field ($kr \ll 1$) there is an opposite situation, and the *reactive* component of the particle velocity is the most significant. In the far acoustic field generated by a point source the relation between the sound pressure and particle velocity phasors, see equation (1.224), is the same as for a plane, progressive wave, travelling away from the source. However, for spherical waves the amplitudes of both the sound pressure and the particle velocity vary in inverse proportion to r, the distance from the point source, whereas for plane waves these amplitudes are unchanged with distance. In the near field, where the sound pressure phasor and particle velocity phasor are nearly in quadrature, the amplitude of sound pressure decreases as $1/r$, but the amplitude of the particle velocity diminishes proportionally to $1/r^2$, see equations (1.168) and (1.185). The amplitude of the (instantaneous) active intensity (see Appendix B) changes with the distance from the point sound source as $1/r^2$, but the amplitude of the reactive intensity as $1/r^3$, hence the reactive intensity is significant only in the near field.

It should be also noted that in the far field the phase velocity of the particle velocity wave $c_{u,ph}$ is equal to the phase velocity for the sound pressure wave $c_{p,ph}$, and to the sound velocity, i.e.,

$$c_{u,ph} = c_{p,ph} = c_0. \tag{1.225}$$

Since

$$\Phi_u(r,t) = \Phi_u(r + \Delta r, t + \Delta t) = \text{const.} \tag{1.226}$$

defines at different instants the surfaces of constant phase for the particle velocity wave, the phase velocity of the particle velocity wave is

$$c_{u,ph} = \frac{dr}{dt} = \frac{\partial \Phi_u}{\partial t} \Big/ \frac{\partial \Phi_u}{\partial r}. \tag{1.227}$$

Remembering that

$$\Phi_u = \Phi_p + \phi = \omega t - kr + \phi, \tag{1.228}$$

where Φ_p is the phase of the sound pressure wave and ϕ is defined by equation (1.177), we have:

$$c_{u,ph} = -\frac{\omega}{-k + \partial \phi / \partial r} \tag{1.229}$$

$$= c_0 \left(1 + \frac{1}{k^2 r^2} \right). \tag{1.230}$$

Hence, for $kr \gg 1$ the equation (1.225) is fulfilled.

In the near field, however, the phase velocity of the particle velocity wave can be much greater than the sound velocity, see equation (1.230). Additionally, from equation (1.175) we can deduce that in the near field of the point source $|\vec{u}_{ro}| \gg p_0/(\rho_0 c_0)$, where $\vec{u}_{ro}$ and p_0 denote (in complex vector formalism) the amplitudes of the particle velocity and sound pressure, respectively.

The sound field from the compact, pulsating sound source of strength Q for $r \gg D$, where D is the dimension of the compact source, is the sound field from the simple source. This sound field possesses all the attributes of the field from the point sound source with real strength $Q_p = Q$, see Appendix C. However, the concept of the simple source excludes from consideration the sound field region situated near the simple source at distances comparable with the dimension of the source D.

The reader should also be aware of a possible different definition of the far and near fields [32].

When sound sources of finite dimensions and various irregular shapes are considered, the specification of only two characteristic regions of the sound field seems to be insufficient. Hence, in general the sound field generated by a sound source in a free field may be divided into three zones, namely, the hydrodynamic near field, the geometric near field and the geometric far field [33].

1.19.1 Hydrodynamic near field

The hydrodynamic near field is adjacent to the vibrating surface of the sound source. In this region, much smaller than one sound wavelength, the fluid behaves like an incompressible medium. Analysing, for example, the formulae (C.6) and (C.9) from Appendix C, we can deduce that in the close vicinity of the sound source, in this case the pulsating sphere, the solution has not the form of a progressive wave. The sound pressure is in phase at all points of the hydrodynamic near field, see equation (C.9). Hence, the motion in this region has the character of an incompressible fluid motion, and the use of the term hydrodynamic near field is justified for the region which is closest to the sound source. In the hydrodynamic near field the phasors of sound pressure and of particle velocity are in general out of phase (nearly in quadrature) and $|\vec{u}_0| \gg p_0/(\rho_0 c_0)$, where $\vec{u}_0$ and p_0 are the amplitudes of the particle velocity and sound pressure phasors, respectively. In addition the reactive component of the particle velocity dominates the active, see Appendix B.

The region adjacent to the hydrodynamic near field is called the geometric near field or Fresnel near field.

1.19.2 Geometric (Fresnel) near field

The angle which the sound source subtends at the observation point in the geometric near field is comparatively large. The *Fresnel zone* is defined as a region where conditions $D^2/r^2 \ll 1$ and $\lambda \ll r$ are fulfilled. The largest linear dimension of the source, visible from the point of observation, is denoted by D, and r is the distance of an observer from the sound source surface under observation. In this region the interference of the sound waves emitted from different points on the sound source surface plays a significant role. Hence, the amplitude of the sound pressure does not change in the geometrical near field inversely with distance from the centre of the sound source. Instead of a monotonical change of the sound pressure amplitude there occur relative minima and maxima. In the geometric near field the particle velocity and the sound pressure phasors may be almost in phase, however, both the active and reactive intensity vectors, as well as the particle velocity vector, are not parallel to each other and not directed radially from the centre of the source. In this region, as in the hydrodynamic zone, the fluid particles do not move along straight lines, but follow elliptical paths.

1.19.3 Geometric far field

The region of the sound field extending beyond the geometric near field to infinity is called the geometric far field [33] or *Fraunhofer zone* [17]. The angle, at which the sound source is visible from the observation point, situated in this zone, is small. The radii connecting the observation point with different points on the sound source surface are almost parallel. More precisely, the Fraunhofer zone is defined as a region where conditions $D/r \ll \lambda/(2\pi D) = 1/(kD)$ and $r \gg \lambda$ are fulfillled [17]. The phasors of the sound pressure and particle velocity are in phase in this zone and the vectors of sound intensity and particle velocity are radially directed.

1.20 Decibel scales

The decibel scale is a logarithmic scale applied in acoustics to scaling the ratio of sound intensities or the ratio of sound pressures. A logarithmic scale is used in acoustics to provide the possibility of covering the wide range of sound pressure amplitudes detected by the human ear — a range of up to 10 million to one is covered by the ear. Thus during the course of a measurement a period of quiet sound, when the sound pressure amplitudes are of the order of, say, 0.01 Pa, may be followed by an intense sound with sound pressure amplitudes of the order of 10 Pa. If the results of measurements in the form of values of the root mean square sound pressure p_{rms} are plotted against time using a linear scale, see Figure 1.22, the details of the quiet sound are lost. On the other hand, a logarithmic scale makes the presentation of the data more compact.

Another reason for using a logarithmic scale is that the human response to external stimuli, for example, sound or light, has a logarithmic character. The human response to sound may be approximated by the two *laws of Weber* and *Fechner* [34-36]. Both are approximate and based on observations of human physiological and psychological reactions to ex-

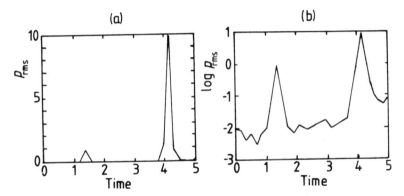

Figure 1.22 Representation of rms sound pressure versus time: (a) linear and (b) logarithmic.

ternal stimuli. Weber's law states that 'the minimum increase of stimulus which will produce a perceptible increase of sensation is proportional to the pre–existent stimulus'. Fechner's law concerns the intensity of the human response to a stimulus and states that the intensity of human sensation changes logarithmically with the energy which causes the sensation.

1.20.1 Sound pressure level

The sound pressure level (SPL) is an important quantity characterising sound and is especially useful as it can be easily measured. Two equivalent equations may be used to define SPL:

$$SPL = 10 \log \frac{p_{rms}^2}{p_{ref}^2} \text{ dB} \tag{1.231}$$

$$SPL = 20 \log \frac{p_{rms}}{p_{ref}} \text{ dB}, \tag{1.232}$$

where p_{ref} is the reference sound pressure and its value is 2×10^{-5} Pa ($20 \, \mu$Pa). The value of the reference pressure is approximately the value of the rms sound pressure corresponding to a very faint sound that can be heard in the mid–frequency range. The rms sound pressure p_{rms} is expressed in Pascals (Pa) and the values of the sound pressure level are given in decibels (dB). The factor of 10 in equation (1.231) was introduced to avoid a scale which is too 'compressed'.

Example 1.1

If the root mean square pressure is 2.5 Pa, what is the sound pressure level?

Solution

$$SPL = 20 \log \frac{2.5}{2 \times 10^{-5}} \text{ dB or}$$
$$SPL = 20 \log 2.5 - 20 \log (2 \times 10^{-5})$$
$$= 20 \log 2.5 - 20 \log 2 + 100 \log 10$$
$$= 7.96 - 6.02 + 100$$
$$= 101.9 \text{ dB}$$

In most cases it would be regarded as sufficient accuracy to quote the result as 102 dB.

1.20.2 Intensity level and sound power level

The intensity level is defined by the equation,

$$IL = 10 \log \left(\frac{\overline{I}}{I_{\text{ref}}} \right) \text{ dB,} \tag{1.233}$$

where $\overline{I}$ is the time–averaged sound intensity and I_{ref} is the reference sound intensity of 10^{-12} W/m². The reference sound intensity, as with p_{ref} in equation (1.231), corresponds approximately to the faintest sound we can hear.

For the far field of an arbitrary sound source, as well as for the sound field generated by a point sound source and for a plane wave, the relation $\overline{I} = p_{\text{rms}}^2/(\rho_0 c_0)$ is always fulfilled, see equations (1.145), (1.192) and in Appendix C, equations (C.9) and (C.15). Consequently, instead of equation (1.233) the following equation may be applied:

$$IL = 10 \log \left(\frac{p_{\text{rms}}^2}{\rho_0 c_0 \times 10^{-12}} \right) \text{ dB.} \tag{1.234}$$

Equation (1.234) may also be presented as

$$IL = 10 \log \left(\frac{p_{\text{rms}}^2}{(2 \times 10^{-5})^2} \right) \times \left(\frac{(2 \times 10^{-5})^2}{\rho_0 c_0 \times 10^{-12}} \right) \text{ dB} \tag{1.235}$$

$$IL = 10 \log \left(\frac{p_{\text{rms}}^2}{(2 \times 10^{-5})^2} \right) + 10 \log \left(\frac{400}{\rho_0 c_0} \right) \text{ dB.} \tag{1.236}$$

Taking into account equation (1.231), we can present equation (1.236) as

$$IL = SPL + 10 \log \left(\frac{400}{\rho_0 c_0} \right) \, dB. \qquad (1.237)$$

The value of the specific acoustic resistance $\rho_0 c_0$ at room conditions (20°C and standard atmospheric pressure) is 415 $kgm^{-2}s^{-1}$ and for this value:

$$IL = SPL - 0.2 \, dB. \qquad (1.238)$$

For many measurements a value of 0.2 dB is not significant, and intensity level (IL) and sound pressure level (SPL) can be regarded as numerically the same. This point is significant because it means that a meter designed to measure the sound pressure level (a sound level meter) also measure directly the intensity level. It should be noted that there are some locations in the world where, due to high altitude, a correction will have to be made if the intensity level is obtained from the SPL.

Finally, *sound power level* L_W of a source of sound power W is defined by the equation,

$$L_W = 10 \log(W/W_{ref}) \, dB, \qquad (1.239)$$

where W_{ref} is the reference sound power and is 10^{-12} W.

1.21 Addition and subtraction of sound pressure levels

In many practical situations the measuement of sound pressure level is performed in a sound field which is the result of the interaction of many sound fields, generated by different sound sources. For such a sound field the sound pressure p at an arbitrary point of a sound field may be presented as the sum of the sound pressures originating from single sound sources. Thus, for n sound sources we have:

$$p = \sum_{i=1}^{n} p_i. \qquad (1.240)$$

Hence,

$$p_{rms}^2 = \overline{p^2} = \overline{\left(\sum_{i=1}^{n} p_i \right)^2}. \qquad (1.241)$$

However, we are interested particularly in the case for which

$$\overline{\left(\sum_{i=1}^{n} p_i \right)^2} = \sum_{i=1}^{n} \overline{p_i^2}, \qquad (1.242)$$

because only when the condition (1.242) is fulfilled can the concept of sound pressure level addition be regarded as useful in analysing sound from many sources. Equation (1.242) is valid for *incoherent* sources. Sound sources are incoherent if their phase differences are random or changing with time. For example, two sound sources which generate harmonic waves of different frequencies, say f_1 and f_2, are mutually incoherent sound sources.

Hence, for n incoherent sound sources we have:

$$\frac{p_{\text{rms}}^2}{p_{\text{ref}}^2} = \sum_{i=1}^{n} \frac{\overline{p_i^2}}{p_{\text{ref}}^2} \qquad (1.243)$$

and consequently from equations (1.231) and (1.243) we obtain:

$$\text{SPL} = 10 \log \sum_{i=1}^{n} \frac{\overline{p_i^2}}{p_{\text{ref}}^2} \text{ dB} \qquad (1.244)$$

$$\text{SPL} = 10 \log \sum_{i=1}^{n} \frac{p_{\text{rms}_i}^2}{p_{\text{ref}}^2} \text{ dB}. \qquad (1.245)$$

Since

$$\text{SPL}_i = 10 \log \frac{p_{\text{rms}_i}^2}{p_{\text{ref}}^2} \text{ dB}, \qquad (1.246)$$

equation (1.245) may be presented in the forms,

$$\text{SPL} = 10 \log \left(\sum_{i=1}^{n} 10^{\text{SPL}_i/10} \right) \text{ dB} \qquad (1.247)$$

$$10^{\text{SPL}/10} = \sum_{i=1}^{n} 10^{\text{SPL}_i/10}. \qquad (1.248)$$

Let us assume that at a certain point in a sound field the values of the rms pressures p_{rms_i} are the same ($p_{\text{rms}_1} = p_{\text{rms}_2} = \ldots p_{\text{rms}_n}$). Formula (1.245) leads in this case to the relations,

$$\text{SPL} = 10 \log \frac{n p_{\text{rms}_1}^2}{p_{\text{ref}}^2} \text{ dB} \qquad (1.249)$$

$$\text{SPL} = \text{SPL}_1 + 10 \log n \text{ dB}. \qquad (1.250)$$

Thus, if there are n machines in a workshop and each produces a sound which can be expressed approximately by the same value of the sound pressure level SPL_1, the value of the sound pressure level from all machines is determined by equation (1.250).

In the general case, when each sound source produces sound with different SPLs, equation (1.247) should be used.

Equation (1.247) applied to two sources is reduced to:

$$SPL = 10\log\left(10^{SPL_1/10} + 10^{SPL_2/10}\right) \text{ dB.} \qquad (1.251)$$

If we have, for example, a situation such that $SPL_1 > SPL_2 > 0$, we can define the difference between the sound pressure levels of the two sources $\Delta SPL = SPL_1 - SPL_2$ and present equation (1.251) in the forms,

$$SPL = 10\log\left[10^{SPL_1/10}\left(1 + 10^{-\Delta SPL/10}\right)\right] \qquad (1.252)$$

$$SPL = SPL_1 + 10\log\left(1 + 10^{-\Delta SPL/10}\right) \text{ dB.} \qquad (1.253)$$

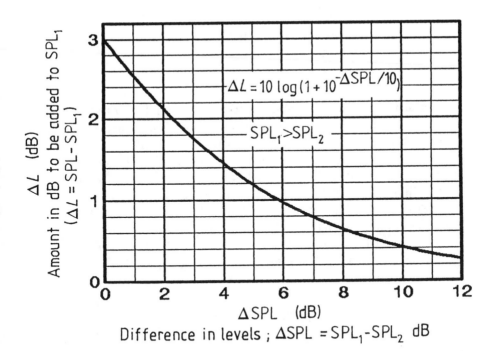

Figure 1.23 Combining decibels.

The relation between $SPL - SPL_1$ and ΔSPL is presented in Figure 1.23. Figure 1.23 may be used to estimate the amount ΔL,

$$\Delta L = 10 \log \left(1 + 10^{-\Delta SPL/10}\right) \text{ dB}, \qquad (1.254)$$

which should be added to the larger sound pressure level SPL_1 to obtain the sound pressure level SPL in the combined field of the two sound sources. The procedure for combining decibels, which makes use of Figure 1.23, may also be applied to n sound sources in the manner illustrated by the following example.

Example 1.2

Three machines independently produce sound pressure levels of 87, 89 and 86 dB. What is the SPL produced by the three machines together?

Solution The value for the SPL, see equation (1.247), is given by

$$\begin{aligned}
SPL &= 10 \log(10^{8.7} + 10^{8.9} + 10^{8.6}) \\
&= 10 \log(5.01 \times 10^8 + 7.94 \times 10^8 + 3.98 \times 10^8) \\
&= 10 \log 10^8 + 10 \log 16.93 \\
&= 80 + 12.3 \\
&= 92.3 \text{ dB}.
\end{aligned}$$

Alternatively, the 87 dB and 89 dB may be combined together with the aid of Figure 1.23 or equation (1.254). Thus, as the difference between 87 and 89 is 2 dB, an amount 2.1 dB has to be added to the largest value, namely 89 dB. The result is $89 + 2.1 = 91.1$ dB. The 91.1 dB is next combined with 86 dB. The difference is 5.1 dB, so the amount to be added to 91.1 is 1.2 dB. The final result is 92.3 dB. The order in which the combination is carried out does not matter. For example, it is possible first to combine 86 and 87 dB — add 2.5 to 87 to give 89.5 dB. Next 89.5 is combined with 89 by adding 2.8 to 89.5 to give 92.3 dB.

If great accuracy is not required it is possible to make use of a simple decibel combination rule which is explained in Table 1.2.

When applied to the data of Example 1.2, the use of Table 1.2 leads to a result of 93 or 92 dB, depending upon the order in which the combination is made.

TABLE 1.2 Approximate scheme for the addition of decibels

$\Delta\,\mathrm{SPL}(=\mathrm{SPL}_1-\mathrm{SPL}_2)$ Difference in levels (dB)	ΔL Amount to be added to larger level SPL_1 (dB)
0 - 1	3
2 - 3	2
4 - 9	1
10 or more	0

1.21.1 Correction for background noise

In many pactical situations there is a problem with background noise and it is necessary to determine the value of the SPL of a sound source alone from measurements of SPL in the sound field caused by the sound source and the background. For example, consider measurements in a factory hall with many machines generating noise. Let us assume that we wish to determine the sound pressure level only for one arbitrarily chosen machine. The sound emitted by all the other machines is described as background noise and is characterised by $\mathrm{SPL}_{\mathrm{bg}}$. From the measurement of SPL, when all the machines are acting, we obtain the value SPL_t. Equation (1.248) can be used to determine the value of the sound pressure level $\mathrm{SPL}_{\mathrm{mach}}$ of the chosen machine. Thus,

$$10^{\mathrm{SPL}_t/10} = 10^{\mathrm{SPL}_{\mathrm{mach}}/10} + 10^{\mathrm{SPL}_{\mathrm{bg}}/10}$$

or,

$$10^{\mathrm{SPL}_{\mathrm{mach}}/10} = 10^{\mathrm{SPL}_t/10}\left\{1 - 10^{(\mathrm{SPL}_{\mathrm{bg}}-\mathrm{SPL}_t)/10}\right\} \qquad (1.255)$$

or,

$$\mathrm{SPL}_{\mathrm{mach}} = \mathrm{SPL}_t + 10\log\left\{1 - 10^{-(\mathrm{SPL}_t-\mathrm{SPL}_{\mathrm{bg}})/10}\right\}. \qquad (1.256)$$

Let the correction C that should be subtracted from the sound pressure level of the total noise be given by:

$$C = -10\log\left\{1 - 10^{-(\mathrm{SPL}_t-\mathrm{SPL}_{\mathrm{bg}})/10}\right\}. \qquad (1.257)$$

Hence, the sound pressure level of the machine may be written as

$$SPL_{mach} = SPL_t - C \text{ dB.} \tag{1.258}$$

The correction C as a function of the difference between the sound pressure levels of the total sound and the background sound is presented in Figure 1.24. The figure applies in general to the 'subtraction' of decibels.

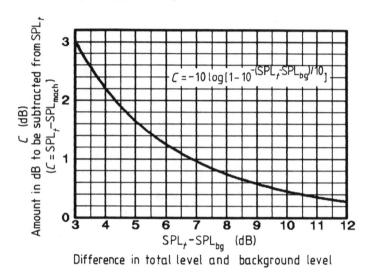

Figure 1.24 Subtraction of decibels.

Example 1.3

The sound pressure level of a machine is measured in a noisy workshop and found to be 92 dB. When the machine is switched off the background noise is 88 dB. What would the sound pressure level of the machine be without the background noise?

Solution $SPL_t = 92 \text{ dB}$, $SPL_{bg} = 88$, dB and from equation (1.257) or Figure 1.24, $C = 2.2 \text{ dB}$. Hence from equation (1.258),

$$SPL_{mach} = 92 - 2.2 = 89.8 \text{ dB or } 90 \text{ dB.}$$

1.22 Closure

For centuries we have been fascinated by wave motion, the essence of acoustics. Chaucer in the 14th century was already aware that sound is a wave motion which is analogous with ripples on the surface of water:

I preve hit thus – take hede now –
By experience; for if that thou
Throwe on water now a stoon,
Wel wost thou, hit wol make anoon
A litel roundel as a cercle,
Paraventure brood as a covercle;
And right anoon thou shalt see weel,
That wheel wol cause another wheel,
And that the thriddle, and so forth, brother,
Every cercle causing other,
Wyder than himselve was;
And thus, fro roundel to compas,
Ech aboute other goinge,
Caused of otheres steringe,
And multiplying ever–mo,
Til that hit be so fer y–go
That hit at bothe brinkes be.

. . . . . .

And right thus every word, y–wis,
That loude or privee spoken is,
Moveth first an air aboute,
Another air anoon is meved,
As I have of the water preved,
That every cercle causeth other.

We define sound as an oscillatory motion of small amplitude in an elastic medium. In the majority of physical problems sound propagation can be treated as an isentropic process. Sound propagates in a compressible medium in the form of waves. It is only in the close proximity of the sound source — in the hydrodynamic zone — that the sound field (vibration field) does not behave like a wave. However, in the case of a noncompact sound source of characteristic size D, when $kD \gg 1$, almost the whole sound field is wave–like.

The most fundamental types of sound wave are the plane wave and the spherical wave from a point source. In the case of a plane wave the amplitude of the wave maintains a constant value during propagation in contrast to the spherical wave for which the wave amplitude decays during propagation from the sound source.

The simple sound source is also a fundamental concept in acoustics which is very useful in noise control. In the case when the sound source can be treated as a simple source the measurement of the time–averaged sound intensity can be reduced to the measurement of the sound pressure. In practical applications it is essential to know how to define the near and far fields of the sound source and to know whether the relation between p_{rms}, the rms sound pressure and the time–averaged sound intensity expressed in the form $\bar{I} = p_{\text{rms}}^2/(\rho_0 c_0)$ is also valid not only in the far field. It should be realised that this relation is not only fulfilled for plane waves but is also applied in the sound fields, represented by a point sound source field, such as the simple source field and the pulsating sphere field (except hydrodynamic zone), as well as in the far field of any arbitrary sound source (see Appendix D).

Section 1.21 is an important practical section dealing with the addition or combination of sound levels in decibels. There are also significant implications for noise control. In Example 1.2 three noise sources of 87, 89 and 86 dB were considered and they produced together a sound of 92.3 dB. If the greatest noise source — of 89 dB — were totally eliminated by noise control techniques, the noise would be 89.5 dB. A reduction of almost 3 dB is not significant and is a small reward for such a great effort. Similarly, a workshop may be filled with 10 machines, each producing 90 dB, so that the total sound is 100 dB. Suppose the necessity arises to replace one machine and that there are two possibilities — to buy a machine similar to the existing or a new design which produces only 80 dB, but which is very much more expensive. Purchase of the new machine will have negligible effect on the noise in the workshop and is difficult to justify to the accountants. But clearly it is the correct decision, because in the long term, when all the machines have been replaced, there is the possibility of a 10 dB reduction. Indeed current legislation is designed to encourage the long term view.

References

[1] G. K. Batchelor (1970). *An Introduction to Fluid Dynamics.* Cambridge University Press, Cambridge.

[2] J. S. Anderson and M. Bratos–Anderson (1987). *Solving Problems in Vibrations.* Longman Scientific, Harlow.

[3] M. W. Zemansky (1968). *Heat and Thermodynamics.* McGraw–Hill Kogakusha, Tokyo.

[4] P. A. Thompson (1972). *Compressible Fluid Dynamics.* McGraw–Hill, New York.

[5] P. M. Morse (1948). *Vibration and Sound.* McGraw–Hill, New York.

[6] L. E. Kinsler and A. R. Frey (1962). *Fundamentals of Acoustics,*

second edition. John Wiley, New York.

[7] L. D. Landau and E. M. Lifshitz (1987). *Fluid Mechanics*, second edition. Pergamon Press, Oxford.

[8] S. Temkin (1981). *Elements of Acoustics*. John Wiley, New York.

[9] J. D. Fast (1970). *Entropy*. Macmillan, London.

[10] S. R. de Groot and P. Mazur (1984). *Non-equilibrium Thermodynamics*. Dover Publications, New York.

[11] J. Lighthill (1978). *Waves in Fluids*. Cambridge University Press, Cambridge.

[12] A. D. Pierce (1989). *Acoustics, an introduction to its physical principles and applications*. Acoustical Society of America, New York.

[13] G. B. Whitham (1974). *Linear and Nonlinear Waves*. John Wiley, New York.

[14] S. N. Rschevkin (1963). *The Theory of Sound*. Pergamon Press, Oxford.

[15] L. D. Landau and E. M. Lifshitz (1986). *Theory of Elasticity*, third edition, Pergamon Press, Oxford.

[16] A. E. H. Love (1944). *Mathematical Theory of Elasticity*. Dover Publications, New York.

[17] I. Malecki (1969). *Physical Foundations of Technical Acoustics*. Pergamon Press, Oxford.

[18] H. Lamb (1960). *The Dynamical Theory of Sound*. Dover Publications, New York.

[19] H. Lamb (1945). *Hydrodynamics*. Dover Publications, New York.

[20] P. M. Morse and H. Feshbach (1953). *Methods of Theoretical Physics*. McGraw-Hill, New York.

[21] L. L. Beranek (1954). *Acoustics*. McGraw-Hill, New York.

[22] Ya. B. Zeldovich and Yu. P. Raizer (1967). *Physics of Shock Waves and High Temperature Hydrodynamic Phenomena*. Academic Press, New York.

[23] K. Rektorys (Ed.) (1969). *Survey of Applicable Mathematics*. Iliffe Books, London.

[24] G. Arfken (1985). *Mathematical Methods for Physicists*, second edition. Academic Press, Orlando.

[25] E. Skudrzyk (1971). *The Foundations of Acoustics*. Springer Verlag, New York.

[26] P. M. Morse and K. U. Ingard (1968). *Theoretical Acoustics*. McGraw-Hill, New York.

[27] R. Aris (1962). *Vectors, Tensors and Basic Equations of Fluid Mechanics*. Dover Publications, New York.

[28] M. E. Goldstein (1976). *Aeroacoustics*. McGraw–Hill, New York.

[29] F. J. Fahy (1989). *Sound Intensity*. Elsevier Applied Science, London.

[30] J. A. Mann, J. Tichy and A. J. Romano (1987). Instantaneous and time–averaged energy transfer in acoustic fields. *Journal of the Acoustical Society of America*, **82** (1), pp 17–30.

[31] D. D. Reynolds (1981). *Engineering Principles of Acoustics*. Allyn and Bacon, Boston.

[32] A. P. Dowling and J. E. Ffowcs Williams (1983). *Sound and Sources of Sound*. Ellis Horwood, Chichester.

[33] D. A. Bies and C. H. Hansen (1988). *Engineering Noise Control; theory and practice*. Unwin Hyman, London.

[34] J. Jeans (1968). *Science and Music*. Dover Publications, New York.

[35] S. S. Stevens and H. Davis (1983). *Hearing, its psychology and physiology*. American Institute of Physics, New York.

[36] G. von Bekesy (1960). *Experiments in Hearing*. Acoustical Society of America, New York.

2 The Measurement and Analysis of Sound

I often say that when you can measure what you are speaking about, you can express it in numbers, you know something about it. But when you cannot measure it, when you cannot express it in numbers, your knowledge is of a meagre and unsatisfactory kind. It may be the beginning of knowledge but you have scarcely, in your thought, advanced to the stage of science whatever the matter may be.

Lord Kelvin, Lecture to the Institution of Civil Engineers, 3 May, 1882

2.1 Introduction

In the measurement of sound the most basic quantity is the sound pressure level. Consequently, the sound level meter, the name given to the instrument which measures the sound pressure level, is the most widely used of sound measuring equipment. The electrical signals from the sound level meter may be processed in different ways; in particular an analysis may be carried out to determine the frequency content of the output signal. Descriptions of the sound level meter and equipment for frequency analysis form the main part of this chapter.

2.2 The sound level meter

The sound level meter (SLM) normally comprises a microphone, preamplifier, an input amplifier and attenuator, weighting networks, an output amplifier and attenuator, an rms detector and an indicating device. The scheme is shown in Figure 2.1. Modern sound level meters may have many

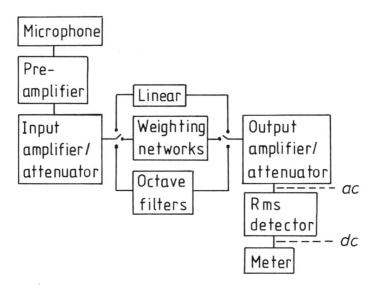

Figure 2.1 Schematic diagram of sound level meter.

more facilities and may have a microprocessor for control and analysis, but the basic system for measuring sound pressure level is as shown.

Sound level meters are built according to certain standards, either national or international. Many national standards, such as the British, are based on the international standard [1]. The international standard specifies four types, from 0, the most accurate, to 3.

2.3 Microphones

There are two types of microphone which are commonly used in sound level meters, namely, the condenser microphone and the electret microphone. Both are capacitive transducers in which the sound signal to electrical signal conversion is based on the phenomenon of the change of electrical capacitance of two conducting plates (electrodes).

2.3.1 Condenser microphones

Condenser microphones are particularly associated with the firm of Brüel and Kjaer who have developed a family of microphones of different sizes [2]. A schematic diagram is shown in Figure 2.2.

The condenser microphone acts as a capacitor (condenser) with a constant charge. The metal diaphragm and rigid backplate, separated by an air gap, constitute the electrodes of the capacitor whose capacitance is

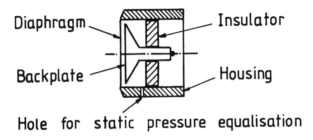

Figure 2.2 Schematic diagram of a condenser microphone.

inversely proportional to the distance between the plates.

During measurements sound waves impinge on the thin, tensioned diaphragm — typically made of nickel and about 2 μm thick. The circular diaphragm, supported around the circumference, is deflected by the sound waves. Thus the width of the air gap between the diaphragm and the rigid backplate is varying and the capacitance of the microphone is also changing. Condenser microphones require a constant charge on the electrodes. The constant charge on the capacitor is maintained thanks to a large *dc polarisation voltage* (normally 28 or 200 V) which is applied between the diaphragm and the backplate from a high resistance source.

The charge on the capacitor is constant and the capacitance is inversely proportional to the voltage across the capacitor. On the other hand, as was already mentioned, the capacitance is inversely proportional to the distance between the capacitor electrodes. Hence, the voltage across the capacitor microphone is proportional to the distance between the electrodes. The increase in voltage is associated with an increase in the distance between the diaphragm and backplate.

It should be noted that the impedance of a condenser microphone is necessarily capacitive and a preamplifier, which is mainly acting as an impedance converter, is required to ensure that the output impedance of the preamplifier is resistive.

Microphone sensitivity The main characteristic of any transducer is its *frequency response* or transfer function which can be presented in the form of a pair of frequency functions, namely, the responses for amplitude and phase. The amplitude response defines the *sensitivity of the transducer*. The open circuit sensitivity of a condenser microphone (unloaded by a preamplifier) is the ratio of the output voltage of the microphone itself to the input sound pressure (ratio of the electrical output to the mechanical input) for a given polarisation voltage and is normally expressed in mV/Pa. The sensitivity increases with the diameter of the microphone diaphragm.

In general the microphone sensitivity depends on frequency, particularly at very high and very low frequencies. However, in the so–called flat response range the microphone sensitivity is constant and is referred to as the *nominal microphone sensitivity*. Hence, the nominal sensitivity of the microphone is the slope of the linear part of the input–output characteristic of the microphone.

The $\frac{1}{2}$ inch microphone is the most commonly used size. The specification may vary, but a typical $\frac{1}{2}$ inch condenser microphone (Brüel and Kjaer type 4133) has an open circuit sensitivity of 12.5 mV/Pa. (Note that a sensitivity of 12.5 mV/Pa is sometimes expressed as -38 dB relative to 1 V/Pa.) A $\frac{1}{2}$ inch microphone with this sensitivity can be used for a range of overall sound pressure levels from 40 to 160 dB. The one inch microphone has a sensitivity of about 50 mV/Pa which enables it to be used over a range of 20 to 145 dB and is thus more suitable for the measurement of sound with low amplitudes. A $\frac{1}{4}$ inch microphone, with a sensitivity of 4 mV/Pa, can be used between 50 and 165 dB.

The upper limit of the operating range of frequencies of condenser microphones is controlled by the size of the microphone and the form of its construction. Condenser microphones can be constructed to have the greatest upper frequency limit either in situations when the sound wave is at normal incidence (microphone pointing at the sound source), or when the sound is reaching the microphone randomly from all directions (as in a reverberant or semi–reverberant room such as a factory hall). Thus the user has the possibility of purchasing two types of microphone, firstly a *frontal incidence microphone* (which is pointed at the source) or secondly a *random incidence microphone* (for use in a reverberant space). The former type is sometimes called a free field microphone and the latter a pressure response microphone.

Pressure response   The frequency response of a microphone may be obtained by using an *electrostatic actuator* which is placed close to the diaphragm of the microphone. A polarisation voltage as well as an ac voltage are supplied between the rigid metallic grid of the actuator and the microphone diaphragm, causing an alternating electrostatic force between the grid and the diaphragm. Consequently, the diaphragm of the microphone is caused to vibrate. Vibration of the diaphragm can be related to the sound pressure changes which would induce the same displacement of the microphone diaphragm.

The ratio of the microphone output voltage (amplitude or rms value) to the input sound pressure (amplitude or rms value) as a frequency function is the frequency response or more strictly the amplitude response of the microphone. When the input sound pressure (amplitude or rms value) is constant over the entire diaphragm of the microphone the frequency response is called the *pressure response*. In the case when the electrostatic actuator is in use the input sound pressure is understood as a hypothetical sound pressure which corresponds to the executed displacement of the microphone diaphragm. The pressure response is indicated in Figure 2.3a as curve (i), where the response at low frequencies is taken to be 1 (or 0 dB).

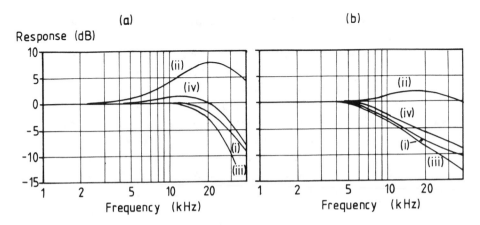

Figure 2.3 Frequency response of $\frac{1}{2}$ inch condenser microphone with protective grid   (a) random incidence or pressure response type of microphone, (b) frontal incidence or free field type;
(i) pressure response, (ii) normal incidence $\theta = 0°$, (iii) grazing incidence $\theta = 90°$ and (iv) random incidence.

Free field response   Another important characteristic of the microphone is the free field response, defined as the ratio of the output microphone voltage to the sound pressure (both in terms of amplitude or rms values), which would exist in the free sound field at the diaphragm position in the absence of the microphone. Hence, the free field response is the frequency response which would be obtained in idealised conditions, namely, in boundary free space where the measured sound field would not be disturbed by reflection, diffraction and interference phenomena caused by the presence of the microphone. Usually, the pressure response as well as the free field response of the microphone refers to the microphone itself (the microphone cartridge) unloaded by the preamplifier, i.e., to open circuit. In a wide frequency range the frequency response of the microphone itself does nor differ from the frequency response of the system, microphone with preamplifier. The difference can be more significant only for very low frequencies. At low frequencies the free field response of the microphone does not differ from the microphone pressure response. However, the difference can be significant for frequencies at which the sound wavelength is comparable with the microphone dimensions.

Let us imagine plane sound waves which are normally incident on the microphone diaphragm. Since the sound waves are partly reflected from the diaphragm, the effective sound pressure increases at the microphone diaphragm. Thus, the free field response can be treated as the sum of the pressure response and a pressure increase correction. The pressure increase correction and, consequently, the free field response — both functions of frequency — depend upon the diameter of the microphone as well as the angle at which the sound wave is incident on the microphone diaphragm.

As already mentioned, condenser microphones can be constructed to possess in specified conditions the best frequency response, i.e., the widest frequency range of the 'flat' response, where the frequency response does not change with frequency. Thus, for example, the microphone of the random incidence (or pressure response) type should be used in situations where the local pressure — not the whole acoustic field which can be disturbed by the presence of the microphone — is of interest. In general microphones of the random incidence type are used in diffuse field measurements.

In contrast to the microphone of the random incidence type the frontal incidence microphone is most useful in the sound field of a single source. The microphones of both types when used in different conditions, not always optimal for their frequency characteristics, show different frequency responses, as illustrated in Figure 2.3.

Figure 2.3a applies to a $\frac{1}{2}$ inch microphone and shows that the frequency response at $\theta = 0°$, curve (ii), increases with frequency and is about +1 dB at 5000 Hz. The same microphone has a different frequency response at grazing incidence ($\theta = 90°$), shown by curve (iii), which is close to the pressure response. The frequency response to sound coming from all directions, the random incidence response, is shown as curve (iv) which differs only slightly from curve (iii). The microphone with the frequency responses illustrated in Figure 2.3a is a microphone with a random incidence response (or a random incidence microphone), because the best frequency response of the microphone occurs for a random incidence. It is the most proper microphone to use in a workshop where sound is reaching the microphone from many machines at different angles of incidence. If this type of microphone is used in the free field of one sound source, the microphone should be almost at grazing incidence ($\theta = 90°$ or even better $\theta = 80°$).

It is possible to modify the internal construction of the microphone so that the best frequency response occurs for normal incidence ($\theta = 0°$). The various frequency responses for a microphone constructed to give the best frequency response for normal incidence are shown in Figure 2.3b and a microphone with this performance is called a frontal incidence microphone. A microphone of this type is for use with one source in a free field, for example, out–of–doors or in an anechoic chamber, and should be pointed at the source. The frequency response curves (ii) to (iv) in Figure 2.3 have been obtained with a protective grid covering the fragile diaphragm. The protective grid has an effect on the directional response of the microphone.

There is no difference between the performance of a frontal incidence microphone and a random incidence microphone at low frequencies. The frequency responses for the two types, at a given angle of incidence of a sound wave, diverge at a certain frequency whose value decreases with the increase in size of the microphone. Thus the difference between a $\frac{1}{4}$ inch frontal incidence microphone and a $\frac{1}{4}$ inch random incidence microphone is hardly important in the audible range of frequencies.

The $\frac{1}{2}$ inch microphone of the frontal incidence type can be used at frequencies up to about 40 kHz when pointed at the sound source; see Figure 2.3b, curve (ii). The $\frac{1}{4}$ inch frontal incidence microphone can be used up to a frequency of 100 kHz. The smaller $\frac{1}{8}$ inch microphone, which has an upper limit of 140 kHz, is used for special purposes, for example, in ducts where the small size is important or in model studies when small scale reconstructions of concert halls or street layouts are made, and measurements have to be carried out at high frequencies in order to maintain similarity with the full scale system.

In some modern sound level meters it is possible to use one type of microphone and change the frequency response of the complete system (microphone and sound level meter) by adding electronically a network in the circuitry of the sound level meter. Thus, with such a type of sound level meter the one microphone is used, but a switch on the front of the meter allows selection of either frontal incidence or random incidence equivalents.

The advantages and disadvantages of condenser microphones can be summarised as follows.

Advantages:—

    (1) stable, i.e., the nominal microphone sensitivity does not change
with time,

    (2) good frequency response,

    (3) adequate sensitivity,

    (4) family of interchangeable microphones of different sizes available,

    (5) low self-noise, hence good dynamic range,

    (6) low sensitivity to mechanical vibration.

Disadvantages:—

    (1) fragile,

    (2) expensive,

    (3) requires a power supply,

    (4) susceptible to humidity.

2.3.2 Electret microphones [3]

A major disadvantage of the condenser microphone is that it requires
a charge from a power supply. The electret microphone overcomes this
drawback. It is well known from electrostatics that certain materials can
be given a charge. For example, amber when rubbed with a cloth will retain a charge for a short period of time sufficient for the amber to attract
small pieces of paper. Some materials, called electrets, when processed in
a certain way can retain their charge for long periods; PTFE (teflon) is an
example of an electret material. The electret microphone is essentially a
variant of the condenser microphone. In one particular design of an electret microphone the diaphragm of the condenser microphone is replaced
by a thin electret foil [4]. An alternative design by Brüel and Kjaer places
the electret on the backplate; electret microphones manufactured under
their name are called *pre-polarised* [5].

2.3.3 Other types of microphones

Piezo-electric microphones [4] have in the past often been used with sound
level meters, particularly for industrial grade meters, but have now been
replaced by electret microphones. Piezo-electric microphones are inexpensive, have a good sensitivity and require no additional power supply.
A disadvantage is that they are sensitive to vibration.

    Moving coil microphones are used in broadcasting studios, but not
with sound measuring equipment.

    Condenser microphones may be modified by the addition of a long,
thin probe to enable measurements of sound pressure to be made in otherwise inaccessible locations [2]. *Probe microphones* are often used where it

is essential to avoid blockage of the sound, as, for example, in the standing wave tube (see Section 4.5). A diagram of a probe microphone is shown in Figure 2.4. Resonances can occur in the probe tube — normally at tube lengths corresponding to a quarter, three quarters of a wavelength, etc., see Section 6.5. The variation of the frequency response of the microphone, caused by the resonances, can be reduced by applying the correct amount of damping in the tube. Correct damping is generally a matter of trial and error, but can be realised by a tuft of steel wool inside the tube. A steel probe, 240 mm long, of outer diameter 4 mm when used with a B & K $\frac{1}{2}$ inch microphone can achieve a flat frequency response from about 15 Hz to 2 kHz.

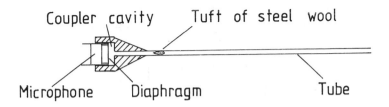

Figure 2.4 Schematic diagram of a probe microphone.

2.4 Weighting networks

For many measurements made in the laboratory the sound level meter will be required to have the best possible linear frequency response over the audible range of frequencies, i.e., a weighting of 1 over all audible frequencies. Sound pressure levels of sound signals obtained by using the linear frequency weighting are sometimes referred to as *overall*, *overall unweighted* or *linear* sound pressure levels; they may be designated 'Lin'.

As we shall see in Section 3.2, the way in which the ear responds to loudness depends upon frequency. Weighting networks are introduced into the sound level meter to take into account the variable response of the ear to loudness. Since these networks depend upon frequency, they are sometimes called frequency or spectral weighting networks. The most common weighting networks are designated A, B and C, and their relative responses in dB are shown as a function of frequency in Figure 2.5. The A weighting network is widely used, particularly for industrial measurements. If the sound level meter does not have a linear frequency response, the C weighting network will give the nearest approximation to a linear one. The results of measurements obtained by the sound level meter with the weighting networks in use are called *weighted sound pressure levels, weighted sound levels* or *sound levels.*

The term relative response of a network is defined in decibels as $20 \log (V_2/V_1)$ dB, where V_1 is the input voltage to the network and V_2 the output voltage.

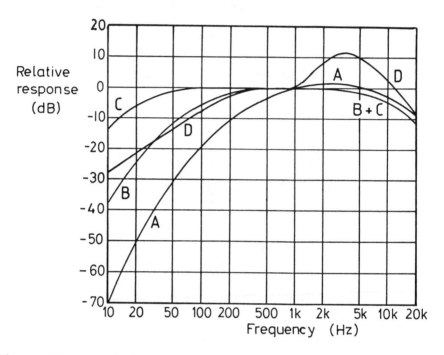

Figure 2.5 The relative frequency responses of weighting networks for sound level meters.

The tolerances on the relative frequency response of the weighting networks depend on the type of the sound level meter. The most accurate, type 0, has tolerances of ±0.7 dB at 1000 Hz, whereas for type 3 the corresponding tolerances are ±2.0 dB. The tolerances for the weighting networks of sound level meters of types 0, 1 and 2 are shown in Table 2.1. Also shown in Table 2.1 is the relative frequency response in dB of the A weighting network; in the table similar information on the A network to that presented graphically in Figure 2.5 is given in a more accurate form.

A weighting network D is sometimes used, particularly for the measurement of aircraft noise; it is shown in Figure 2.5.

The weighting networks are discussed further in Sections 3.4 and 3.6.

TABLE 2.1 Relative frequency response (in dB) of A weighting network and tolerances on the response (in dB) for sound level meters of types 0, 1 and 2

Nominal frequency (Hz)	A weighting response (dB)	Tolerances Type 0 (dB)	Type 1 (dB)	Type 2 (dB)
10	−70.4	+2;−∞	+3;−∞	+5;−∞
12.5	−63.4	+2;−∞	+3;−∞	+5;−∞
16	−56.7	+2;−∞	+3;−∞	+5;−∞
20	−50.5	±2	±3	±3
25	−44.7	±1.5	±2	±3
31.5	−39.4	±1	±1.5	±3
40	−34.6	±1	±1.5	±2
50	−30.2	±1	±1.5	±2
63	−26.2	±1	±1.5	±2
80	−22.5	±1	±1.5	±2
100	−19.1	±0.7	±1	±1.5
125	−16.1	±0.7	±1	±1.5
160	−13.4	±0.7	±1	±1.5
200	−10.9	±0.7	±1	±1.5
250	−8.6	±0.7	±1	±1.5
315	−6.6	±0.7	±1	±1.5
400	−4.8	±0.7	±1	±1.5
500	−3.2	±0.7	±1	±1.5
630	−1.9	±0.7	±1	±1.5
800	−0.8	±0.7	±1	±1.5
1000	0	±0.7	±1	±1.5
1250	+0.6	±0.7	±1	±1.5
1600	+1.0	±0.7	±1	±2
2000	+1.2	±0.7	±1	±2
2500	+1.3	±0.7	±1	±2.5
3150	+1.2	±0.7	±1	±2.5
4000	+1.0	±0.7	±1	±3
5000	+0.5	±1	±1.5	±3.5
6300	−0.1	+1;−1.5	+1.5;−2	±4.5
8000	−1.1	+1;−2	+1.5;−3	±5
10000	−2.5	+2;−3	+2;−4	+5;−∞
12500	−4.3	+2;−3	+3;−6	+5;−∞
16300	−6.6	+2;−3	+3;−∞	+5;−∞
20000	−9.3	+2;−3	+3;−∞	+5;−∞

2.5 Measurement of root mean square of acoustic quantities

For a sound level meter to specify the sound pressure level the rms value of the sound pressure must first be determined (see equation (1.21)). The rms sound pressure can be obtained approximately by means of a detector which comprises a full wave bridge rectifier, a squaring network and a lowpass filter. The lowpass filter consists of a resistance R and a capacitance C, as shown schematically in Figure 2.6. An rms detector of this kind is used as an element in the traditional sound level meter and is said to make use of an *RC averager*.

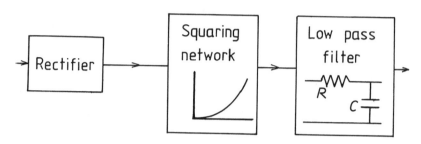

Figure 2.6 Schematic diagram of rms detector based on RC averaging.

The time constant of the lowpass filter or RC network affects the rate at which the needle moves, i.e., the effective damping of the needle of the meter. Two damping settings are usually available — *slow* and *fast*. The time constant RC for the fast setting is of the order of 125 ms and 1 s for slow. The fast and slow damping settings are sometimes called the fast and slow *time weightings*. The way in which the meter and its needle respond is not in fact defined in terms of time constants, but in terms of the response to pulses. For example, in the international standard [1] it is stated that a sound level meter of type 1 with the fast time weighting, when subjected to a sinusoidal pulse at 2000 Hz of 0.2 s duration, should give a maximum reading which is between 0 and −1 dB, relative to that of the meter reading for a steady, continuous signal of the same frequency and amplitude.

Averaging with RC averagers is not true averaging. The equivalent true averaging time is twice the RC time constant for a random sound signal which lasts for a long time [6,7].

In the RC averager type of detector the logarithm of the sound pressure is generally achieved by correct markings on the meter scale. This means that a 2 dB variation, for example, covers a large amount of the scale at the right–hand end, but a similar 2 dB is confined to a smaller range at the left.

In many modern sound level meters more complicated circuits are used. These circuits may have variable square networks and logarithmic mean square detectors.

Many sound level meters have outputs referred to as *ac* and *dc*. The ac output occurs before the rms detector and gives the instantaneous sound pressure as a function of time. The dc output leads from the rms detector and is similar to the variation shown on the meter by the needle. The ac and dc outputs are indicated in Figure 2.1.

2.6 Calibration of sound level meters

A sound level meter must be calibrated, preferably before and after each measurement session. For a sound level meter in good condition and in regular use with the same microphone, the calibration should essentially be a process of checking. It is likely that something is wrong, if adjustment has to be continually made.

Calibrators for sound level meters are devices that fit around the microphone and emit a pure tone of known intensity. The most accurate type is called a *pistonphone*. It emits a tone at 250 Hz with a sound intensity level of 124 dB. It is a mechanical device in which two opposed pistons are driven by a rotating cam. A schematic diagram is shown in Figure 2.7. As the pistons move in and out, the volume of the space between the pistons and the microphone changes. The pistonphone should preferably only be used for sound level meters with a linear frequency response. If the sound level meter has no linear response, but only an A weighting network, a correction of −8.6 dB should be added to the value 124 dB, i.e., the meter should give a reading of 115.4 dB. If the atmospheric pressure is different from the normal, an additional small correction to the meter reading should be taken into account.

A less expensive, electrically–driven calibrator produces a sound intensity level of 94 dB at 1000 Hz. In this type an oscillator drives on a bridge a piezo–electric crystal which causes a diaphragm to vibrate near the microphone. Since the frequency of the sound produced is 1000 Hz, the calibrator can be used with either the linear response or the A weighting network without the need for a correction. There is a small free field correction which depends upon the size of the microphone; e.g., the correction amounts to 0.2 dB for $\frac{1}{2}$ inch microphones.

The pistonphone should be checked every year in an approved laboratory. In the United Kingdom the calibration can be carried out at the National Physical Laboratory, Teddington, or at 'Aquila', the Ministry of Defence's Standards Laboratory at Bromley, Kent.

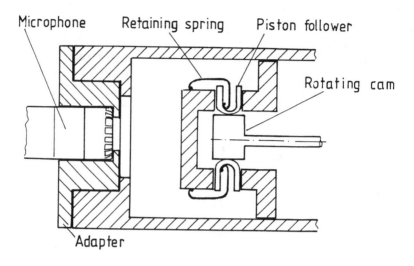

Microphone   Retaining spring   Piston follower

Rotating cam

Adapter

Figure 2.7 Schematic diagram of pistonphone.

2.7 Points to observe when using a sound level meter

2.7.1 Fast and slow time weightings of the meter; reading the meter

The traditional type of sound level meter with *RC* averager has a meter
with a needle which fluctuates as the noise varies. Later designs of SLM
have a digital readout, possibly with a bar display where the length of
the bar increases with the sound pressure level. The former type has some
advantages in that the oscillating movement of the needle of the meter can
be related by the observer to the audible fluctuations of the noise. Visual
averaging of the position of the needle during fluctuations is relatively
easy. On the other hand digital readouts are particularly suitable for
integrating meters which give an average result after a certain period of
time (see Section 2.8).

When measurements are made according to a particular standard,
guidance is given on whether to use the fast or slow time weightings. In
general the slow setting is used for fluctuating sound. About 2 s is required
for the averaging process to be completed. The fast setting can be used
for steady sound when a quick reading is required.

The following comments refer to a sound level meter with a conventional meter and provide instructions on how to interpret the fluctuations
of the needle. Further information is given by Bruce [8].

If the noise gives rise to a *steady* displacement of the needle of the meter, where steady means that the fluctuations of the needle range over less than about 4 dB with the slow setting, measure the average position of the needle on the meter with either the fast or slow settings.

If the noise causes the needle displacement to be *non–steady* but *continuous* (and the fluctuations are between 4 and 10 dB on slow) use the slow setting and read the upper and lower limits of the needle positions together with the average position of the needle. The true reading will be in excess of the average position, because of the nonlinearity of the scale, so either add 1 dB to the average, or use a more complicated scheme described by Steele [9]. For fluctuations greater than 10 dB regard the noise as intermittent.

For *intermittent* noise record with the meter setting on slow the average position of the needle of the meter during periods when the noise is on and off.

With *repeated impulsive noise*, if the repetition rate is greater than about 10 per second, the noise can be treated as quasi-steady and the slow time weighting should be used.

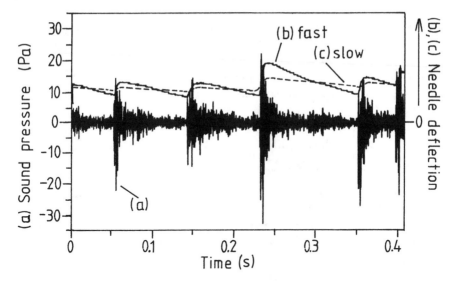

Figure 2.8 Impulsive noise from a tapping machine (a) instantaneous sound pressure; deflection of needle of meter on (b) fast and (c) slow settings.

The effect of the fast and slow settings on the needle displacement of a SLM with a conventional meter is illustrated in Figure 2.8, where the noise record is shown as a repeated impulse from a tapping machine. (A tapping machine is used as a noise source in the measurement of structure–borne sound.) Shown in Figure 2.8 are (a) the instantaneous sound pressure against time, (b) the needle deflection in the fast setting and (c) the approximate needle displacement with the slow setting. The instantaneous trace is obtained from the ac output of the sound level meter (see Figure 2.1), whilst (b) and (c) are derived from the dc output. With the slow setting the needle is displaced only a small amount from a mean.

If *standing waves* are present in a room where measurements are taken, the sound level meter with the slow setting selected should be continuously moved through at least half a wavelength — so that maximum and minimum values of the sound pressure level are obtained — and the average position of the needle on the meter noted, i.e., the meter reading is averaged over distance rather than time. See Section 4.4 for a discussion of standing waves.

2.7.2 Electrical noise level of the SLM and dynamic range

The amplifiers and the other units which are parts of the meter generate their own electrical noise. When measurements of low intensity sound are made care should be taken to ensure that it is not the electrical noise which is being measured. The electrical base noise can be measured by covering the microphone with the pistonphone calibrator and taking a reading of SPL without the calibrator in operation. In this way the pistonphone is providing sound insulation so that the effect of the background noise is excluded.

The highest sound pressure levels that can be obtained depend upon the nonlinearity of the amplifiers or the distortion limit of the microphone.

2.7.3 Frequency response of sound level meter

The frequency response of the system of the microphone and the amplifiers must be considered. In general the manufacturer's information has to be relied upon for the system response. Information about the phase response may not be available and is not usually significant. The ear is not sensitive to phase and as long as rms measurements of noise are taken phase distortion is not important.

2.7.4 Position of the observer when making measurements

If the observer is positioned close to the microphone, sound is reflected from the observer's body on to the microphone and the readings are in error. The SLM should be either placed on a tripod or held as far away as possible from the body of the observer. It is important to avoid crowding around the microphone.

2.7.5 Environmental conditions

Atmospheric pressure affects the calibration made by using the piston-phone. A means of performing the correction, which results from the deviation of the atmospheric pressure from the normal, is supplied with the pistonphone. Atmospheric pressure, in so far as it affects density, controls the relationship between SPL and IL (see equation (1.238)).

Low temperature reduces the effectiveness of the batteries that provide the power for the portable SLM.

Humidity can have serious effects on condenser microphones. If moisture is trapped in the air space between the diaphragm and the backplate then sharp pulses in output occur.

When outdoor measurements are taken wind will be the most serious factor. Noise caused by wind can be as much as 100 dB at a wind velocity of 40 km/hr, and is particularly significant at low frequencies. Specially constructed wind shields are available to reduce this wind noise and should be used for all outdoor measurements. Even indoors, in the absence of air flow, wind shields can be useful, since they provide protection against accidental damage to the microphone. However, for precision measurements indoors the wind shield should be removed.

2.8 Integrating sound level meters

The basic sound level meter is essentially an instrument for measuring the rms sound pressure over a short period of time; it works well as long as the rms sound pressure is not changing too much with time. However, in many industrial and urban environments the noise fluctuates considerably with time. In such circumstances what is required to be measured is a quantity which is proportional to the time average of the total energy detected by the meter. The average total energy is proportional to the true mean square of the sound pressure $p(t)$, obtained after averaging over time T. The time T over which $p^2(t)$ is averaged in general does not denote the period of a harmonic wave. It refers in this case to the measurement (record) time. The product of 10 and the logarithm of the ratio of the mean square pressure to the square of the reference pressure $(2 \times 10^{-5} \mathrm{Pa})$ is defined as the *equivalent continuous sound level*. It is often designated by the symbol L_{eq}. Since the parameter T is important in the measurement of L_{eq}, a more precise designation of the equivalent continuous sound level is $L_{eq(T)}$ or $L_{eq,T}$ (or $L_{Aeq,T}$, if it is wished to emphasise the A weighting).

Hence,

$$L_{eq,T} = 10 \log \left[\frac{1}{T} \int_0^T \frac{p^2(t)dt}{(2 \times 10^{-5})^2} \right] \text{dB.} \tag{2.1}$$

The definition of the equivalent continuous sound level given by equation (2.1), although written differently, is essentially the same as the definition of sound pressure level in equation (1.231). The difference between the designation of sound pressure level and equivalent continuous sound level is really a matter of the difference in the measurement processes. The sound pressure level is a quantity obtained by a sound level meter which is an instrument with an RC averager or a circuit that simulates an RC averager. The mean square value of the sound pressure is obtained after averaging over a short time (about $\frac{1}{4}$ s on the fast setting and 2 s on slow); the sound pressure level which is indicated on the meter is a continuous update which is based on the previous $\frac{1}{4}$ or 2 s of signal.

The equivalent continuous sound level is determined by an instrument called an *integrating sound level meter* [10]. Once a measurement is initiated with an integrating sound level meter (or L_{eq} meter), the L_{eq} indicated on the meter, or on the digital display, is continuously updated, but the update is based upon the square of the sound pressure which is averaged over the elapsed time, i.e., the time from initiation of the measurement to the time when the reading on the meter is observed. Many integrating meters are preset to stop after a specified elapsed time of 5 s, 60 s, 8 h, 24 h, etc.

There are various design principles for integrating sound level meters [11]. The integration of the square of the sound pressure in time, as required by equation (2.1), is not generally carried out. One commonly used procedure is to sample the output of an RC averager which has been passed through a square root circuit and a logarithmic circuit. The sampled level is L_n and for N samples the L_{eq} is given by:

$$L_{eq} = 10 \log \frac{1}{N} \sum_{n=1}^{N} 10^{L_n/10} \text{dB.} \tag{2.2}$$

Hedegaard [11] refers to an instrument in which the RC detector fast (with a time constant of 125 ms) is used and the output from the detector is sampled 128 times per second. If this type of circuitry is used, a conventional sound level meter and an integrating sound level meter can be easily combined in one hand–held unit.

2.9 Tape recorders

For a permanent record of sound signals magnetic tape or cassette recorders may be used. For recording signals in the audio–frequency range the *direct record* principle is sufficient [12]. A good quality direct record system should have a linear frequency response from 25 Hz to 20 kHz ± 1 dB at a tape speed of $7\frac{1}{2}$ in/s, and a signal to noise ratio better than 50 dB.

For the measurement of low frequency signals, i.e., infra–sound or mechanical vibrations, a *frequency modulated* (FM) tape recorder is required. FM recorders have a linear frequency response from dc (0 Hz) to an upper limit which depends upon the tape speed. Typical frequency ranges are: dc to 5 kHz at a tape speed of 15 in/s and dc to 20 kHz at 60 in/s. Electrical noise in tape recorders may interfere with the signal. The signal to noise ratio depends upon the quality of the recorder, but may be as low as 40 dB.

Some large, expensive FM tape recorders are multi–channel (up to 14 tracks) and they use 1 inch wide tape. These often conform to the IRIG standard (inter–range instrumentation group). It is possible to record a signal on one tape recorder and playback on a second, provided both conform to the IRIG standard.

During the recording of sound on a tape recorder it is essential to register a calibration signal and the test signal on the same tape. It is preferable to keep the line input or microphone input controls at the same position for both calibration and test signals.

2.10 Measurement of impulsive sound

When impulsive noise occurs it is often necessary to know the peak sound pressure during the transient process. Some sound level meters have a 'peak' facility which will immediately give the value (in Pa) of the peak of the sound pressure. To be sure that the result is correct, and also to obtain full information how the sound pressure varies with time, it is advisable to obtain a trace of the signal on an oscilloscope, preferably of the storage or digital type.

When measurements are made outside the laboratory it may be necessary to take a tape recording of the signal for subsequent playback in the laboratory. Care is required to ensure that the amplifiers of the tape recorder are not overloaded. Normally, the needle on the meter of the monitoring device is only designed to respond accurately to steady signals. With a transient sound the impulsive displacement of the needle on the monitor may be 10 dB less than the true peak.

An example of the measurement of a transient sound from a diesel-driven pile driver is provided in Figure 2.9. A pistonphone calibration signal and the transient sound were first recorded on magnetic tape. Because of the different sound intensities of the calibration signal and the test signal, the amplifier/attenuator settings of the sound level meter have to be changed. During the calibration of the sound pressure of the test signal with the aid of the pistonphone, it is important to be aware that the pistonphone sound pressure level of 124 dB corresponds to an rms value of 31.7 Pa or an amplitude of 44.8 Pa.

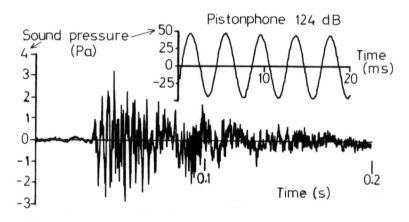

Figure 2.9 Measurement of peak instantaneous sound pressure of a diesel–driven pile driver; pistonphone calibration signal also shown.

As an example, the noise signal from a pile driver is presented in Figure 2.9. The pistonphone was used for calibration. During the calibration the sound level meter (B & K type 2203) was set with the amplifier/attenuator setting on '120 dB'. Hence, for a correctly adjusted SLM the position of the needle on the meter should be 4 dB. In the case of the measurement of the pile driver noise the amplifiers were set to 90 dB to account for the lower level of the signal. Thus the calibration factor obtained from the pistonphone test (part of the pistonphone signal is shown in Figure 2.9) has to be reduced by 31.6 (note that $20\log 31.6 = 30$dB) when applied to the pile driver signal. The peak instantaneous sound pressure from the pile driver in this case is 3.15 Pa, see Figure 2.9.

Some sound level meters have an 'impulse' mode setting with a capacity for holding the needle at the position of maximum displacement. In the impulse mode the time constant of the RC averager is 35 ms. Thus the sound pressure level registered on the meter will be greater than if the slow or fast time settings had been used. However, the peak instantaneous sound pressure is not registered.

2.11 Frequency analysis of sound

It is often useful to know how the energy content of a sound signal varies with frequency. Frequency analysis is carried out by analogue techniques with electrical filters, by digital techniques or by a combination of analogue and digital methods. Analogue techniques deal with continuous signals whilst digital analysis is concerned with discrete values of the signals.

Electrical filters can be classified as *lowpass, highpass, bandpass* or *bandstop*. The first three are illustrated in Figure 2.10. A lowpass filter passes only low frequency signals. Ideally there is a cutoff frequency above which no signal is transmitted out of the filter. In reality the cutoff does not occur discontinuously. There is a gradual decrease in the relative response and the cutoff frequency is that frequency at which the response is 3 dB below the response in the pass band. The decrease in the response is often called the rolloff or the rejection of the filter. Good lowpass filters have a rolloff of 72 dB per octave (i.e., per doubling of frequency); values of 120 dB per octave are possible.

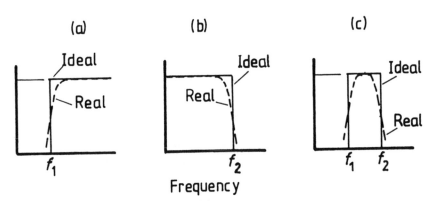

Figure 2.10 Relative frequency response of (a) highpass, (b) lowpass and (c) bandpass filter.

A highpass filter only passes a signal above the cutoff frequency.

A bandpass filter is a combination of a lowpass and a highpass filter. Signals with frequencies in the pass band formed by the lower cutoff (lower limit) and the upper cutoff (upper limit) are transmitted. Bandpass filters are used in frequency analysis. The *centre frequency* and the *pass band* or *bandwidth* are the important parameters with this type of filter. Bandpass filters are either *constant bandwidth* or *constant percentage*.

A constant bandwidth filter has a bandwidth or pass band which is independent of the centre frequency. In a constant percentage filter the bandwidth B is a fixed fraction of the centre frequency f_m, i.e.,

$$\frac{B}{f_m} = k_f, \tag{2.3}$$

where k_f is a constant.

In acoustical analysis the most commonly used filters are one octave and one third octave filters; both are constant percentage filters. Constant bandwidth filters are used for the detection of pure tones in sound signals.

2.12 Octave filters

The term octave is used in music. Two pure tones with frequencies which are related by a factor of two, for example, 100 Hz and 200 Hz, are an octave apart. In an octave filter the frequency of the upper band limit is twice the frequency of the lower band limit.

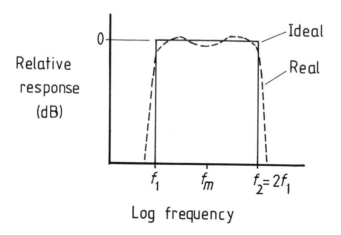

Figure 2.11 Relative frequency response of octave filter.

The relative frequency response of a real and an ideal octave filter is shown in Figure 2.11. The lower band limit is f_1 and the upper band limit f_2. The bandwidth B is given by:

$$B = f_2 - f_1. \tag{2.4}$$

For an octave filter

$$f_2 = 2f_1. \tag{2.5}$$

Octave filters are specified by their centre frequencies [13]. The centre frequency f_m for an octave filter is the geometric mean of the lower and upper band limits:

$$f_m = \sqrt{f_1 f_2}. \tag{2.6}$$

For the case when a logarithmic scale is used for the abscissa, and when the centre frequency f_m is the geometric mean (see equation (2.6)), f_m appears midway between f_1 and f_2, as shown in Figure 2.11. Taking into account the relations (2.4), (2.5) and (2.6), we find that

$$\frac{B}{f_m} = \frac{f_2 - f_1}{\sqrt{f_1 f_2}} = \frac{2f_1 - f_1}{\sqrt{f_1 \cdot 2f_1}} = \frac{1}{\sqrt{2}} = 0.707. \tag{2.7}$$

Thus the ratio B/f_m is a constant, confirming that equation (2.3) is satisfied and that the octave filter is a constant percentage filter. An octave filter is a 70.7% filter. In terms of the centre frequency the expressions for the lower and upper band limits are:

$$f_1 = f_m/\sqrt{2} \tag{2.8}$$

and

$$f_2 = \sqrt{2}f_m. \tag{2.9}$$

The nominal centre frequencies form a preferred number series. The centre frequencies of octave filters which cover the audio range are: 31.5, 63, 125, 250, 500 Hz and 1, 2, 4, 8 and 16 kHz. The exact centre frequencies are given by the formula,

$$f_m(n) = 1000 \times 10^{3n/10}, \tag{2.10}$$

where n is an integer. The band limits are obtained from equations (2.8) and (2.9), and from the equation (2.10) which determines the exact centre frequencies.

A set of octave filters with centre frequencies from 31.5 Hz to 16 kHz can be attached to the sound level meter to form a single hand–held unit. The filters are generally connected electrically between the input and output amplifiers/attenuators in place of the weighting networks, see Figure 2.1. Each filter is switched into operation in turn and for each filter a reading in dB (re 2×10^{-5} Pa) is obtained and this reading is called an *octave band pressure level* (BPL). The octave BPL can be plotted against f_m on a logarithmic scale in the form of a histogram plot, as shown in Figure 2.12a, or in the form of a curve joining up the values of the BPLs at the centre frequencies (Figure 2.12b). The former is preferable, but the latter more usual. As f_m is expressed on a logarithmic scale, the separation between two subsequent centre frequencies corresponds to an equal interval on the abscissa.

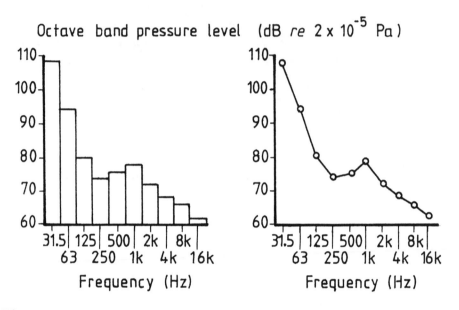

Figure 2.12 Presentation of octave band pressure levels with respect to the centre frequency f_m as (a) histogram and (b) curve.

When measuring octave band pressure levels in the field the operator should also measure the A–weighted sound level and the linear sound pressure level, see Section 2.4. If the operator is recording the results by hand, the measurements will take at least 2 minutes for a noise which does not cause the needle of the meter to fluctuate more than, say, 4 dB. A check should be carried out to see if the results have been recorded correctly, or to assess whether the noise has changed during measurement. The octave band pressure levels can be 'added' according to the decibel addition rule to obtain the overall or linear sound pressure level. The octave BPLs may be summed directly, according to equation (1.247), or the graphical method outlined below may be used. In the example, illustrated below, the values 61, 68... to 68 refer to the octave band pressure levels (in dB) obtained by using filters with centre frequencies from 31.5 Hz to 8 kHz. In the graphical method each pair of octave BPLs is added with the help of Figure 1.23, see also equation (1.253).

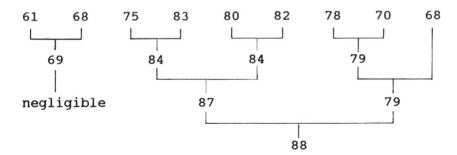

Hence, the linear sound pressure level is 88 dB.

A similar procedure may be used to obtain the A-weighted sound level. The relative response of the A weighting network (in dB) is added to the band pressure levels. From Table 2.1 it is seen that at 31.5 Hz the relative response of the A weighting network is approximately −39 dB; at 63 Hz it is −26 dB, etc.

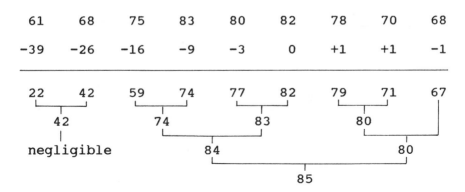

The A–weighted sound level is 85 dB(A). This result should agree with the measurement of the A–weighted sound level.

In carrying out the above procedure it does not matter which pairs of results are taken first, although more accuracy is achieved if large values or identical values are paired off first.

2.13 One third octave filters

For a more detailed frequency analysis one third octave filters can be used. With one third octave filters the frequency range covered by an octave filter is divided into three parts, as shown in Figure 2.13. The ideal filters, shown in Figure 2.13, have centre frequencies of f_{m1}, f_{m2} and f_{m3}. The centre frequencies are the geometric means of the pass band limits. Since $f_a/f_1 = f_b/f_a = f_2/f_b = \sigma$, where σ is the common ratio for the geometric series: f_1, f_a, f_b, f_2, and since $f_2 = 2f_1$, we obtain $\sigma = 2^{1/3}$.

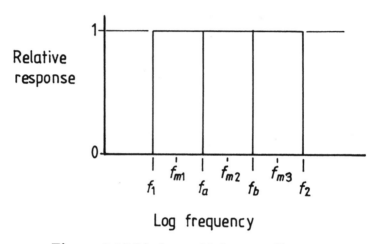

Figure 2.13 Ideal one third octave filters.

Consequently, the sequence of the pass band limits of the one third octave filters can be presented in an explicit form as the geometric series: $f_1, f_1\sigma, f_1\sigma^2, f_1\sigma^3$, etc.. Similarly, the centre frequencies of the one third octave filters f_{mi}, where $i = 1, 2, 3$, etc., also creates a geometric series with the common ratio $\sigma = 2^{1/3}$, namely, $f_{mi} = f_{m1}\sigma^{i-1}$, where $i = 1, 2, 3$, etc. For the one third octave filter the ratio k_f of the bandwidth to the centre frequency, see also equation (2.3), is given as follows:

$$k_f = \frac{f_a - f_1}{f_{m1}} = \frac{f_1(\sigma - 1)}{f_1\sqrt{\sigma}} = \frac{\sigma - 1}{\sqrt{\sigma}} = 2^{1/6} - 2^{-1/6} = 0.23156.$$

Hence, a one third octave filter is a 23% constant percentage filter.

For a one third octave filter the lower band limit is given by $0.8909 f_m$ and the upper band limit by $1.1225 f_m$, where f_m is the centre frequency. The exact centre frequencies are quoted as $1000 \times 10^{n/10}$, where n is an integer [14].

2.13.1 $(1/q)$th octave filters

The frequency range of the octave filter can be completely covered by q filters. These are $(1/q)$th octave filters, where q denotes a natural number.

Let $f_{m,\text{oct}}$ denote the centre frequency of the octave filter covered by the $(1/q)$ th octave filters and f_{mi} denote the centre frequency of the i th filter in the sequence of $(1/q)$ th octave filters, see Figure 2.14.

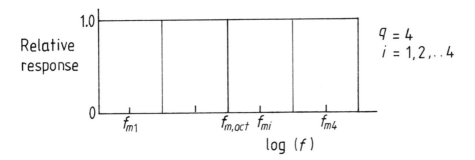

Figure 2.14 Sequence of $(1/q)$ th octave filters.

The centre frequencies of the $(1/q)$ th octave filters formulate a geometric series with the common ratio $\sigma = 2^{1/q}$, namely,

$$f_{mi} = f_{m,\text{oct}} \times 2^{(1/q)(i - \frac{q+1}{2})}, \tag{2.11}$$

where $i = 1, \ldots q$, and the lower and upper band limits of the $(1/q)$ th octave filters are, respectively,

$$f_{li} = f_{mi} \times 2^{-1/2q}, \tag{2.12a}$$
$$f_{ui} = f_{mi} \times 2^{1/2q}. \tag{2.12b}$$

For the $(1/q)$ th octave filter the ratio of the bandwidth to the centre frequency is expressed by the formula

$$\frac{B}{f_{mi}} = 2^{1/2q} - 2^{-1/2q}. \tag{2.13}$$

In certain problems it is essential to design filters in such a manner that the arbitrarily chosen frequency, e.g., 1000 Hz, is the centre frequency of the filter, as opposed to having an octave filter covered by q filters. The formula

$$f_{mn} = 1000 \times 2^{n/q}, \tag{2.14}$$

where $n = -N, -(N-1), \ldots, 0, \ldots, M$, and M and N are natural numbers, gives the centre frequency of the filters in the sequence of $(1/q)$ th octave filters.

The lower and upper limits of the filters, f_{ln} and f_{un}, respectively, are defined by equations (2.12). Hence, taking into account also equation (2.14)

$$f_{ln} = f_{mn} \times 2^{-1/2q} = 1000 \times 2^{(n/q-1/2q)}$$

and

$$f_{un} = f_{mn} \times 2^{1/2q} = 1000 \times 2^{(n/q+1/2q)}.$$

Equation (2.14) can be presented in another form

$$\log f_{mn} = 3 + \frac{n}{q} \log 2. \tag{2.15}$$

Since $\log 2 \simeq 0.3$, we obtain from equation (2.15)

$$\log f_{mn} = 3 + \frac{3n}{10q} \tag{2.16}$$

or

$$f_{mn} = 10^3 \times 10^{3n/10q} \tag{2.17}$$

which is an approximation of equation (2.14). For octave filters ($q = 1$) equation (2.17) is reduced to equation (2.10). For one third octave filters $q = 3$.

2.13.2 International standard on octave and one third octave filters

Numerical values of the exact centre frequencies and the band limits for one third octave and octave filters are included in Table 2.2. Details of the octave filters are given in heavy type. The table is based upon the procedures laid down in the international standard [14]. In the standard there is reference to both a nominal and an exact centre frequency. The exact centre frequency is given by equation (2.17) with $q = 1$ for octave and $q = 3$ for one third octave filters. For example, the octave filter with both a nominal and exact centre frequency of 1000 Hz has a lower band limit of 708 Hz and an upper band limit of 1413 Hz.

Tolerances on the frequency response of the octave and one third octave filters are given in the international standard [14]. The bounds within which the octave and one third octave filters should be designed to satisfy the international standard are shown in Figure 2.15 for the upper part of the filter frequency responses. The requirements for the relative response (or in the terminology of the standard the attenuation Δ

TABLE 2.2 Centre frequencies and band limits for octave and one third octave centre frequencies

Nominal centre frequency (Hz)	Exact centre frequency f_m (Hz)	Approximate lower band limit f_1 (Hz)	Approximate upperband limit f_2 (Hz)
12.5	12.59	**11.3**	14.1
16	**15.85**	14.1	17.8
20	19.95	17.8	**22.4**
25	25.12	**22.4**	28.2
31.5	**31.62**	28.2	35.5
40	39.81	35.5	**44.7**
50	50.12	**44.7**	56.2
63	**63.10**	56.2	70.8
80	79.43	70.8	**89.1**
100	100.0	**89.1**	112.2
125	**125.9**	112.2	141.2
160	158.5	141.2	**177.8**
200	199.5	**177.8**	223.9
250	**251.2**	223.9	281.8
315	316.2	281.8	**354.8**
400	398.1	**354.8**	446.7
500	**501.2**	446.7	562.3
630	631.1	562.3	**708.1**
800	794.3	**708.1**	891.3
1000	**1000.0**	891.3	1122.0
1250	1258.9	1122.0	**1413.0**
1600	1584.9	**1413.0**	1778.0
2000	**1995.3**	1778.0	2239.0
2500	2511.9	2239.0	**2818.0**
3150	3162.3	**2818.0**	3548.0
4000	**3981.1**	3548.0	4467.0
5000	5012.9	4467.0	**5623.0**
6300	6309.6	**5623.0**	7080.0
8000	**7943.3**	7080.0	8913.0
10000	10000	8913.0	**11220.0**
12500	12590	**11220.0**	14126.0
16000	**15850**	14126.0	17846.0
20000	19950	17846.0	**22390.0**

expressed in dB) are shown in tabular form in Table 2.3 for both octave and one third octave filters.

Most manufacturers will aim to design their filters so that the relative

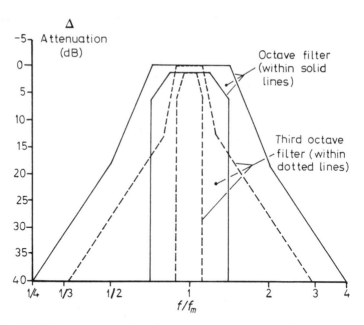

Figure 2.15 Tolerances on the frequency responses of octave and one third octave filters, as required by international standard [14].

frequency response is 3 dB down at the band limits f_1 and f_2. However, the response could be 6 dB down and still satisfy the international requirements, see Figures 2.11 and 2.15.

2.14 Band pressure level (BPL) and spectrum level (SL)

The output from a bandpass filter depends upon the sound which is being measured and the filter characteristics. For a broad band random sound the larger the bandwidth of the filter the greater is the *band pressure level.* In general at a certain location in a sound field the signal may be defined by the sound pressure as a function of time. Knowing $\overline{p^2(t)}$, we can determine the sound pressure level from equation (1.231). In performing a spectral analysis of the sound signal, we are interested in the filtered signal, i.e., in the primary signal component $p_{f_m,B}(t)$, the frequency content of which is within the bandwidth B and of centre frequency f_m. The band pressure level is defined as

$$\mathrm{BPL} = 10\log\left[\frac{\overline{p^2_{f_m,B}(t)}}{(2 \times 10^{-5})^2}\right] \mathrm{dB}, \qquad (2.18)$$

TABLE 2.3 Attenuation Δ of octave and one third octave filters with respect to centre frequency f_m

Octave filters	One third octave filters	Attenuation (dB)
From $0.8409 f_m$ to $1.1892 f_m$	From $0.9439 f_m$ to $1.0595 f_m$	$-0.5 \leq \Delta \leq 1$
From $0.7071 f_m$ to $1.4142 f_m$	From $0.8909 f_m$ to $1.1225 f_m$	$-0.5 \leq \Delta \leq 6$
At $0.5 f_m$ and $2 f_m$	—	$\Delta \geq 18$
—	At $0.7937 f_m$ and $1.2599 f_m$	$\Delta \geq 13$
Less than $0.25 f_m$ and greater than $4 f_m$		$\Delta \geq 40$
—	Less than $0.25 f_m$ and greater than $4 f_m$	$\Delta \geq 50$
Less than $0.125 f_m$ and greater than $8 f_m$	Less than $0.125 f_m$ and greater than $8 f_m$	$\Delta \geq 60$

where the extended bar indicates time average. Consequently, $p_{f_m,1}(t)$ denotes the primary signal component with a frequency content within the bandwidth $B = 1$ Hz and of centre frequency f_m. Both signal components can be simply related to each other, provided that the noise contains no pure tone and is fairly uniform over the bandwidth B, by the expression,

$$B\left[\overline{p_{f_m,1}^2(t)}\right] = \overline{p_{f_m,B}^2(t)}. \tag{2.19}$$

Defining a *spectrum level* (SL) as

$$SL = 10 \log\left[\frac{\overline{p_{f_m,1}^2(t)}}{(2 \times 10^{-5})^2}\right] \text{dB} \tag{2.20}$$

we obtain

$$SL = BPL - 10 \log B \text{ dB}. \tag{2.21}$$

The spectrum level characterises a property of the noise and is independent of the type of filter used in carrying out the frequency analysis, see definition (2.20).

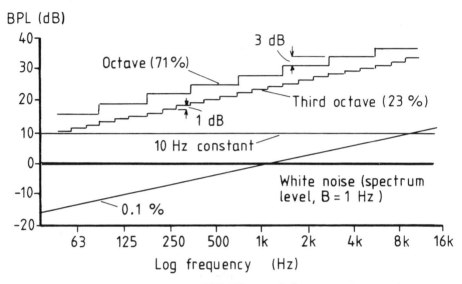

Figure 2.16 Comparison of band pressure levels of white noise as measured with different bandpass filters.

For a given noise the band pressure levels vary for measurements with different filters. In Figure 2.16 the results are shown for white noise measured by analysers with different kinds of filters. *White noise* is a random noise for which the spectrum level is constant and in the case presented in Figure 2.16 is 0 dB. When the white noise is measured with a filter with a constant bandwidth of 10 Hz, the band pressure level is $10 \log 10$ or 10 dB greater than the spectrum level (see equation (2.21)). In the case of a constant percentage filter the bandwidth increases with frequency. For a 0.1% constant percentage filter the bandwidths are 0.5, 1 and 2 Hz at centre frequencies of 500, 1000 and 2000 Hz, respectively. Thus the band pressure levels are -3 dB, 0 dB and $+3$ dB at these frequencies. If the centre frequency is changed continuously, the BPL against frequency appears as a straight line of slope 3 dB/octave.

Analysis by octave and one third octave filters is normally carried out by a set of contiguous filters that cover the entire frequency range; the centre frequency is changed in a discrete manner rather than continuously. Thus the band pressure levels from the octave and one third octave filters increase with frequency in steps at a rate of 3 dB/octave. In the case, for example, of the octave filter with centre frequency 1000 Hz the band pressure level is $10 \log 707$ or 28.5 dB greater than the spectrum level.

A noise source is often required which when measured by constant percentages filters provides a band pressure level which is constant with frequency. Random noise which produces a constant BPL with octave or one third octave filters is called *pink noise*. The spectrum level of pink noise is of the form

$$SL = K - 10 \log f \text{ dB}, \qquad (2.22)$$

where K is a constant and f is the frequency.

2.15 Power spectral density function

The power spectral density function (PSD) is a quantity of considerable theoretical significance which is formally defined in terms of an auto–correlation function [15]. The concept of PSD is used to analyse the spectral properties of signals such as the energy characteristics of signals in the frequency domain and is specifically applicable to the analysis of random signals. Signals are specified by time–varying quantities, which appear in the formalism describing such phenomena as sound, vibration, surface roughness, etc. In particular a sound signal is defined as the sound pressure dependence on time.

The *one–sided power spectral density function* $\mathcal{G}(f)$ is a function of frequency and is a characteristic property of a signal. The way in which $\mathcal{G}(f)$ might vary with frequency is shown in Figure 2.17. The mean–squared sound pressure $\overline{p^2}$ for the entire signal is equal to the area under the curve, i.e.,

$$\overline{p^2} = \lim_{T \to \infty} \left[\frac{1}{T} \int_0^T p^2(t)dt \right] = \int_0^\infty \mathcal{G}(f)df, \qquad (2.23)$$

where T denotes the time of the record (observation time).

The units of the power spectral density function in this case are Pa^2/Hz. The function could be in terms of, e.g., sound intensity, in which case the units would be $Wm^{-2}Hz^{-1}$. Hence, the units of the power spectral density function depend upon the type of signal being analysed.

The one–sided power spectral density function $\mathcal{G}(f)$ is defined for frequencies $(0, \infty)$. However, the concept of the *two–sided power spectral*

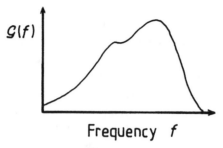

Figure 2.17 Power spectral density function.

density function $S(f)$, which is defined in terms of the Fourier transform of the auto–correlation function [15], formally introduces the range of frequencies $(-\infty, \infty)$. The one–sided spectral density function $G(f)$ is related to the two–sided spectral density function $S(f)$ as follows: $G(f) = 2S(f)$. The plot of power spectral density function versus frequency f is called the *power spectrum* of the signal. It should be noted that the same name is used for the integral of the power spectral density function as a function of frequency, see, for example, Sections 2.21 and 2.22.

The mean–squared value for the primary signal component $p_{f,B}(t)$ with frequency content in the bandwidth B and of centre frequency f can be expressed by the power spectral density function $G(f)$, as follows:

$$\overline{p_{f,B}^2(t)} = \lim_{T\to\infty} \left[\frac{1}{T} \int_0^T p_{f,B}^2(t)dt \right] = G(f,B),\qquad(2.24)$$

where

$$G(f,B) = \int_{f-B/2}^{f+B/2} G(f_0)df_0.\qquad(2.24a)$$

For sufficiently small bandwidth B

$$\overline{p_{f,B}^2} = G(f,B) \simeq G(f) \times B.\qquad(2.25)$$

The formulae (2.24) and (2.25) imply that the power spectral density function can be precisely defined as:

$$G(f) = \lim_{B\to 0} \frac{1}{B} \left[\lim_{T\to\infty} \frac{1}{T} \int_0^T p_{f,B}^2(t)dt \right] = \lim_{B\to 0} \frac{\overline{p_{f,B}^2(t)}}{B}.\qquad(2.26)$$

Combining formulae (2.18) and (2.24) leads to the relation between $G(f, B)$ and the band pressure level in the bandwidth B:

$$\text{BPL} = 10 \log \left[\frac{G(f, B)}{(2 \times 10^{-5})^2} \right] \text{dB.} \qquad (2.27)$$

Note that $G(f, B)$ denotes the area under the curve of power spectrum $\mathcal{G}(f)$ between the frequencies $f - B/2$ and $f + B/2$, see Figure 2.17. Thus, for example, the band pressure level in the octave band of centre frequency 1000 Hz is given by

$$\text{BPL} = 10 \log \left[\frac{\int_{708}^{1413} \mathcal{G}(f) df}{(2 \times 10^{-5})^2} \right] \text{dB.}$$

In general if the power spectral density function is fairly constant in the pass band of the filter:

$$\text{BPL} = 10 \log \left[\frac{\mathcal{G}(f) \times B}{(2 \times 10^{-5})^2} \right] \text{dB}$$

or,

$$\text{BPL} = 10 \log \mathcal{G}(f) + 10 \log B + 94 \text{ dB.} \qquad (2.28)$$

Comparison of equations (2.21) and (2.28) shows that

$$\text{SL} = 10 \log \mathcal{G}(f) + 94 \text{ dB.} \qquad (2.29)$$

The concept of power spectral density function can be a useful aid in interpreting the results from analysis with different bandpass filters. The following example provides an illustration.

Example 2.1

The noise from a machine is mainly random in nature and was measured at a given point by a one third octave analyser with centre frequencies of 800, 1000 and 1250 Hz; the band pressure levels were 72, 74 and 76 dB, respectively. At a subsequent date noise control measures were carried out on the machine. Unfortunately, the only analyser available had a filter with a constant bandwidth of 125 Hz. The analyser was of the type in which the centre frequency of the filter could be continuously changed. Band pressure level readings were taken at 100 Hz intervals from

700 to 1500 Hz. The levels were 60, 58, 60, 61, 63, 64, 61, 60 and 60 dB. What is the attenuation achieved by the noise control measures in the octave band of centre frequency 1000 Hz?

Solution

Of course, this is no way to carry out a test, but it illustrates the point we wish to make. Table 2.4a shows the stages in obtaining the spectrum level SL and the power spectral density function $\mathcal{G}(f)$ for the unsilenced machine. Similar results are shown in Table 2.4b for the silenced machine. The power spectral density function for the unsilenced machine is shown in Figure 2.18a. The frequency is presented on a linear scale and the PSD function is shown in the form of a histogram the (the bandwidths can be obtained from Table 2.2). In the case of the silenced machine, as there is some overlap in the BPL measurements, the PSD function is plotted as a continuous curve which joins up the measurement points. See Figure 2.18b.

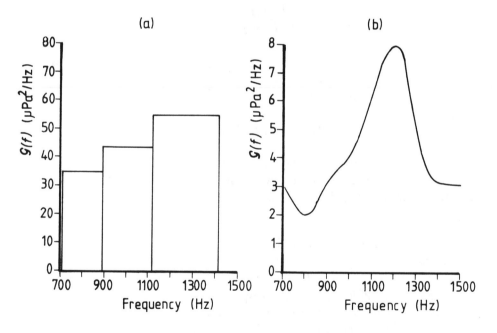

Figure 2.18 Power spectral density function (Example 2.1) (a) unsilenced machine (b) silenced machine.

2 THE MEASUREMENT AND ANALYSIS OF SOUND 119

TABLE 2.4a Example 2.1: power spectral density function for unsilenced machine

One third octave centre frequency (Hz)	BPL (dB)	Bandwidth B (Hz)	10logB	SL (dB)	$\mathcal{G}(f)$ (Pa²/Hz)
800	72	183.3	22.6	49.4	3.48×10^{-5}
1000	74	230.7	23.6	50.4	4.39×10^{-5}
1250	76	291.0	24.6	51.4	5.52×10^{-5}

TABLE 2.4b Example 2.1: power spectral density function for silenced machine

Constant bandwidth centre frequency (Hz)	BPL (dB)	Bandwidth B (Hz)	10log B	SL (dB)	$\mathcal{G}(f)$ (Pa²/Hz)
700	60	125	21.0	39.0	3.18×10^{-6}
800	58	125	21.0	37.0	2.00×10^{-6}
900	60	125	21.0	39.0	3.18×10^{-6}
1000	61	125	21.0	40.0	4.00×10^{-6}
1100	63	125	21.0	42.0	6.34×10^{-6}
1200	64	125	21.0	43.0	7.98×10^{-6}
1300	61	125	21.0	40.0	4.00×10^{-6}
1400	60	125	21.0	39.0	3.18×10^{-6}
1500	60	125	21.0	39.0	3.18×10^{-6}

The area under the PSD function between frequencies of 708 and 1413 Hz provides the mean–squared sound pressure in that bandwidth. In the case of the unsilenced machine the area can be determined easily as $(183.3 \times 3.48 + 230.7 \times 4.39 + 291.0 \times 5.52)10^{-5}$ Pa² and finally amounts to

$$3.26 \times 10^{-2}\,\text{Pa}^2.$$

For the silenced machine the area can be estimated with sufficient accuracy by a graphical method. The area is

$$3.06 \times 10^{-3}\,\mathrm{Pa}^2.$$

The octave band pressure level of the unsilenced machine, see formulae (2.27) and (2.28), is

$$\mathrm{BPL} = 10\log\left[\frac{3.26 \times 10^{-2}}{(2 \times 10^{-5})^2}\right] = 79.1\,\mathrm{dB}$$

and for the silenced machine:

$$\mathrm{BPL} = 10\log\left[\frac{3.06 \times 10^{-3}}{(2 \times 10^{-5})^2}\right] = 68.8\,\mathrm{dB}.$$

Thus the attenuation in the octave band is 10.3 dB.

Note that the octave band pressure level for the unsilenced machine could have been more easily obtained directly from the values of the one third octave band pressure levels by addition of the mean–squared sound pressures p_{800}^2, p_{1000}^2, p_{1250}^2 in the one third octave bands with centre frequencies 800, 1000 and 1250 Hz, respectively. Thus the octave band pressure level of the unsilenced machine is

$$10\log[(1.585 + 2.512 + 3.981) \times 10^7] = 79.1\,\mathrm{dB}.$$

In the example the one third octave levels are 72, 74 and 76 dB, corresponding to the 800, 1000 and 1250 Hz one third octaves. If the levels had been 76, 74 and 72 dB in the 800, 1000 and 1250 one third octave bands, respectively, the same octave band pressure level of 79.1 dB would have been obtained.

2.16 Narrow band analysis using analogue techniques [16]

Many manufacturers make hand–held sound level meters to which an octave or one third octave filter box can be attached. These boxes generally contain active filters with operational amplifiers, so that the final combined unit is still light enough to be hand–held for field use.

Analysis with filters narrower than one third octave is generally confined to the laboratory, if analogue devices are used. (With digital signal processing equipment it is possible to construct equipment which is small and light, suitable for field use). Analogue techniques require a long analysis time and bulky equipment. The narrower the filter the longer is the time required to analyse the sound. The *normalised standard error* ϵ (ratio of standard deviation to the mean square value) in the power spectral density function of a random signal is given by [17]

$$\epsilon = 1/\sqrt{BT_A}, \tag{2.30}$$

where B is the bandwidth and T_A is the averaging time. A typical wave analyser incorporating a narrow band filter is a swept device and the scan rate S_r must not be too great and should satisfy the inequality:

$$S_r < B/T_A. \tag{2.31}$$

Thus, if $B = 2\,\text{Hz}$ (constant bandwidth) and $T_A = 2\,\text{s}$, the error in PSD is 50% and the scan rate should be less than 1 Hz/s. To scan frequencies from 20 to 2000 Hz will take 1980 s, or over 30 minutes, and even then the accuracy will be poor. Equation (2.30) applies to a random signal; discrete tones can be analysed more quickly.

2.17 Digital signal analysis

Narrow band analysis of sound by analogue methods can take a long time. Analogue techniques are mostly confined to octave and one third octave analysis in which stepped filters are used. A more detailed frequency analysis generally makes use of digital techniques. Digital signal analysers are used because they are able to process acoustical signals — or any other signal — much quicker than analogue equipment. Also digital analysers are often linked to a small computer, or have capabilities similar to that of a computer; consequently it is possible to store and process data, thus giving great flexibility to digital analysis.

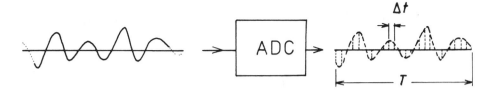

Figure 2.19 The sampling process.

Analogue equipment deals with a continuous signal from a sound level meter or tape recorder. In the case of digital analysis a signal corresponding to a finite time is analysed. This signal must be sampled or digitised by an *analogue to digital converter* (ADC). The sampling process is illustrated in Figure 2.19. The output is a series of numbers which should accurately represent the section of the analogue signal which was digitised. The output is, for example, a series of N numbers from p_0 to p_{N-1}, which represents the sound pressure signal. If the interval of time between any two data points is Δt, the time T of the record is given by:

$$T = N\Delta t. \qquad (2.32)$$

2.18 The discrete Fourier transform (DFT)

In this section Fourier transform theory [6,18,20] is applied to determine the monochromatic components of a signal. In the case when the signal is finite and sampled in time the finite and discrete Fourier transform must be used. The time series $\{p_n\}$, which is the discrete time representation of the signal $p(t)$ after its segmentation and sampling, can be transformed from the discrete time domain to the discrete frequency domain by using the *discrete Fourier transform* (DFT). In other words by applying the DFT we can obtain the frequency domain representation of the primary signal of finite length T. The time sequence $\{p_n\}$ is assumed to be periodic in the time domain with period T. The discrete Fourier transform $\{\mathcal{F}_j\}$ of the time series $\{p_n\}$ is defined in the form of a frequency domain series (series of Fourier components $\mathcal{F}_j$):

$$\mathcal{F}_j = \mathcal{F}(j, T) = \frac{T}{N} \sum_{n=0}^{N-1} p_n e^{-i2\pi jn/N}, \qquad (2.33)$$

where $i = \sqrt{-1}$ and where $j = 0, 1, \ldots, N-1$. The *inverse discrete Fourier transform* (IDFT) of the series $\{\mathcal{F}_j\}$ is given by the time series $\{p_n\}$, namely,

$$p_n = \frac{1}{T} \sum_{j=0}^{N-1} \mathcal{F}_j e^{i2\pi jn/N}, \qquad (2.34)$$

where $n = 0, 1, \ldots\ldots, N-1$. The finite time interval of the signal record is denoted by T and p_n refers to the nth data point in the sampled signal, or in other words to the nth sample of the time series $\{p_n\}$. $N = T/\Delta t$, where Δt denotes the time interval between any two data points p_n. Both n and j are integers. The integer n corresponds to time and the integer

j to frequency, since $t = n\Delta t$ and $\omega_j = 2\pi f_j = 2\pi j/T = 2\pi j\Delta f$. The reciprocal of T is the spectrum resolution Δf:

$$\Delta f = 1/T. \tag{2.35}$$

The Fourier component $\mathcal{F}_j$ is called also the j th coefficient of the DFT.

In general to have a scale factor equal to 1 in front of the summation sign in equation (2.34) a 'new' Fourier component F_j may be defined as

$$F_j = \mathcal{F}_j/T \tag{2.36}$$

and this modification leads to the other form of the discrete Fourier transform pair, namely:

$$F_j = \frac{1}{N}\sum_{n=0}^{N-1} p_n e^{-i2\pi jn/N}, \tag{2.37}$$

where $j = 0, \ldots, N-1$ and

$$p_n = \sum_{j=0}^{N-1} F_j e^{i2\pi jn/N}, \tag{2.38}$$

where $n = 0, \ldots, N-1$.

The discrete Fourier transform pair $\{F_j\}$ and $\{p_n\}$, defined by equations (2.37) and (2.38), is usually quoted in the literature [6,21,22] and often used in spectral analysis.

The Fourier components F_j are complex quantities. The plot presenting values of $|F_j|$ versus frequencies f_j has a repeatable pattern, namely, values of $|F_j|$ for $f_j > F_{\max}$, where

$$F_{\max} = \frac{N}{2T} = \frac{N}{2}\Delta f, \tag{2.39}$$

are repetitions of those at frequencies below $F_{\max}$, see also Section 2.21. Consequently, the correct Fourier components in equation (2.38) are only those referring to frequencies $f_j \leq F_{\max}$. Hence, the maximum value of integer j reached in the summation in equation (2.38) should be $N/2 - 1$, instead of $N - 1$.

The plot showing the distribution of the $N/2$ values of the moduli of the Fourier components $|F_j|$ in the discrete frequency domain is a *discrete* or *line spectrum*. The name line spectrum is also used for the plots of

$\text{Re}(F_j)$ and $\text{Im}(F_j)$ versus f_j frequencies. The line spectrum of the signal is a representation of the discrete Fourier transform in the frequency domain and is visualised in the form of spectral lines with a resolution between lines of Δf. The resolution Δf is the reciprocal of the time of record T and can be compared with the bandwidth B of the analogue filter.

From equation (2.37) we can conclude that

$$F_0 = \frac{1}{N} \sum_{n=0}^{N-1} p_n \tag{2.40}$$

is the mean value of the series $\{p_n\}$. The Fourier component F_1 corresponds to a frequency $f_1 = \Delta f$, i.e., it represents the amplitude and phase of the harmonic component of the signal with a period equal to T, the time of record. It should be noted that any signal spectrum obtained using the DFT method for any arbitrary signal is discrete and has finite resolution since the time interval T is finite.

Example 2.2

As an illustration of the discrete Fourier transform consider a discrete time series $\{p_n\}$ of a sound pressure signal, for which $N = 16$, as shown in Figure 2.20. Obtain the DFT.

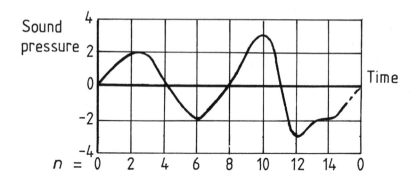

Figure 2.20 Discrete time series for Example 2.2.

Solution

The Fourier component F_0, corresponding to $j = 0$, is given by

$$F_0 = \frac{1}{16} \sum_{n=0}^{15} p_n$$

since the exponential term is unity. Thus, approximately,

$$F_0 = (0 + 1 + 2 + 2 + 0 - 1 - 2 - 1 + 0 + 2$$
$$+ 3 + 0 - 3 - 2 - 2 - 1)/16$$
$$= -0.125 .$$

The Fourier component F_0 is the mean value or dc value.

Normally, for a time of recording T which is sufficiently long and for a time interval Δt sufficiently small the mean value is zero, as sound pressure is a variation of air pressure relative to the atmospheric pressure. With a short sample a mean value which deviates from zero is possible.

The Fourier component F_1 is given by:

$$F_1 = \frac{1}{16}\left[p_0 + p_1 e^{-i2\pi \times 1 \times 1/16} + p_2 e^{-i2\pi \times 1 \times 2/16}\right.$$
$$\left. + \ldots + p_{15} e^{-i2\pi \times 1 \times 15/16}\right].$$

Using Euler's formula $e^{-i\theta} = \cos\theta - i\sin\theta$, we have:

$$F_1 = \frac{1}{16}\left[\left(\cos\frac{\pi}{8} - i\sin\frac{\pi}{8}\right) + 2\left(\cos\frac{\pi}{4} - i\sin\frac{\pi}{4}\right)\right.$$
$$\left. + 2\left(\cos\frac{3\pi}{8} - i\sin\frac{3\pi}{8}\right) + \ldots - \left(\cos\frac{15\pi}{8} - i\sin\frac{15\pi}{8}\right)\right]$$
$$= \frac{1}{16}\left[(0.924 - i0.333) + 2(0.707 - i0.707)\right.$$
$$\left. + 2(0.383 - i0.924) + \ldots - (1.248 - i0.105)\right]$$
$$= \frac{1}{16}(-1.243 - i1.682)$$
$$= -0.078 - i0.105 .$$

The values of the other Fourier components F_2 to F_7 may be calculated in a similar manner. The line spectra of the signal as the real and imaginary parts of the DFT in the frequency domain are shown separately in Figure 2.21 as (a) and (b). Information about the frequency content of the signal is only provided at discrete frequencies at the lines corresponding to $j = 1, 2$, etc. It is always advisable to be aware that the discrete Fourier transform $\{F_j\}$ has a line spectrum. The discrete spectrum is a consequence of the method rather than of the type of signal. Thus, if one is

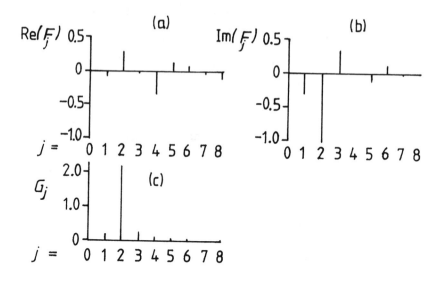

Figure 2.21 Line spectra (Example 2.2), (a) real part of DFT (b) imaginary part of DFT (c) power spectrum.

attempting to determine a tone in a signal which has a frequency not corresponding to the lines of the spectrum of the signal, a correct result cannot be obtained. In this simple example it is clear that the spectra obtained by the DFT method are discrete. In a real measurement, where N is sufficiently large, the lines of the frequency spectrum will be situated very close together. Consequently, the spectrum presented on a screen or by a plotter can be seen as continuous.

2.18.1 Power spectrum

Most commercially available instruments that are based on the discrete Fourier transform provide as, the final output, a power spectrum instead of spectra for the real and imaginary parts of the transform.

The plot of power spectral density function against frequency is the power spectrum (auto–spectrum) of the signal, see [6, 12] and Section 2.15. The 'power spectrum' or 'auto–spectrum' presented as an output from digital instruments is often the distribution of $G(f, \Delta f)$ in the frequency domain, see also equation (2.24a). Thus the function $G(f, \Delta f)$ defined as

$$G(f, \Delta f) = \int_{f-\Delta f/2}^{f+\Delta f/2} \mathcal{G}(f_0) df_0 \qquad (2.41)$$

will be also called the *power spectrum* in the frequency domain. The function $\mathcal{G}(f)$ is the power spectral density function, see Section 2.15. The function $G(f, \Delta f)$ denotes the *power of the signal harmonics* with frequencies within the limits $f - \Delta f/2$ and $f + \Delta f/2$ and is equal to the mean–squared value of $p(t)$, namely, $G = p_{f, \Delta f}^2(t)$. The power spectrum can be obtained directly using the DFT method. From Parseval's theorem [23] it can be proved that the discrete Fourier transform defined by equation (2.33) relates to the power spectral density function $\mathcal{G}(f)$ as

$$\mathcal{G}_j = \mathcal{G}(f_j) = \lim_{T \to \infty} \frac{2}{T} |\mathcal{F}_j|^2. \qquad (2.42)$$

Consequently, for a sufficiently small resolution, see equation (2.41), we have for G (and $\mathcal{G}$) estimate:

$$G_j = G(f_j, \Delta f) = \mathcal{G}(f_j) \Delta f \simeq \frac{2}{T^2} |\mathcal{F}_j|^2, \qquad (2.43)$$

where G_j denotes the power of the signal harmonic of frequency j/T Hz. On the other hand since $F_j = \mathcal{F}_j/T$, see equation (2.36), we obtain:

$$G_j \simeq 2|F_j|^2 = 2F_j F_j^*. \qquad (2.44)$$

Hence, the power spectrum is a sequence $\{G_j\} = \{2F_j F_j^*\}$ in the frequency domain, where F_j and F_j^* denote the Fourier component and its complex conjugate, respectively. Since the Fourier components are complex numbers, they can be presented as:

$$F_j = a_j + ib_j \quad \text{and} \quad F_j^* = a_j - ib_j,$$

where a_j and b_j are the real and imaginary parts of the complex quantity F_j. Consequently, the power spectrum is a sequence of real numbers $\{G_j\} = \{2(a_j^2 + b_j^2)\}$ in the frequency domain. Note that $\mathcal{G}_j = G_j \times T$. The power spectrum $\mathcal{G}(f)$ (power spectral density function in the frequency domain) relates to the power spectrum $G(f)$ via the constant factor T.

The power spectrum $G(f)$ of the signal (Example 2.2) shown in Figure 2.20 is presented in Figure 2.21c as a sequence of real numbers versus frequency. The power spectrum is a much more convenient representation of the frequency content of the signal. However, the phase information relating to the signal components has been lost and it is impossible to recover the original time data by an inverse transform.

In Example 2.2 the sampling time Δt for the signal shown in Figure 2.19 is 1 ms. Thus the time of record T, see equation (2.32), is given by:

$$T = 16 \times 10^{-2}\,\text{s}.$$

Hence, the resolution between lines, see equation (2.35), is

$$\Delta f = \frac{1}{16 \times 10^{-3}} = 62.5\,\text{Hz}.$$

Putting the values for Δf into equation (2.39), we obtain for the value of the maximum frequency F_{max}

$$F_{\text{max}} = \frac{8}{16 \times 10^{-3}} = 500\,\text{Hz}.$$

Note that, for example, the frequency corresponding to $j = 3$ is 187.5 Hz.

2.19 The fast Fourier transform (FFT)

The discrete Fourier transform can be calculated by using a computer program which realises equations (2.33) or (2.37). A program written in what might be described as the direct method requires N^2 multiplications, if the signal (time series) has N samples. Considering Example 2.2, one can see that the values of many of the sine and cosine terms are the same. For example:

$$\cos\frac{\pi}{8} = \cos\frac{15\pi}{8} = \sin\frac{3\pi}{8} = \sin\frac{5\pi}{8}\,,\text{etc.}$$

Various algorithms are available which make use of these equivalences and other factors to enable the discrete Fourier transform to be computed more quickly. The DFT calculated by these algorithms is known as the *fast Fourier transform* or FFT [22,24,25]. The FFT algorithm results in the reduction of the number of multiplications from N^2 to approximately $N \log_2 N$. For example, for $N = 1024$ the computation is speeded up by a factor of more than 100. Use of the fast Fourier transform enables frequency analysis by means of equation (2.33) to be realised in practice.

In order to carry out frequency analysis by means of the FFT the investigator can write his own program or make use of commercially available programs. The use of software in this way provides considerable flexibility in analysis. However, most people make use of special instruments, available from many manufacturers. The instruments may be computers with ADCs and programs with FFT algorithms, or they have special purpose hardware for performing the FFT procedure.

Most of the machines which use the FFT are adapted to input N data points in such a manner that $N = 2^m$, where m is a natural number, i.e., the FFT procedures are based on radix 2. Typically, $m = 10$, so that $N = 1024$. Once N is chosen, either Δt or F_{max} must be selected according to:

$$T = N\Delta t,$$
$$\Delta f = 1/T, \tag{2.45}$$
$$F_{max} = \frac{N}{2}\Delta f.$$

Thus, for example, if $N = 1024$ and $\Delta t = 50 \times 10^{-6}$ s:

$$T = 51.2 \times 10^{-3} \text{ s},$$
$$\Delta f = 19.53 \text{ Hz},$$
$$\text{and} \quad F_{max} = 1000 \text{ Hz}.$$

In order to increase the value of F_{max} a smaller sampling time can be chosen, say, 25×10^{-6} s. In this case $F_{max} = 20$ kHz, but the resolution Δf is increased to 39.06 Hz. To increase F_{max} to 20 kHz and to maintain the value of the resolution at 19.53 Hz, the number of data samples N must be increased to 2048.

Analysers based on the FFT principle are equivalent to a set of parallel filters with a constant bandwidth equal to the resolution Δf.

2.20 Ensemble averaging

A random signal, in contrast to a deterministic signal, cannot be determined explicitly, i.e., its time–dependent function which represent the signal cannot be presented in analytic form. The random time history records (random signals) obtained from a series of experiments performed under identical conditions form an ensemble. Each time history record is one of many possible representations of the random time–dependent phenomenon.

The statistical values characterising the random signals are determined by *ensemble averaging*. The ensemble average, such as the mean value at a specific instant in time t_s, is obtained by summing the instantaneous values of the records of the ensemble at time t_s and dividing the sum by the number of records [6,23]. The above definition concerns averaging over the ensemble with the number of records approaching infinity. As this requirement is obviously not fulfilled in practice, the ensemble average can only be estimated with a certain accuracy, never calculated exactly.

The ensemble averaging procedure is applicable in the case of *stationary random signals*. A random signal is defined as *stationary* when the statistical properties, such as ensemble averages, do not depend on time. When the ensemble averages depend on time, the random signal is said to be *non–stationary*. For stationary random data the ensemble averages at all times can be determined from the ensemble averages at a single time.

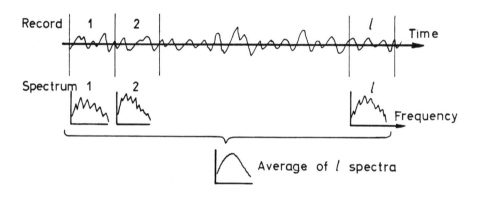

Figure 2.22 Segmentation of data for ensemble averaging.

In the case of stationary random signals the ensembles of records can be created by continuous segmentation (sampling) in time of a single primary record. Every segment of the primary signal is also a stationary random signal of equal duration, say, T (see Figure 2.22). For a stationary random signal the ensemble–averaged power spectrum should not be sensitive to the segmentation procedure. Thus for a stationary random signal the ensemble average of the power spectrum should not be dependent on the choice of the starting time of the segments. Obviously for a non–stationary random signal the value of the power spectrum changes from segment to segment much more significantly than for a stationary random signal.

Let us analyse the ensemble averaging procedure for the power spectra of stationary signals in a more systematic manner. The random stationary signal (primary data in the time domain) is divided into a sequence of l segments or records of time length T, as it is shown in Figure 2.22. For each segment of the signal the discrete Fourier transform is determined and from equation (2.44) the power spectrum $\{G_j\}$ is obtained. The values of the power spectrum $\{G_j\}$ in the frequency domain (as well as the values of the power spectral density function $\{\mathcal{G}_j\}$), see equation (2.42), are different for each segment (sample), although the time of the record for all segments is the same, namely, T. This difference is a manifestation of the *random error* of the estimate.

The random error is defined by the standard deviation of the estimate. It is a measure of the spread of a set of estimated values about the mean. The *normalised standard deviation* or the *normalised standard error* is a more convenient definition of the random error. Namely, it is defined as the ratio of the standard deviation of the estimate to the value of the quantity being estimated.

The segments of the stationary random signal, as well as the power spectra obtained from different segments, form ensembles, each of l elements, see Figure 2.22. The final ensemble average for the power spectrum $\{\langle G_j \rangle_l\}$ is obtained by averaging values of $|F_j|^2$, over the number of spectra l, namely,

$$\langle G_j \rangle_l = (2/l) \sum_{m=1}^{l} |F_j|^2{}_m .\qquad (2.46)$$

The normalised standard error ϵ for the estimate of the power spectrum $\{\langle G_j \rangle_l\}$ is given by the expression [27]

$$\epsilon = \sqrt{1/l},\qquad (2.47)$$

where l denotes the number of elements in the power spectrum ensemble. The effect of ensemble averaging is illustrated in Figure 2.23. The estimate of power spectra of band–limited noise from a noise generator has been obtained for (a) one record and (b) an ensemble of 100 records.

The variability of the estimate is considerably reduced in the case shown in Figure 2.23b, as would be expected from equation (2.47). For $l = 1$ the normalised standard error is equal to 1, whilst for $l = 100$ the error amounts to 0.1.

Power spectrum

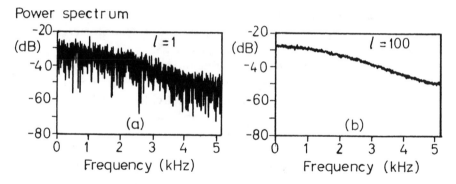

Figure 2.23 Estimates of power spectra of band–limited noise obtained from (a) one record and (b) ensemble of 100 records (FFT analysis: $N = 2048$, $T = 200\,\text{ms}$).

In Figure 2.22 it is implied that the segments are contiguous and that no data is lost in the analysis. This is only possible if the FFT calculation is carried out in a time which is less than or equal to that required for the data to be read into the processor via the ADC. In other words the analyser or computer is capable of working in *real time*. If the computation of the FFT takes more time than is needed for the data input, the segments of time length T will not be contiguous and data between the segments will be lost. For stationary processes the loss of data is unlikely to matter. However, there is always the possibility of retaining all information by recording the data first onto a disc store.

For a discrete Fourier transform performed on 1024 numbers a time of, say, 100 ms may be needed. If $T = 100\,\text{ms}$, $N = 1024$, then $F_{\text{max}} = 5120$ Hz. Thus real time operation for the transform is possible at frequencies less than 5120 Hz. However, if we wish to obtain the ensemble–averaged power spectrum as well as the DFT, real time operation will only be possible for signals with components within a smaller frequency bandwidth than 5120 Hz.

2.21 Aliasing errors

In general the data points in the time domain (time series) obtained after sampling of the continuous signal can describe different continuous signals, not only the one from which they were selected. In other words many continuous signals can possess the same representation in the form of a discrete time series. The indistinguishability of continuous signals in this sense is a source of aliasing error [28].

The sampling rate of the analogue to digital converter must be sufficiently large so that an accurate representation of the analogue signal is obtained for the computer to process. For example, if a high frequency signal is sampled at too low a rate, entirely false, low frequency signal occurs in the DFT output. This type of aliasing, related to the appearance of low frequency signals, is illustrated in Figure 2.24.

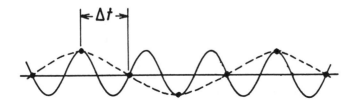

Figure 2.24 Origin of aliasing errors.

A sine wave with frequency f_{sig}, shown in Figure 2.24, is sampled at a rate of F_s, where $F_s = 1/\Delta t$. In this example, $F_s = (4/3)f_{\text{sig}}$. The dotted curve constructed through the sampled data is the aliased sine wave with frequency

$$f_{\text{alias}} = F_s - f_{\text{sig}}. \tag{2.48}$$

Thus, if $f_{\text{sig}} = 150$ Hz and the sine wave with frequency f_{sig} is sampled at a rate of 200 Hz, the aliased frequency f_{alias} is 50 Hz.

As was already mentioned the sampled data can represent not only the real signal but also signals with aliased frequencies. The sine wave with frequency f_{sig}, presented in Figure 2.34, can be confused not only with the aliased sine wave whose frequency is given by equation (2.48) but also with other aliased sine waves. The aliased frequencies formulate a sequence and are related to the sampling rate F_s and the frequency of the signal f_{sig} in the following manner:

$$(F_s \pm f_{\text{sig}}), (2F_s \pm f_{\text{sig}}), (3F_s \pm f_{\text{sig}}) \cdots \tag{2.49}$$

At least two samples of the original signal per cycle are required to define a frequency component of the original signal. Increasing the sampling rate F_s allows the possibility of detecting from the discrete time series the higher frequency components of the original signal. The fequency F_{max}, where

$$F_{\text{max}} = \frac{1}{2\Delta t} = \frac{F_s}{2}$$

denotes the maximum frequency which can be detected in the sampled data after sampling the original signal at the rate F_s. In other words all frequency components of the original signal of frequency $f_{sig,c}$, fulfilling the condition $0 \le f_{sig,c} \le F_{max}$, can be identified in the frequency spectrum, whilst those frequency components for which $f_{sig,c} > F_{max}$ are not distinguishable.

If the signal possesses high frequency components, i.e., $f_{sig,c} > F_{max} = F_s/2$, we can deduce from equation (2.48) that even in the range of lower frequencies $(0 - F_{max})$ the aliased frequency components appear. On the other hand if all frequency components are such that $0 < f_{sig,c} < F_{max}$, then in this range of frequencies there are no aliased frequency components. Note, that the frequency F_{max} can be defined as the lowest frequency which coincides with one of the aliased frequencies, see equation (2.49).

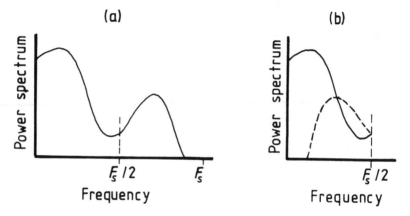

Figure 2.25 Illustration of aliasing (a) true power spectrum of signal to be measured (b) power spectrum from DFT (output) showing folding about frequency $F_s/2 (= F_{max})$.

The consequences of aliasing are further illustrated in Figure 2.25, which presents (a) the true power spectrum of a signal and (b) the aliased power spectrum. The signal component with the greatest frequency occurs at a frequency just less than the sampling rate F_s. The aliased power spectrum, obtained from the sampled signal using the DFT method and shown in Figure 2.25b, is a result of the folding of the higher frequency part of the true power spectrum $(f_{sig,c} > F_{max} = F_s/2)$ onto the lower frequency part of the same power spectrum. The folding occurs about the frequency $F_{max} = F_s/2$ and finally results in false values of the power

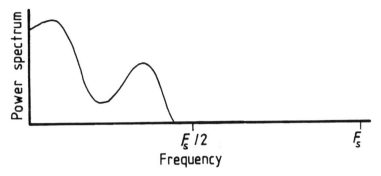

Figure 2.26 Avoidance of aliasing errors by doubling the sampling rate F_s.

spectrum. Hence, the frequency $F_{max} = F_s/2$ is also called the *folding frequency* or the *Nyquist frequency*.

Aliasing cannot occur if the sampling rate F_s is greater than twice the highest frequency of the signal spectrum. Thus one method of eliminating aliasing is to increase the sampling rate F_s. In the case of the signal presented in Figure 2.25a, doubling the sampling rate allows all the frequencies in the spectrum to be less than $F_s/2$, and no aliasing is possible, see Figure 2.26.

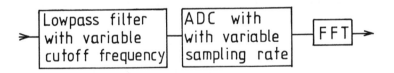

Figure 2.27 Lowpass filter to prevent aliasing.

It is often not possible to increase the sampling rate. The most common way of reducing the effect of aliasing is to include a lowpass filter before the ADC, as shown in Figure 2.27. An ideal lowpass filter with a cutoff at $F_s/2(= F_{max})$ would eliminate any possibility of aliasing, see Figure 2.28a. However, filters are never ideal and with a real filter there will always be some possibility of aliasing, if the original signal has frequency components with frequencies greater than $F_s/2$, as shown in Figure 2.28b. In practice it is advisable to set the cutoff at a frequency less than F_{max}. Of course, the use of the lowpass filter destroys the high frequency part of the signal, and if these components of the signal are important the sampling rate must be increased.

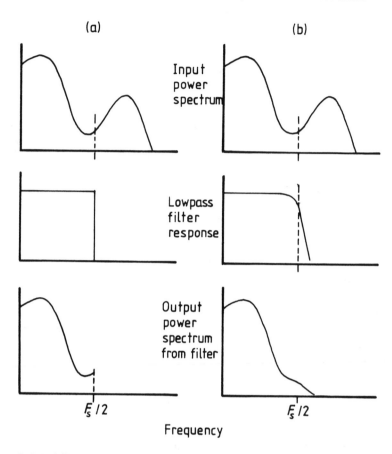

Figure 2.28 Aliasing prevented by (a) ideal lowpass filter and reduced by (b) real lowpass filter.

Lowpass filters which are used to reduce the effect of aliasing are referred to as anti–aliasing filters. In many commercially available analysers the anti–aliasing filters are built in and are automatically adjusted to a suitable cutoff frequency which depends upon the values of F_{max} or Δt selected by the operator. Usually the analysers are constructed according to the scheme illustrated in Figure 2.29. A lowpass filter with a fixed cutoff frequency is placed in line before an analogue–to–digital converter with a fixed sampling rate. The fixed sampling frequency is the greatest rate available for the analyser and the cutoff frequency of the lowpass filter is fixed at half or less than half of the sampling frequency. A digital filter follows the ADC. A digital filter is a filter which operates on sampled time data and simulates the more usual type of filter which handles continuous data. The digital filter, shown in the scheme in Figure 2.29, is called a decimating filter, because it not only provides lowpass filtering but also reduces the sampling rate to give the required value of F_{max}.

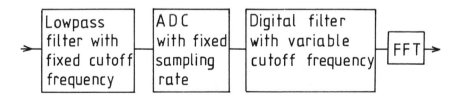

Figure 2.29 Scheme to reduce aliasing by digital filtering of the sampled data.

Usually the computation of the discrete Fourier transform (or FFT) is terminated at a frequency F_{max}, i.e., for $j = N/2 - 1$. However, if the computation is extended beyond the frequency F_{max}, the DFT coefficients have repeatable values. Thus the resulting aliased power spectrum has also a repeatable pattern. It can be presented as a superposition of the true power spectrum and the power spectra which result from aliasing. The aliased spectra are situated around frequencies F_s, $2F_s$, $3F_s$, etc., see also expressions (2.49). Let us assume that, for example, the power spectrum of the signal $G(f)$ under analysis has the shape of the hypotenuse AB of the triangle AOB, as shown in Figure 2.30, and that for the signal component with the highest frequency $f_{\mathrm{sig},c}$ the condition $F_s/2 < f_{\mathrm{sig},c} < F_s$ is satisfied. In this example the true power spectrum (in the frequency range OB) is not significantly aliased, since OB is only slightly greater than $F_s/2$.

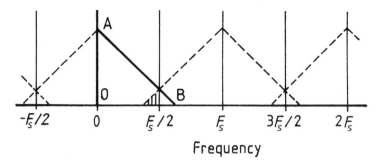

Figure 2.30 Example of aliasing: $F_s/2 < f_{\mathrm{sig},c} < F_s$.

For a spectrum extending to higher frequencies, as in the example shown in Figure 2.31, where $OB' > F_s$, aliasing takes place over OB', the entire useful range of the signal power spectrum; at low frequencies there is double aliasing.

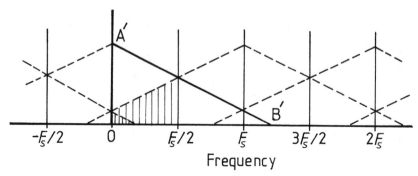

Figure 2.31 Example of aliasing: $F_s < f_{\text{sig},c} < 3F_s/2$.

The effect of aliasing is also illustrated in the example of the power spectrum presented in Figure 2.32. In this case the input signal being analysed is a sine wave at 4 kHz from an oscillator (generator). The FFT computation is performed for $N = 1024$ and $F_{\max} = 5$ kHz. It corresponds to $\Delta t = 100\,\mu s$ and $\Delta f = 9.77$ Hz. The tone at 4 kHz is clearly visible in the power spectrum which was obtained after ensemble averaging an ensemble with number of elements $l = 100$.

When the sampling time Δt is doubled to 200 μs, i.e., when the maximum frequency $F_{\max}$ is reduced to 2.5 kHz, the sine wave with frequency 4 kHz cannot, of course, be detected. The power spectrum shown in Figure 2.32b indicates the existence of a peak at a frequency of 1 kHz, identical to the peak at 4 kHz in the power spectrum presented in Figure 2.32a. The appearance of this peak (at 1 kHz) is a result of aliasing caused by folding of the 4 kHz peak, situated in the higher frequency part of the spectrum, onto the lower frequency part of the power spectrum ($f_{\text{sig},c} < 2.5$ kHz). Folding occurs about the frequency 2.5 kHz. Both power spectra (a) and (b), presented in Figure 2.32, possess the peak at frequency 2 kHz. This is also an aliased result which arises from folding the real peak, existing at 8 kHz, about the frequency 5 kHz. The true peak at 8 kHz refers to the second harmonic of the fundamental oscillator tone. As has already been mentioned and shown in Figures 2.30 and 2.31, there is a repetitive pattern in the aliased power spectra, and 5 kHz — as well as 2.5 kHz — can act as a folding frequency.

A signal, such as a square wave, which has harmonics with large

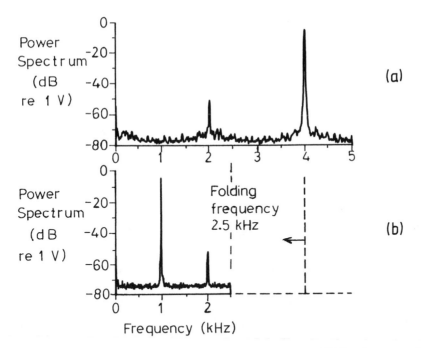

Figure 2.32 The effect of aliasing; power spectrum $G(f)$ of a sampled sine wave of frequency 4 kHz (a) with $F_{max} = 5$ kHz ($\Delta t = 100\,\mu s$) and $N = 1024$, and (b) with $F_{max} = 2.5$ kHz ($\Delta t = 200\,\mu s$) and $N = 1024$.

amplitudes in the low frequency range is likely to be the source of aliasing errors unless an appropriate lowpass filter is used. It is very important that aliasing is prevented, otherwise in the analysis of an unknown signal discrete tones may be detected which could be fallaciously ascribed to some real source.

2.22 Window functions

In the case of spectrum analysis with analogue equipment a continuous signal is analysed. On the other hand digital data analysis is performed for records with finite length. In general the records with finite length T are obtained from a continuous signal by segmentation. The segmentation procedure can be understood as the selection of the part of the continuous signal which is 'seen' through a unity amplitude *rectangular time window* of length T. The segment (sample) of the signal $p_T(t)$ is equivalent to the product of the continuous signal $p(t)$ and the rectangular time window $w(t)$, where $w(t)$ is a time–dependent function such that $w(t) = 1$ for

$0 \leq t \leq T$ and $w(t) = 0$, otherwise, see Figure 2.33. Thus,

$$p_T(t) = w(t)p(t). \tag{2.50}$$

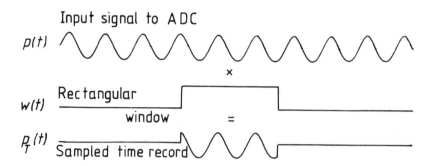

Figure 2.33 Effect of a rectangular window on sampled time data.

The Fourier transform of the segment of record $p_T(t)$ is the convolution of $W(f)$ and $P(f)$, the Fourier transforms of functions $w(t)$ and $p(t)$, respectively. The *Fourier transform pair for continuous signals* is defined as

$$P(f) = \int\limits_{-\infty}^{+\infty} p(t)e^{-i2\pi ft} dt \tag{2.51}$$

$$p(t) = \int\limits_{-\infty}^{+\infty} P(f)e^{i2\pi ft} df,$$

where $P(f)$ exists if $\int_{-\infty}^{\infty} |p(t)| dt < \infty$ and where $-\infty < f < \infty$, $-\infty < t < \infty$ and $i = \sqrt{-1}$.

The *convolution* of the Fourier transforms $W(f)$, $P(f)$ is defined by the equation:

$$C(f) = \int\limits_{-\infty}^{\infty} W(\alpha)P(\alpha)d\alpha \tag{2.52}$$

and is equal to $P_T(f)$, the Fourier transform of the segment of the signal $p_T(t)$. Similarly, the inverse transform of the product $W(f) \cdot P(f)$ is the convolution of the function $w(t)$ and $p(t)$.

The distribution of the modulus of $W(f)$, namely $|W(f)|$, in the frequency domain is called the *spectral window*. Since the Fourier transform of the rectangular time window function $w(t)$ is

$$W(f) = T\frac{\sin \pi f T}{\pi f T}e^{-i\pi f T} \qquad (2.53)$$

the distribution $|W(f)|$ has the form presented in Figure 2.34.

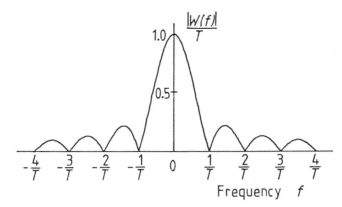

Figure 2.34 Spectral window for the rectangular time window of length T.

We can deduce from equation (2.52) that the spectrum $P_T(f)$ of the segment of the signal is clearly different from the spectrum of the primary continuous signal. Since in practice only finite records are analysed, it is important to know to what degree the true spectrum of the continuous signal is affected by segmentation, or in other words to what degree the spectral window changes the properties of the continuous signal spectrum.

The phenomena which result from segmentation and which are responsible for the difference between the spectrum of the continuous signal and the spectrum of the segment of the signal are called *leakage* and *picket-fence* effects [22].

Figure 2.35 Segment of signal (time record).

Leakage takes place when the primary continuous signal is not periodic (with period T) in the rectangular time window, of length T, as illustrated in an example with a sinusoidal signal shown in Figure 2.35. In this case during segmentation a discontinuity occurs at every interval of time T. In practice the condition of periodicity of the primary signal in the rectangular time window is rarely satisfied. The spectrum of the unity amplitude rectangular time window has zero crossings at 0, $\pm 1/T$, $\pm 2/T$, etc., see Figure 2.34. If the continuous signal is periodic in the time window of length T, then its power spectrum has spectral lines at frequencies $1/T$, $2/T$, etc., i.e., zero crossings of the Fourier transform of the time window function coincide with positions of the spectral lines of the continuous signal spectrum. The convolution of the Fourier transforms $W(f)$ and $P(f)$ results in a spectrum with the main lobes or spectral lines situated at the same positions in the frequency domain as the spectral lines of the continuous signal spectrum, see Figure2.36b. Hence, in the digital approach the resulting spectrum of $p_T(t)$ the segment of the signal is the same as that of the continuous signal.

In the case when the continuous signal is not periodic in the time window the line spectrum of the segment of the signal has additional spectral lines, as shown in Figure 2.36a. This phenomenon is called *leakage*. The term leakage implies that the power spectrum of the finite time record (segment of the signal) can be interpreted as a continuous signal spectrum which 'leaks' into the spectral lines distributed around the spectral lines of the continuous signal spectrum. Leakage can be reduced by the suitable choice of a *time window function* (sometimes called a *lag window*) which replaces the rectangular window.

The effect of leakage can be explained by an example. Consider a sine wave with a frequency of 97.5 Hz which is viewed in a rectangular window of length 200 ms. The segment of time record is shown in Figure 2.36 as (a). As the number of periods of the sine wave is not an integer in the range of the time window, there is a discontinuity of the type shown in Figure 2.35. The power spectrum of the segment of the time record, shown also as (a) in Figure 2.36A, has a peak at 100 Hz, but instead of a sharp spike at that frequency there is some leakage of power into the adjacent lines. This smearing of the spectrum could destroy low level results at frequencies close to 100 Hz.

If the sine wave possesses an integer number of periods in the rectangular time window, there would be no leakage losses, as indicated in (b) of Figure 2.36B, where a sine wave of exactly 100 Hz is shown with exactly 20 periods in the time window. In this case there is no discontinuity and

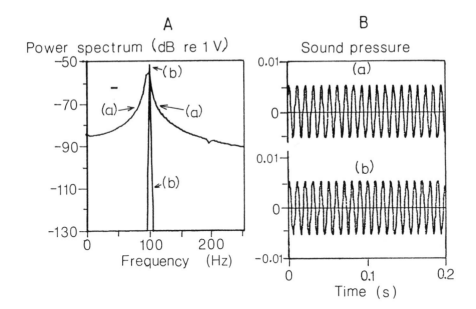

Figure 2.36 Leakage losses for a sine wave; (A) power spectrum, (B) segment of the signal (time domain); (a) non-integer number of periods in 0.2 s time window, (b) integer number of periods in the time window.

all the power in the 100 Hz sinusoidal signal refers to one line in the output spectrum.

Leakage can be alleviated by the use of a time window function such as the *Hanning window*. There are many time window functions, but the Hanning — sometimes called a cosine bell — is one of the most widely used. The Hanning function W_H is defined as a function of time t:

$$W_H(t) = \frac{1}{2}\left(1 - \cos\frac{2\pi t}{T}\right), \qquad (2.54)$$

where T is the length of the Hanning time window. The function is shown in Figure 2.37.

An illustration of the effect of using the Hanning function is shown in Figure 2.38, where it is applied to the sinusoidal signal with a non–integer number of periods in the time window of length T, see (a) in Figure 2.36 as well as (B) in Figure 2.38. The effect of the Hanning function on the signal (in the time domain) as well as on the signal power spectrum is shown in Figure 2.38b. In the time domain the signal is reduced to zero at the

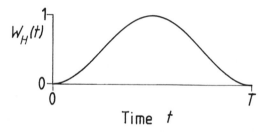

Figure 2.37 Hanning function.

beginning and end of the time window and thus the discontinuity shown
in Figure 2.35 cannot exist. The advantage of the Hanning function is
clearly seen in Figure 2.38, where the leakage is reduced to such an extent
that a harmonic of the signal — a sine wave with some distortion — can
be seen at 200 Hz.

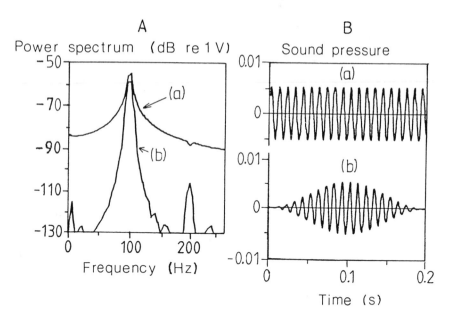

Figure 2.38 Elimination of leakage using the Hanning window function;
(A) power spectrum (B) segment of the time record; (a) rectangular time
window (b) Hanning time window.

The signals denoted by (a) in Figure 2.36 and in Figure 2.38, and
by (b) in Figure 2.38 are non–periodic in the time window and originate

from the same sound source. In the case of the signal shown as (b) in Figure 2.38 the Hanning function is applied to a continuous signal in the time domain. In all of these signals the component with frequency close to 100 Hz dominates; the signals were generated by an oscillator. The signal denoted as (b) in Figure 2.36 is a pure tone with a frequency of exactly 100 Hz; it was artificially generated within a Fourier analyser to ensure that it was exactly periodic in the rectangular time window.

The result obtained using the Hanning function is not, of course, as good as the spectrum obtained for the signal with an integer number of periods in the time window (see Figure 2.36b). Compared with this result, use of the Hanning function leads to the effect of broadening the apparent bandwidth of the spectrum and reducing the level of the peak. (Note that a direct comparison of Figures 2.36b and 2.38b cannot be made as the frequencies of the main tone are not exactly the same.) Application of the Hanning function to the sine wave with an integer number of periods in the time window reduces the value of the power spectrum at 200 Hz by a factor of 4 (or 6 dB). (The reduction is less in Figure 2.38, since the considered sine wave is not exactly with an integer number of periods in the Hanning window.) In the case of broad band random sound the power spectrum should be multiplied by a correction factor of 8/3 when Hanning is used [26].

Window functions are also able to reduce an error known as the *picket-fence effect* [22]. This effect can be described with reference to Figures 2.36A and 2.38A. When a power spectrum is obtained by means of the DFT, values of the power spectrum are only obtained at the frequencies of the discrete spectral lines ($j = 1$, 2, etc.). No information is available from DFT results at frequencies between the spectral lines. In the case of the sine wave shown as (b) in Figure 2.36 the frequency coincides exactly with a spectral line obtained by the DFT method and the power spectrum obtained is correct. However, with the sine wave shown as (a) in Figure 2.36 the frequency of the wave at 97.5 Hz lies half way between two spectral lines ($j = 19$ and $j = 20$). Since the output spectrum can only be filled in at spectral lines, the values between the spectral lines cannot be correctly estimated. The term picket–fence refers to the analogy between the output resulting from the DFT and the performance of a set of bandpass filters. The response of the bandpass filters, as shown in Figure 2.39, are said to be similar in shape to the top of a picket fence. Each filter response is centred round a spectral line. In the case of the rectangular time window the response for each filter is 0.637 at frequencies half way between the spectral lines. The Hanning window reduces the

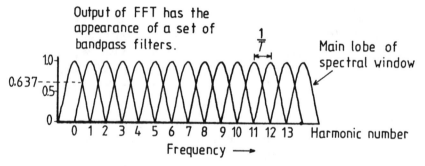

Figure 2.39 Picket-fence effect.

picket–fence effect.

The picket–fence effect is noticeable, if analysis is carried out of a tone whose frequency is varying whilst its amplitude remains constant, e.g., the tone from a hydraulic motor under variable load. If the frequency of the tone varies sufficiently so that it covers several spectral lines, the output spectrum will exhibit several peaks with the shape of a picket–fence.

2.23 Sound intensity analyser

The properties of sound intensity has already been explained in Section 1.13 (and Appendix B), where it was shown that the instantaneous sound intensity is a vector which is the product of the sound pressure and the particle velocity. The relation between the particle velocity and the sound pressure gradient ∇p is specified by the equation (B.4), see also equations (B.14) and (1.15). The pressure gradient can be measured approximately by two microphones separated by a distance d, as shown schematically in Figure 2.40.

Thus, if the sound pressure measured by microphone A is p_A, and p_B is the sound pressure at the second microphone, the pressure gradient ∇p is approximately defined by the equation

$$\frac{p_B - p_A}{d} = \vec{e}_d \cdot \nabla p.$$

Note that the directional derivative of p in the direction $\vec{e}_d$ is:

$$\frac{dp}{d\vec{e}_d} = \vec{e}_d \cdot \nabla p.$$

For example, in the case of the measurement performed in the sound field of a simple sound source:

$$\frac{p_B - p_A}{d} \simeq \vec{e}_d \cdot \vec{e}_r \frac{\partial p}{\partial r},$$

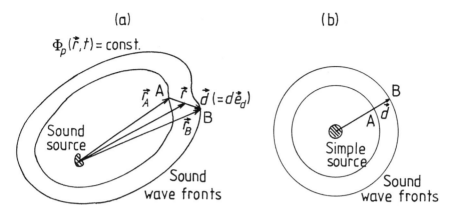

Figure 2.40 Estimation of the sound pressure gradient by using two microphones (A,B) in the sound field from (a) arbitrary sound source and (b) simple sound source.

where $\vec{e}_r = \vec{r}/|\vec{r}|$ and for the radial configuration of two microphones, shown in Figure 2.40b, we have $\vec{e}_d = \vec{e}_r$.

Hence, in this case

$$\frac{p_B - p_A}{d} \simeq \frac{\partial p}{\partial r}$$

and

$$\nabla p \simeq \frac{p_B - p_A}{d} \vec{e}_r.$$

The two microphones are built into a probe, as shown in Figure 2.41. It is possible for microphone A to face away from the sound source and still give a correct reading of sound pressure, because the microphone can be designed to be omni–directional up to a certain frequency (see Section 2.3.1 and Figure 2.4). An alternative arrangement is possible in which the two microphones are side by side with their axes parallel, but the microphone positions shown in Figure 2.41 are preferable, because the sound pressure is the same over the whole diaphragms when the probe is pointing at the noise source.

One way in which the time–averaged sound intensity measurement can be realised is shown schematically in Figure 2.42. It is important that all components in the two channels, and particularly the microphones, have identical characteristics with respect to amplitude and phase. The sound intensity is particularly sensitive to differences in phase response of the microphones.

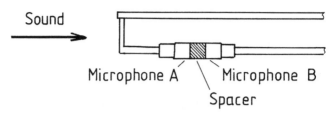

Figure 2.41 Sound intensity probe.

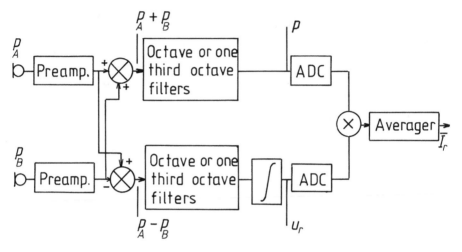

Figure 2.42 Schematic diagram of sound intensity meter.

For the arrangement shown in Figure 2.42 to be able to measure the pressure gradient with sufficient accuracy the separation d has to be small compared with the wavelength of sound. Sound intensity meters operate successfully within a range of frequencies in which the upper limit of the frequency range is determined by the separation d and the lower limit is controlled by the phase difference of the measurement channels. In order to achieve an accuracy of ± 1 dB in the time–averaged sound intensity with $\frac{1}{4}$ inch microphones the frequency ranges should be [29]

$$250 \text{ Hz to } 10 \text{ kHz} \qquad \text{for d} = 6 \text{ mm}$$
$$\text{and} \quad 125 \text{ Hz to } 5 \text{ kHz} \qquad \text{for d} = 12 \text{ mm.}$$

For $\frac{1}{2}$ inch microphones the frequency ranges should be

$$125 \text{ Hz to } 5 \text{ kHz} \qquad \text{for d} = 12 \text{ mm}$$
$$\text{and} \quad 31.5 \text{ Hz to } 1.25 \text{ kHz} \quad \text{for d} = 50 \text{ mm.}$$

The sound intensity meter is used for measuring sound power and for locating noise sources (see Section 8.6.8). In the latter application the time–averaged sound intensity is a maximum when the probe is pointing at the source and is zero when the axis of the probe is at right angles to the direction of the sound. Sound intensity analysers have found wide applications [30].

2.24 Closure

There are two basic types of test in which sound measurement equipment is used. One form of measurement involves the acquisition of data in a factory or in the environment. Once acquired the data is compared with, say, the criterion for hearing loss. The end product of the measurement is a series of numbers in a notebook or results on a paper chart. For this type of simple analysis relatively simple sound measurement equipment is needed; a sound level meter, octave filters and a chart recorder are sufficient. The second type of test is diagnostic in nature and may involve a great deal of equipment, such as narrow band frequency analysers and sound intensity analysers, which is used for the detection of specific noise sources. As a consequence of the information gained, remedial measures may be taken to reduce the noise and further measurements are taken to confirm the improvement. The experimenter and the equipment form a part of a feedback process. There is always a danger of obtaining too much information which is difficult to interpret. This is particularly the case when narrow band analysers are used, and there is a temptation to select a very small frequency resolution in the hope that more discrete tones will be detected. However, variations in speed or unsteadiness in a process may mean that the quest for more information is futile.

The equipment needed for the collection of basic data may only be a sound level meter. Some modern sound level meters can be very complicated, with switches which have dual functions and with interchangeable modules which are capable of statistical analysis, frequency analysis, etc. The alternative to a complicated sound level meter is to use a basic meter interfaced with a microcomputer. The computer–based approach is more flexible, but may often involve the experimenter in some programming.

Several advanced measurement techniques, such as acoustic holography [31] and the use of acoustic telescopes to determine directional characteristics [32], have not been discussed here. These techniques require many microphones and the considerable cost of good quality microphones is clearly a deterrent to the widespread adoption of the methods. Silicon sensors may in the future provide an alternative [33]. Silicon sensors or integrated sensors are fabricated with the aid of the same technology as that used in the production of integrated circuits. It is possible to combine the sensor and the integrated circuits in the same chip. In one form of silicon sensor, used as a pressure transducer, a silicon chip is etched to form a thin diaphragm. In future microphones may be manufactured in a similar manner.

Narrow band analysis of sound has become dominated by analysers based upon the fast Fourier transform method. The output from an FFT analyser is equivalent to the results obtained by a series of constant band-width filters. Analysis with octave band, one third octave or equivalent constant percentage filters can be obtained from the FFT analyser by post–processing of the output. Constant percentage analysis is better performed by analysers based on digital filters.

A further problem with FFT analysers arises in connection with transients. Suppose we have a very short transient of length 1 ms which just fits into the rectangular time window of the analyser and that the transient is defined by 1024 time data points. As $N = 1024$ and $T = 1 \times 10^{-3}$s, we obtain from equations (2.45) $\Delta f = 1000$ Hz, $F_{max} = 512000$ Hz. The resolution Δf may be too coarse and the maximum frequency F_{max} too great. If T is increased to 100 ms, so that the transient occupies only a small part of the time window, we have $\Delta f = 10$ Hz, $F_{max} = 5120$ Hz. The resolution between spectral lines is narrower, but the maximum frequency is reduced and may exclude frequencies of interest (remember that anti–aliasing filters will be needed). As the transient is now defined by only about 100 data points, the accuracy of the analysis is reduced.

In the analysis of transients the set of equations (2.39) imposes limitations in choosing parameters for optimum results. The analysis can be improved by increasing the number of data points N.

The increasing use of microprocessors for collection of data gives the user greater flexibility in selecting the analysis technique. Spectrum analysis can be achieved not by the FFT method but also by using the direct transform of the autocorrelation function [15]. Provided the user is prepared to undertake some programming a wide range of techniques is available, some of which are able to overcome the problem with the transient signal outlined above.

We opened this chapter with a quotation by Lord Kelvin and it seems appropriate to include in the closing comments some further opinions from him. Lord Kelvin was an early and enthusiastic champion of Fourier whose famous treatise on 'The Analytical Theory of Heat' had been ignored for 14 years before its rediscovery by Bessel. Kelvin wrote his first scientific paper, as a young man of seventeen, on the subject of Fourier series and throughout his life was active in extending the applications of Fourier's work to many branches of science. As quoted by Bracewell [20], Kelvin's view was that:

> Fourier's theorem is not only one of the most beautiful results of modern analysis, but it may be said to furnish an indispensable instrument in the treatment of nearly every recondite question in modern physics.

References

[1] International Electrotechnical Commission (1979). Sound level meters. IEC 651:1979 (Also British Standards Institution BS 5969:1981 and AMD 5787, March 1989, implementation of CENELEC HD 425.)

[2] Brüel and Kjaer (1982). Condenser Microphones: data handbook. Brüel & Kjaer, Naerum, Denmark.

[3] G. M. Sessler (Ed.) (1987). *Electrets.* Topics in Applied Physics no 33. Springer Verlag, Berlin.

[4] E. A. Starr (1971). Sound and vibration transducers, in *Noise and Vibration Control* (ed. L. L. Beranek). McGraw–Hill, New York.

[5] E. Frederiksen, N. Eirby and H. Mathiasen (1979). Prepolarised condenser microphones for measurement purposes. *Brüel & Kjaer Technical Review*, no 4.

[6] J. S. Bendat and A. G. Piersol (1971). *Random Data: analysis and measurement procedures*, pp 260–263. Wiley–Interscience, New York.

[7] C. G. Wahrmann and J. T. Broch (1975). On the averaging time of rms measurements. *Brüel & Kjaer Technical Review*, no 2.

[8] R. D. Bruce (1971). Field measurements: equipment and techniques, in *Noise and Vibration Control* (ed. L. L. Beranek). McGraw–Hill, New York.

[9] D. Steele (1971). Data analysis, in *Noise and Vibration Control* (ed. L. L. Beranek). McGraw–Hill, New York.

[10] International Electrotechnical Commission (1985). Integrating — averaging sound level meters. IEC 804:1985 (Also British Standards Institution BS 6698:1986.)

[11] P. Hedegaard (1983). Design principles for integrating sound level meters. *Brüel & Kjaer Technical Review*, no 4.

[12] K. G. Beauchamp (1973). *Signal Processing: using analogue and digital techniques.* Allen and Unwin, London.

[13] International Organization for Standardization (1975) Preferred frequencies for acoustical measurements. ISO R266:1975.

[14] International Electrotechnical Commission (1966). Octave, half-octave and third–octave band filters intended for the analysis of sounds and vibrations. IEC 225:1966 (Also British Standards Institution BS 2475:1964.)

[15] G. M. Jenkins and G. D. Watt (1968). *Spectral Analysis and its Applications.* Holden–Day, San Francisco.

[16] R. B. Randall (1977). *Application of B & K Equipment to Frequency Analysis.* Brüel & Kjaer, Naerum, Denmark.

[17] Reference 6, chapter 6.

[18] D. E. Newland (1975). *An Introduction to Random Vibrations and Spectral Analysis.* Longman, London.

[19] S. L. Marple (1987). *Digital Spectral Analysis.* Prentice–Hall International, London.

[20] R. N. Bracewell (1978). *The Fourier Transform and its Applications,* second edition. McGraw–Hill, New York.

[21] J. W. Cooley, P. A. W. Lewis and P. D. Welch (1969). The finite Fourier transform. *IEEE Transactions on Audio and Electroacoustics,* **AU–17**, pp 77–85. (Also in Digital Signal Processing (eds L. R. Rabiner and C. M. Rader), IEEE Press, 1972.)

[22] G. D. Bergland (1969). A guided tour of the fast Fourier transform. *IEEE Spectrum,* **6**, pp 41–52. (Also in Digital Signal Processing (eds L. R. Rabiner and C. M. Rader), IEEE Press, 1972.)

[23] G. Arfken (1985). *Mathematical Methods for Physicists,* second edition. Academic Press, Orlando.

[24] J. W. Cooley and J. W. Tukey (1965). An algorithm for the machine calculation of complex Fourier series. *Mathematics of Computation,* **19**, no 90, pp 297–301. (Also in Digital Signal Processing (eds L. R. Rabiner and C. M. Rader), IEEE Press, 1972.)

[25] E. O. Brigham (1974). *The Fast Fourier Transform.* Prentice Hall, Englewood Cliffs, New Jersey.

[26] J. S. Bendat and A. G. Piersol (1980). *Engineering Applications of Correlation and Spectral Analysis.* John Wiley, New York.

[27] Reference 6, p 191.

[28] Reference 6, p 228.

[29] S. Gade (1982). Sound intensity (Part 1, theory). *Brüel & Kjaer Technical Review,* no 3.

[30] F. J. Fahy (1989). *Sound Intensity.* Applied Science Publishers, London.

[31] J. Hald (1989). STSF — a unique technique for scan–based near–field acoustic holography without restriction on coherence. *Brüel & Kjaer*

Technical Review, no 1.

[32] M. A. Boone and A. J. Berkhout (1984). Theory and applications of a high-resolution synthetic acoustic antenna for industrial noise measurements. *Noise Control Engineering Journal*, **23** (2), pp 60–68.

[33] S. Middelhoek and S. A. Audet (1989). *Silicon Sensors*. Academic Press, London.

3 Noise Scales, Indices and Rating Procedures

We have a nice little back garden which runs down to the railway. We were rather afraid of the noise of the trains at first, but the landlord said we should not notice them after a bit, and took £2 off the rent. He was certainly right; and beyond the cracking of the garden wall at the bottom, we have suffered no inconvenience.

George and Weedon Grossmith, *The Diary of a Nobody*

3.1 Introduction

The sound level meter is used to provide a physical measurement of the sound pressure. When the linear frequency response is used it gives the overall sound pressure level. The subjective response of a person does not correlate well with the objective measurement of the sound level meter. A noise may be assessed for loudness, which is the subjective measure of the intensity of a sound. The other main factor, apart from sound intensity, which has an influence on loudness is the frequency of the sound. Loudness is expressed in the units of sones. Loudness level refers to loudness on a logarithmic scale and is expressed in units of phons. The loudness level of a given sound is defined as numerically equal to the sound pressure level of a pure tone of 1000 Hz, if a listener (a young person with good hearing) estimates that the loudness of the given sound is equal to the loudness of a reference sound (pure tone). The reference point on the loudness scale is one sone. This reference point refers to 40 phons on the loudness level scale. On the other hand the reference loudness level of 40 phons corresponds to 40 dB on the sound pressure level scale. In other words the loudness level for the pure tone of 1000 Hz and of sound pressure level of 40 dB is assumed to be 40 phons. The loudness scale and loudness level scale are examples of *noise scales*. Other examples are the equivalent continuous sound level (L_{eq}) and perceived noise level scales.

The Noise Advisory Council of Great Britain [1] distinguished between a noise scale and a *noise index*. For the latter, not only are physical measures considered, but also an adjustment is made to take into account the subjective response to a particular environmental situation. Thus, an index may incorporate a correction to account for tonal content or impulsiveness of the sound, if those factors are known to be disturbing. A noise index is usually based on a noise scale with a correction for the reference point of the scale. Introduction of the correction occurs because, for example, noise at night is more disturbing. As an example, we can consider the noise and number index (see Section 3.12) which is based on the perceived noise level scale and also takes into account the number of aircraft movements. Two constants factors are introduced in the formula defining the noise and number index to obtain the appropriate correlation with the annoyance due to aircraft noise. Other examples of indices are the traffic noise index and the articulation index.

A *noise rating procedure* is often defined in terms of a noise index. The aim of a rating procedure is to decide whether the noise from a certain source or from sources in general is acceptable for a particular activity. An important procedure concerns the rating of community noise. Noise in the community can come from aircraft, road traffic, industrial premises or neighbours. Such noise can be a nuisance. In common law in Britain a nuisance is defined as something that interferes with a person's enjoyment of his land [2]. Noise is one form of a nuisance; dust or smells are others. For a certain class of noise source, such as industrial premises, we need a scale or index which is a subjective descriptor of disturbance to residents from this type of source. A criterion should be formed in the same scale. If the noise causes the value on the index scale or noise scale to exceed the criterion, the noise is a nuisance and complaints may be expected.

The details of a noise rating procedure can vary. Sometimes a noise index is used as the descriptor and comparisons are made with a fixed criterion. Alternatively, a noise scale, based on measurements of the noise, is used and this noise scale is compared with a criterion which can be adjusted according to environmental conditions or the type of noise. The procedure based on noise rating (NR) curves (see Section 3.15) is of the latter type.

3.2 Loudness and loudness level (sones and phons)

Loudness is a subjective assessment of the intensity of a sound. Loudness depends upon both intensity and frequency. Equal loudness contours, as a function of intensity and frequency, were first devised in the 1930's, and more recently they have been brought up to date by Robinson and Dadson [3,4] whose curves have been accepted as an International Standard [5]. These curves for equal loudness or loudness level are shown in Figure 3.1.

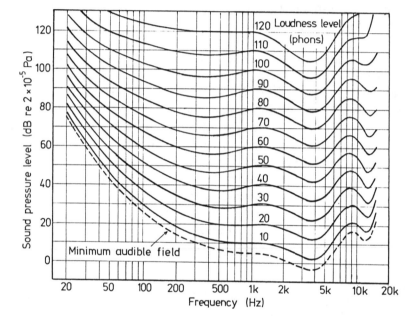

Figure 3.1 Equal loudness level contours for pure tones.

The curves are obtained by submitting people with normal hearing to pure tones from a loudspeaker in a free field room (anechoic chamber). For example, a reference tone with a sound pressure level (SPL) of 40 dB at 1000 Hz is played to the listener. Later a tone at 100 Hz with a sound pressure level of 50 dB is played, and the subject may say that the sound is as loud as the reference tone. If the listener is next played a 1000 Hz tone with an SPL of 50 dB, he or she may say that the tone is now twice as loud as the reference. In this way an equal loudness contour is built up. A large number of subjects are, of course, required to produce the equal loudness curves.

The unit of loudness is the *sone*; one sone is the loudness of a pure tone of 1000 Hz with an SPL of 40 dB. A sound which is three times as loud as this reference is said to have a loudness of 3 sones. As the ear responds logarithmically to stimuli (see Section 1.20) such as sound intensity, a *loudness level* was devised (in *phons*) such that a loudness of 1 sone is equivalent to 40 phons, 2 sones to 50 phons, 4 sones to 60 phons, etc. Doubling the loudness is equivalent to an increase of loudness level of 10 phons. The mathematical relationship between loudness (S) in sones and loudness level (P) in phons is given below:

$$S = 2^{(P-40)/10}. \tag{3.1}$$

The relationship can also be obtained from Figure 3.2.

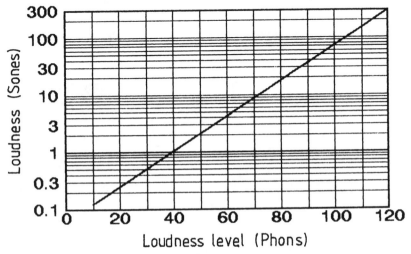

Figure 3.2 Relationship between sones and phons.

The equal loudness level contours indicate how the sensitivity of the ear varies with frequency for different intensity levels. The ear has greatest sensitivity to sound intensity at 4 kHz. At low frequencies and low intensities the ear has a poor sensitivity compared with that at 4 kHz. The sensitivity can decrease by as much as 75 dB from 4 kHz to 20 Hz. At high intensities the sensitivity of the ear varies much less with frequency.

The *minimum audible field*, shown in Figure 3.1, represents the average threshold of audibility for a large sample of people with normal hearing, exposed to sound in a free field.

3.3 Sound levels (the weighting networks A, B and C)

The equal loudness level contours show that as far as loudness is concerned the ear is not uniformly sensitive throughout its frequency range.

In general the ear is most sensitive around the middle frequencies, and the sensitivity falls away at the lower and higher frequencies. It immediately became obvious to the early originators of these curves that an electrical filter could be devised with a frequency response similar in shape to an equal loudness contour 'turned upside down', and if this filter were inserted between the microphone and the meter readout, the meter would give a reading which would be a measure of the loudness of the sound. The sensitivity to loudness depends upon the intensity level, so different filters are really required for different intensities. Originally three electric filters were constructed and called the A, B and C *weighting networks* — the A to cover low intensity sounds, the B medium intensities and the C for high intensities. The A, B and C networks are based on the 40, 70 and 90 phon equal loudness level contours, respectively. The network responses are shown in Figure 2.5.

The A weighting in particular has been frequently used and the results obtained have been found to give a surprisingly good correlation with not only loudness, but also other subjective attributes of noise. The advantages of the *A–weighted sound level* are that it is a simple, single value assessment and that it can be obtained with a relatively inexpensive sound level meter.

The A–weighted sound levels have been shown to be a good descriptor for the noise from motor vehicles [6], even though vehicle noise is of quite high intensity. The correlation of the subjective response with the A–weighted measurements is shown in Figure 3.3 for petrol–driven vehicles.

3.4 Loudness of broad band sound

Figure 3.1 refers to pure tones, whereas many sounds are random noises covering a broad frequency range. The following text outlines a procedure, applicable to broad band sound (sometimes called complex sound in this context), for determining the loudness level (Stevens's mark VI method), which has been adopted as a British and International Standard [7]. The method is applicable when the sound field is diffuse, and the sound spectrum is relatively smooth without any pronounced pure tones. The sound is first measured in octave (or one third octave) bands using an omnidirectional microphone. The chart in Figure 3.4 is then used to determine the *loudness index* for each octave band. For example, for a band pressure level at 500 Hz of 60 dB the loudness index is 4. The total loudness S_t in sones (OD) — where OD means octave diffuse — is obtained from

$$S_t = S_m + F\left(\sum S - S_m\right),$$
(3.2)

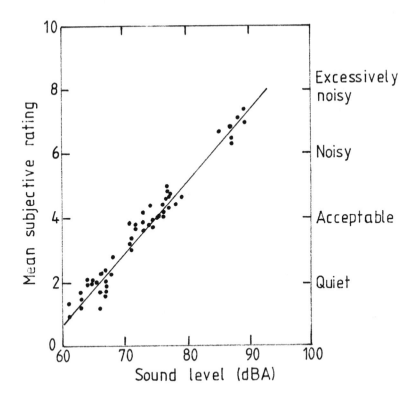

Figure 3.3 Correlation of subjective response with A–weighted sound levels of petrol–driven vehicles (adapted from [6]).

where S_m is the maximum loudness index,
         $F = 0.3$ for octave analysis, 0.15 for one third octave analysis
and $\sum S$ is the sum of all the loudness indices.

The total loudness S_t can be converted into a *calculated loudness level* in phons (OD) by using equation (3.1) or Figure 3.2.

Example 3.1

The band pressure levels in octave bands with centre frequencies from 63 Hz to 8 kHz are listed in Table 3.1. Determine for broad band sound the calculated loudness level in phons by Stevens's method.

TABLE 3.1 Estimation of loudness level and perceived noise level. (Also required in Example 3.2.)

Octave centre frequency (Hz)	Band pressure level (dB)	Loudness index (sones)	Perceived noisiness (noys)
63	50	1.0	0.26
125	55	1.5	1.38
250	60	3.5	3.26
500	65	5.0	5.66
1k	70	9.0	8.00
2k	70	10.0	13.8
4k	65	8.0	12.0
8k	50	4.5	3.02
		$\sum S = 42.5$	$\sum N = 47.38$

From Table 3.1 we note that

$$\sum S = 42.5 \quad \text{and} \quad S_m = 10.0.$$

From equation (3.2)

$$S_t = 10.0 + 0.3(42.5 - 10)$$
$$= 19.75 \, \text{sones.}$$

19.75 sones is equivalent to a calculated loudness level of 83 phons, see equation (3.1).

A second method for calculating loudness is described in the British Standard and is due to Zwicker. Physiological studies of the ear have shown that the inner ear behaves like a series of bandpass filters with bandwidths similar to those of one third octave filters. Thus in Zwicker's method the sound is measured by a one third octave analyser. The procedure for calculating loudness by Zwicker's method is more complicated than that of Stevens and gives a value about 5 phons greater than Stevens's.

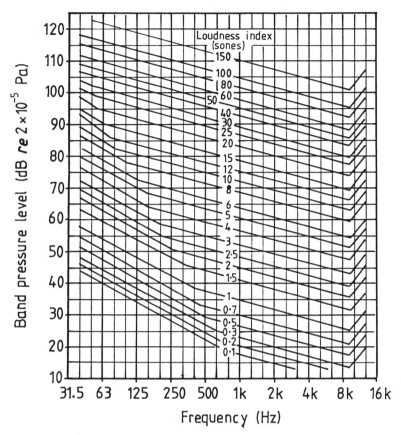

Figure 3.4 Contours of equal loudness for broad band noise (complex sound).

3.5 Perceived noise level (L_{PN})

Loudness is only one of many ways of judging a sound. In addition to loudness a sound can be judged according to its 'unwantedness', 'unacceptableness', 'objectionableness' or 'noisiness'. The concept of *perceived noisiness* was introduced by Kryter [8] to assess all these factors. Perceived noisiness is often taken to be synonymous with *annoyance*, although annoyance implies that certain psycho–social variables — such as one's personality — are taken into account additionally in judging the noise.

Contours of equal perceived noisiness have been constructed from responses of panels of subjects listening to a variety of real–life noises. These contours are similar in shape to equal loudness contours and are shown in Figure 3.5.

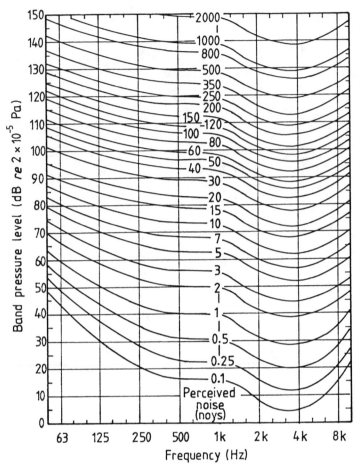

Figure 3.5 Contours of equal perceived noisiness [13].

The unit for perceived noisiness is the *noy* (cf. sone) and for *perceived noise level* (L_{PN}) we have the *perceived noise decibel* PNdB (cf. phon). The perceived noise index for each octave (or one third octave) band is obtained from Figure 3.5 and the total perceived noise N_t is determined according to Stevens's rule (see equation 3.2) which, in the case of perceived noise, is

$$N_t = N_m + F\left(\sum N - N_m\right), \qquad (3.3)$$

where F is 0.3 for octave analysis and 0.15 for one third octave analysis. The total perceived noise is converted into a perceived noise level in PNdB by using the following relationship,

$$N_t = 2^{(L_{PN}-40)/10}.$$
                                                                (3.4)

The whole procedure is very similar to that adopted for the calculation of the loudness of complex sounds according to Stevens's method.

Example 3.2

Consider the same band pressure levels as used in the example in Section 3.4. Obtain the perceived noise level.

Solution   The values of perceived noisiness, in noys, are listed in Table 3.1 along with the values of loudness index.

$$\sum N = 47.38 \quad \text{and} \quad N_m = 13.8$$
$$N_t = 13.8 + 0.3(47.38 - 13.8)$$
$$= 23.9 \quad \text{noys}.$$

From equation (3.4) the value of the perceived noise level is found to be 86 PNdB.

The concept of perceived noisiness, although of general value, has found particular application in the assessment of the noise of aircraft. For example, at London Airport (Heathrow) at certain locations on the perimeter of the airport at the edge of the runway the noise must not exceed 110 PNdB by day or 102 PNdB by night.

Measurements in PNdB are relatively complicated since an octave band analysis of the sound has first to be made. Measurements [9] have shown that there is a correlation between calculated perceived noise level in PNdB and the A–weighted sound level in dB(A) in the case of aircraft. Addition of 13 dB to the A–weighted sound level L_A gives an approximate perceived noise level. The relationship depends upon the type of aircraft. More recently Rice and Walker [10] have suggested a large tolerance on the relationship, namely,

$$L_{PN} = L_A + 13 \pm 5\,\text{dB}.$$

An approximation to the perceived noise level can also be obtained by using the D weighting network, available on some sound level meters, and adding 7 dB to the single figure reading obtained (in dB(D)) [11]. The D weighting network is shown in Figure 2.7 and its frequency response is similar in shape to the inverted 40 noy contour of Figure 3.5.

3.6 Effective perceived noise level (L_{EPN})

A modified form of the perceived noise level, called the *effective perceived noise level* and measured in units of EPNdB, has been developed, mainly by Kryter [12]. This scale is used particularly for aircraft flyover noise, and takes into account the duration of the noise and the spectral content. A complete description of the method of calculation may be found in the International or British Standard [13]. The expression for the effective perceived noise level L_{EPN} is

$$L_{EPN} = L_{TPNmax} + D, \tag{3.5}$$

where L_{TPNmax} is the maximum value during the aircraft flyover of the tone–corrected perceived noise level L_{TPN} and $L_{TPN} = L_{PN} + C$; C is a tonal correction to allow for the case where a one third octave band level may be much greater than the adjacent values.

$$D = 10\log(\tau/\tau_{ref}), \tag{3.6}$$

where τ_{ref} is 10 s [13] and τ is defined so that

$$\tau \times 10^{L_{TPNmax}/10} = \int_{-\infty}^{+\infty} 10^{L_{TPN}/10} dt. \tag{3.7}$$

In the above method the tonal correction is applied to L_{PN} during the stages of the flyover (in, say, 0.5 s intervals) to obtain L_{TPN}. The effective perceived noise level is given by:

$$L_{EPN} = 10\log\left[\frac{1}{\tau_{ref}} \int_{-\infty}^{+\infty} 10^{L_{TPN}/10} dt\right],$$

where $\tau_{ref} = 10s$. In practice the integration is not carried out between infinite time limits, but for a time during which the value of L_{TPN} is greater than the value of $L_{TPNmax} - 10$ dB.

The effective perceived noise level scale is important because it has been adopted by the International Civil Aviation Organization (ICAO) and the Federal Aviation Administration (FAA) of the USA for aircraft noise certification purposes [14]. According to the FAA the noise from Concorde at a point under the flight path at 3.5 nautical miles from the start of take–off roll was 119.4 EPNdB [15]. The corresponding levels for a Lockheed Tristar L–1011 with Rolls Royce RB–211–22 engines and a Boeing 'Jumbo' B–747–200 with Pratt and Whitney JT9D–7 engines are 97 and 108 EPNdB, respectively [16]. See Smith [17] for a review of aircraft noise and a history of legislation.

3.7 The equivalent continuous sound level ($L_{\mathrm{Aeq},T}$)

The equivalent continuous sound level, $L_{\mathrm{Aeq},T}$, or L_{eq}, which has already been described in Section 2.8, is in its A–weighted form one of the most widely used noise scales for describing environmental and industrial noise. If $p_{\mathrm{A}}(t)$ is the instantaneous A–weighted sound pressure, then the A–weighted equivalent continuous sound level is defined as

$$L_{\mathrm{Aeq},T} = 10 \log \frac{\frac{1}{T} \int\limits_{0}^{T} [p_{\mathrm{A}}(t)]^2 \, dt}{(2 \times 10^{-5})^2} \ \mathrm{dB(A)}. \tag{3.8}$$

Different integration times T are required for different noise rating procedures. For environmental noise the equivalent continuous sound level measured over a 24 hour period is often used, i.e., $L_{\mathrm{Aeq,24h}}$. Several noise indices are based on $L_{\mathrm{Aeq,24h}}$. For example, in the United States of America for community noise the Environmental Protection Agency (EPA) developed an index L_{dn}, *the day–night equivalent sound level*, defined as

$$L_{\mathrm{dn}} = 10 \log \left\{ (1/24) \left[15 \left(10^{L_{\mathrm{d}}/10} \right) + 9 \left(10^{(L_{\mathrm{n}}+10)/10} \right) \right] \right\} \ \mathrm{dB(A)}, \tag{3.9}$$

where L_{d} is the daytime equivalent continuous sound level, from 7.00 to 22.00, and L_{n} is the night–time equivalent continuous sound, from 22.00 to 7.00. The night–time levels are subjected to a 10 dB penalty since noise at night is much more disturbing than noise during the day.

Note that, as defined in equation (3.8), $L_{\mathrm{Aeq},T}$ is a noise scale, whereas L_{dn} is a noise index since it has a night–time weighting factor.

3.8 The single event sound exposure level (SEL)

The *single event sound exposure level* SEL is intended to be used with the rating of noise from single events such as transient sound from an aircraft flyover. The idea behind the SEL is similar to that for the flyover time correction for the effective perceived noise level (see equation (3.7)). The SEL of the transient sound signal is defined as the sound level in dB(A) of a hypothetical signal of duration 1 second, which has sound energy equal to the A–weighted sound energy of the noise from the single event under consideration. The definition is

$$\text{SEL} = 10 \log \frac{1}{T_{\text{ref}}} \int_{-\infty}^{\infty} \frac{[p_A(t)]^2 \, dt}{(2 \times 10^{-5})^2} \ \text{dB(A)}, \qquad (3.10)$$

where T_{ref} is the reference time of 1 second. In practice the integration is carried out over a time interval T during which the sound level of the transient sound achieves values greater than the maximal value (during the event) minus 10 dB. The SEL is a measure of the A–weighted acoustic energy of the transient sound.

The A–weighted equivalent continuous sound level $L_{\text{Aeq},T}$ is the average over a time T, and, when the noise derives only from one event, is related to the SEL by the expression,

$$L_{\text{Aeq},T} = \text{SEL} - 10 \log (T/T_{\text{ref}}) \ \text{dB(A)}, \qquad (3.11)$$

with $T_{\text{ref}} = 1$ s.

Examples of two transient noises, caused by a motor cycle and a bus, are shown in Figure 3.6. The maximum sound level for the motor cycle is 97 dB(A) and 81 dB(A) for the bus. The single event sound exposure levels are 96.1 and 84.2 dB(A) for the motor cycle and bus, respectively. The difference between the maxima is 16 dB(A), but only 11.9 dB(A) in the SEL values because the bus passes by more slowly and generates more sound energy than its maximum level would indicate. The relationship between L_{eq} and SEL is

$$\begin{aligned} L_{\text{Aeq},15s} &= \text{SEL} - 10 \log 15 \\ &= \text{SEL} - 11.8 \ \text{dB(A)}. \end{aligned}$$

Since the SEL represents sound energy, for several independent events the total A–weighted sound energy can be obtained by addition of the SEL values for individual events according to the decibel addition rule (see Section 1.21). Consequently, the A–weighted equivalent continuous sound level over a period T for a noise containing n single events with single event sound exposure levels SEL_i is obtained from the expression,

$$L_{\text{Aeq},T} = 10 \log \left(\sum_{i=1}^{n} 10^{\text{SEL}_i/10} \right) - 10 \log (T/T_{\text{ref}}) \ \text{dB(A)}, \qquad (3.12)$$

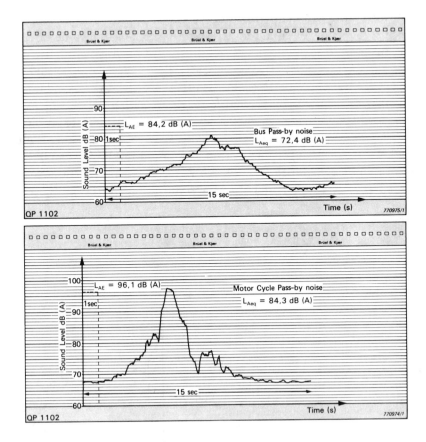

Figure 3.6 Comparison of single event sound exposure levels of bus and motor cycle noise (courtesy of Brüel & Kjaer).

where SEL_i denotes the single event sound exposure level for the ith event. For example, during 5 minutes there are 10 buses and 1 motor cycle with pass–by noises similar to those shown in Figure 3.6. Thus,

$$L_{\text{Aeq},300s} = 10\log(10 \times 10^{8.42} + 10^{9.61}) - 10\log 300$$
$$= 98.3 - 24.8$$
$$= 73.5\,\text{dB(A)}.$$

3.9 Percentile levels L_{10}, L_{30}, L_{50}, etc.

In many cases it is important to have a noise scale that takes into account intermittency of the noise. The equivalent continuous sound level L_{eq} is one solution to the problem. However, the noise caused by several events, each with a low value of sound level, could have the same value of L_{eq} as a single, intense noise. If the low level noises are not disturbing then L_{eq} fails as a satisfactory subjective noise scale. Types of noise scale that have been designed to overcome this disadvantage are the L_{10}, L_{30}, L_{50}, etc., levels, sometimes called the *percentile levels* L_N, where N represents the percentage of time during which the level is exceeded. For example, the L_{10} level is the sound level that is exceeded for 10% of the time. Thus, if the L_{10} level is 70 dB(A) it means that the sound level exceeds 70 dB(A) for 10% of the time. More exactly the A–weighted percentile levels are designated $L_{AN,T}$, where A denotes A weighting and T the measurement duration.

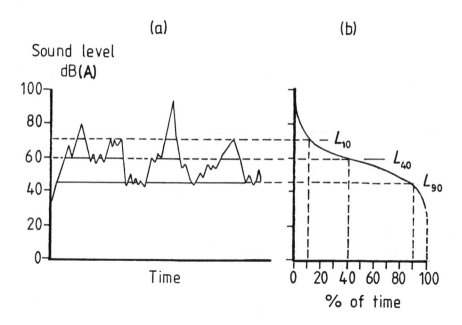

Figure 3.7 (a) illustration of L_{10}, L_{40} and L_{90} levels (b) cumulative probability distribution function of sound levels.

The idea of the L_{10} level is illustrated in Figure 3.7a where the sound level in dB(A) is shown against time. The variation of the dB(A) level is obtained by sampling the sound level at suitable intervals from, for example, the dc output of a sound level meter; Figure 3.7a does not show the instantaneous A–weighted sound pressure which would be obtained from the ac output. The sound level exceeds the 70 dB(A) line for about

10% of the time and exceeds 45 dB(A) for 90% of the time. In this way a cumulative probability distribution function, shown in Figure 3.7b, can be built up.

The A–weighted L_{10} level is extensively used in Great Britain for measurement and prediction of road traffic noise [18] and in noise legislation relating to traffic noise. Typically, traffic noise generates high noise levels for about 10% of the time, so the L_{10} level is a good discriminator for traffic noise. Special purpose microprocessor or computer–based equipment is required, if the L_{10} level is to be conveniently measured. Typical outdoor, urban L_N levels are shown in Figure 3.8 [19].

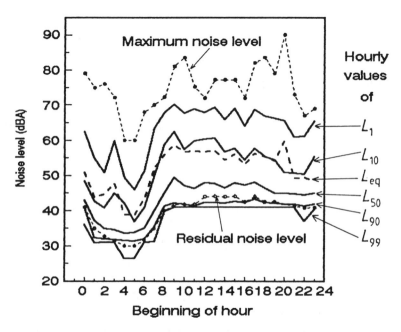

Figure 3.8 Hourly values of L_1, L_{10}, etc., taken during the day for a typical outdoor noise (adapted from [19]).

3.10 Noise pollution level (L_{NP} or NPL)

It must be clear by now that there is a great proliferation of noise scales, and it would be very convenient to have a universal scale capable of being used in all situations. One such scale, proposed by Robinson [20], is called the *noise pollution level*, and is defined as follows:

$$L_{NP} = L_{eq} + 2.56\,\sigma \text{ dB(NP)}, \qquad (3.13)$$

where L_{eq} is the A–weighted equivalent continuous sound level measured over a sufficiently large sampling period, and σ is the sample standard deviation for the A–weighted sound level.

The quantity σ has to be obtained from a statistical distribution analysis and involves relatively expensive equipment. The term in equation (3.13) involving σ has been introduced to take into account the fact that annoyance increases with the variability of the noise, even when the total energy is constant.

An alternate expression for L_{NP} is possible [11]:

$$L_{NP} = L_{eq} + (L_{10} - L_{90}).\tag{3.14}$$

In the search [1] for a unified noise scale, noise pollution level is regarded favourably, but probably loses to L_{eq} alone on grounds of simplicity.

3.11 The traffic noise index (TNI)

The TNI was formulated by Griffiths and Langdon [21] as an index for traffic noise that would correlate well with the dissatisfaction of residents. The TNI is defined as:

$$\text{TNI} = 4(L_{10} - L_{90}) + L_{90} - 30.\tag{3.15}$$

With the TNI there is a possibility that urban background noises other than those from road traffic would have an effect. For this reason the L_{10} alone has been preferred in the UK for traffic noise rating.

3.12 The noise and number index (NNI)

Aircraft noise requires a different type of index from traffic noise, because disturbance from aircraft can be treated as a series of separate noise events, whereas traffic noise, although fluctuating in level, is mostly continuous. The *noise and number index* was introduced in the Wilson report [6] as an index suitable for rating aircraft noise. The NNI is defined in terms of the perceived noisiness, which is known to be a good measure for the assessment of a single aircraft noise, and the number of aircraft movements.

$$\text{NNI} = \langle L_{PNmax} \rangle + 15 \log N - 80,\tag{3.16}$$

where

$$\langle L_{PNmax} \rangle = 10 \log \left\{ \frac{1}{N} \sum_{i=1}^{N} 10^{L_{mi}/10} \right\}$$

is the average maximum perceived noise level and $L_{mi}(i = 1, \ldots N)$ denotes the maximum perceived levels attained by the N aircraft movements during the period from 6.00 to 18.00 hrs. An aircraft must have a maximum perceived noise level of at least 80 dB to qualify. The social survey on which the NNI was based gained insufficient information to extend the concept to night–time use.

In sites where aircraft noise is a problem the NNI has been used as a basis for planning applications in the Department of Environment circular on 'Planning and Noise' [22]. This document sets down various criteria for control of development, and advises, for example, that planning permission for housing should not be given if the NNI exceeds 40. However, as an index for assessing the disturbance from aircraft noise it is expected that the NNI will be superseded by the more universal scale, the A–weighted equivalent continuous sound level, $L_{\mathrm{Aeq}, T}$, where T covers the period from 7.00 to 23.00 hours [22].

3.13 Speech interference level (SIL)

The speech interference level is a noise scale which can be used to estimate the effect of noise on speech communication. The most important frequencies for speech are the mid–frequencies, so in order to assess the disturbing effect of background noise on speech the octave band pressure levels at centre frequencies 500, 1000, 2000 and 4000 Hz are measured. The arithmetic average of these band pressure levels is called the *speech interference level* (SIL) [23]. As the speech interference level increases, the intelligibility of speech decreases in a given situation. Background noises with the same SIL value, but different spectrum shapes, have the same capacity to interfere with speech [23].

Measurement of the SIL, when used in conjunction with Figure 3.9, allows the prediction of the type of speech required for reliable face–to–face communication between two people. Thus, if the SIL is 40 dB and the two people are 8 m apart, they are able to converse with a normal voice. In a noisy environment there is a natural tendency of speakers to raise their voices automatically. The shaded triangular region in Figure 3.9 indicates the range of expected voice levels due to the normal raising of a speaker's voice in the presence of noise. If the speakers are 4 m apart and the SIL is 52 dB, they will not be able to converse in a normal voice. However, they would understand each other's speech, if they used raised voices. In this situation most people would automatically increase their voice effort.

Figure 3.9 is primarily to be used for situations out of doors, but can also be used indoors provided the enclosed space is not too reverberant,

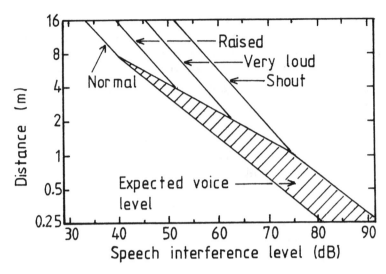

Figure 3.9 Talker–to–listener distances for just reliable communication, modified from [23].

i.e., the reverberation time should be less than 1.5 s. (See Chapter 5 for a discussion of reverberation and reverberation time.) When applied in a typical office the distances for just reliable communication may be greater than those shown in Figure 3.9.

3.14 Noise criteria curves (NC)

The *noise criteria* (NC) curves provide a rather different way of rating noise. First of all the octave band levels of the noise under consideration are measured, and then compared with the NC curves. These are shown in Figure 3.10. Superimposed on this figure is a spectrum of noise which just falls below the NC 40 curve; thus the noise has a rating of NC 40.

The noise rating for a particular noise can then be compared with recommended levels. See Table 3.2. If the noise rating exceeds the recommended NC values the environment is unlikely to be acceptable from the point of view of noise. Thus an NC level of 40 would be acceptable for a restaurant, but not for a theatre. It should be pointed out that Table 3.2 is not incorporated into any standard or legislation, but is one of many such tables, based on practical experience, which are available. A more comprehensive table is provided by Smith [24].

Noise criteria curves are extensively used in the mechanical services industry for rating noise from ventilating systems in offices, departmental

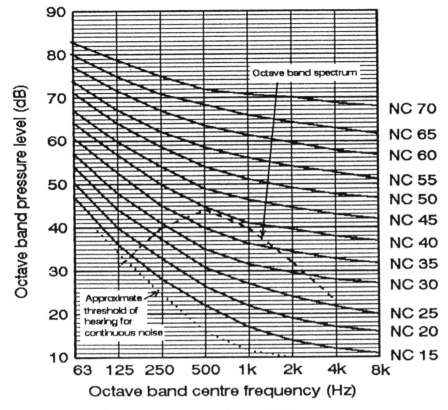

Figure 3.10 Noise criteria (NC) curves.

stores, ships etc. Before a building is constructed a noise criterion (NC) is assigned to a room in the building. A silencer is introduced in the ducting between the fan and the ventilation outlet in order to ensure that the design NC value is attained.

3.15 Community response to noise; noise rating procedures

The previous sections described some important scales and indices. Some noise procedures were also outlined. For example, in Section 3.13 a simple procedure was described for estimating whether intelligible speech is possible in the presence of noise. In Section 3.12 a planning procedure was briefly described for the situation where aircraft noise is a problem. In noise rating procedures a noise index is compared with a criterion which depends upon people's location, where the noise occurs, etc. (In simple procedures a noise scale, such as the speech interference level, is compared

TABLE 3.2 Recommended NC levels for various environments

Environment	Range of NC levels likely to be acceptable
Factories (heavy engineering)	55 – 75
Factories (light engineering)	45 – 65
Kitchens	40 – 50
Swimming baths and sports areas	35 – 40
Department stores and shops	35 – 45
Restaurants, bars, cafeterias and canteens	35 – 45
Mechanised offices	40 – 50
General offices	35 – 45
Private offices, libraries, courtrooms and schoolrooms	30 – 35
Homes, bedrooms	25 – 35
Hospital wards and operating theatres	30 – 35
Cinemas	30 – 35
Theatres, assembly halls and churches	25 – 30
Concert and opera halls	20 – 25
Broadcasting and recording studios	15 – 20

with a criterion.) In the case of the rating of community noise the criteria should incorporate the fact that cities are always going to be noisy to some extent, whereas quiet in the countryside is a reasonable expectation. The criteria may take into account the old legal statement that 'what would be a nuisance in Belgrave Square would not necessarily be so in Bermondsey.'

Noise criteria are often set after the analysis of responses to social surveys in which people have been asked about the degree of their dissatisfaction. This dissatisfaction will depend upon:

(1) damage to health and general well–being,
(2) interference with work,
(3) interference with leisure activities, such as watching TV,
(4) interference with speech communication,
(5) interference with sleep,
(6) general annoyance,

(7) feeling of fear.

There are also other more subtle and less easily definable factors at work. For example, people accept that certain noises, such as from aircraft, cannot be completely silenced. People who live an outdoor life are said to accept noise more than people who are mostly indoors, or at least inhabitants of southern regions are more tolerant of noise than less outgoing northerners.

Comments of the kind made above should indicate that setting criteria is a complex task which is often in the realm of the psychologist. Predicting the response, should the noise exceed the criteria in a given situation, is equally difficult. However, in a noise rating procedure the approach that is adopted is generally the same; first a noise index is assigned to the offending noise, a noise criterion is established for the particular situation, and finally the community response is predicted.

Road traffic gives rise to more complaints than any other source of noise. The results of the London Noise Survey of 1961–62 showed that 36% of people were disturbed by road traffic noise [6]. An American survey of 1977 showed that as many as 50% of the population could be disturbed by road traffic noise in densely populated areas. Road traffic is continually increasing and it is likely that the percentage of people disturbed is now even greater. People have little redress against traffic noise; traffic can be re–routed, barriers can be built, but in general the noise can only be reduced by legislation, enforced by the state, which limits noise made by individual vehicles [25].

TABLE 3.3 Community noise complaints [26]

Light industry	32%
Heavy industry	21%
Transport	14%
Agriculture	8%
Entertainment	8%
Construction	7%
Retail trade	4%
Offices and public buildings	4%
Defence activities	2%

Noise from industrial premises is a more tractable problem in the sense that much of it may be unnecessary and complaints by the public can have an effect. A group of noise consultants in Britain investigated the number of complaints they had to deal with, and found that the distribution was as shown in Table 3.3 [26]. Only 14% of the complaints concern transport, because people do not complain to consultants about cars or aircraft. The same survey found that of the industrial noise sources a fan was the origin of the disturbance in 28% of the cases; a further 25% of the cases were ascribed to process plant and equipment.

3.16 The law relating to noise [27–29]

In Britain a system of *common law* prevails, as opposed to the *civil law* which pertains on the continent of Europe. Common law is based on judgments passed down by judges over many years. Civil law is codified according to rules laid down by governments. In Britain anyone who is disturbed by noise from a neighbour or industrial plant can bring a civil action for private nuisance at common law in a County Court or High Court. A discussion on the law of nuisance, as it applies to noise, has been given by Cronin [2].

Noise is just one of the many forms of nuisance that are dealt with in the courts. A *private nuisance* is generally connected with land and with a person's use or enjoyment of that land. What constitutes a nuisance was established in common law in 1851, and the original ruling, often requoted by judges ever since, is as follows:

> every person is entitled as against his neighbour to the comfortable and healthy enjoyment of the premises occupied by him and in deciding whether, in any particular case, his right has been interfered with and a nuisance thereby caused, it is necessary to determine whether the act complained of is an inconvenience materially interfering with the ordinary physical comfort of human existence not merely according to elegant or dainty modes and habits of living but according to plain and sober and simple notions obtaining among English people.

In a particular case a plaintiff who has served an action for nuisance will be requiring an injunction to stop the noise, damages, or both.

The defendant has to establish that a nuisance was not committed at all. If a nuisance is established then 'best practical means' cannot be proposed as a defence. In other words the defendant cannot claim that he is carrying out an important process which is, of necessity, noisy and that he has done all that is practicable to reduce the noise. The inability to use this type of defence means that action for nuisance is a powerful means of remedy.

3.16.1 The Control of Pollution Act 1974

The noise sections of this Act of the British Parliament came into force on 1st January 1976, and deal with:

>   noise nuisance,
>   control of noise from construction sites,
>   noise in streets,
>   noise abatement zones,
>   noise from plant or machinery.

The Control of Pollution Act is a wide–ranging Act of Parliament only part of which concerns noise. Of the sections dealing with noise only the part concerned with noise nuisance is discussed here. A nuisance can be designated a statutory nuisance by a local authority. If a local authority is satisfied that a noise nuisance exists, or is likely to occur or recur, it can serve a notice requiring the abatement of the nuisance (or prohibiting or restricting its occurrence or recurrence) and specifying, if necessary, any modifications that must be carried out. The person served with the notice can appeal to a Magistrates' Court within 21 days. Potential nuisances that have not yet occurred, but are likely to occur, can be dealt with. The Act is mainly for use by local authorities but it is possible for one aggrieved occupier of premises to take action on his or her own account.

When considering the noise nuisance aspects of the Control of Pollution Act it is important to realise that 'best practicable means' is a legitimate form of defence. In other words it can be claimed in defence that the best possible engineering practice has been applied and that no further practicable means can be taken to reduce the noise. In an action for nuisance at common law this is not necessarily regarded as a defence.

The *Environmental Protection Act 1990* does not introduce any significant changes to the way in which noise is dealt with as a statutory nuisance. The Act provides an integrated system for the control of statutory nuisances such as noise, smoke, fumes, dust, dangerous animals, etc.

3.16.2 The working of the Control of Pollution Act

The usual procedure is for some householder to make a complaint to a local authority, and the Environmental Health Department of that authority will investigate. In assessing the noise the inspectors will generally rely upon guidance laid down by the British Standard BS 4142 (see Section

3.18). If it is decided that there are justifiable grounds for complaints, the inspectors will contact the management of the factory — if it is a factory that is causing annoyance — and they will discuss together how the noise can be dealt with. Generally the factory will take measures to reduce the noise at this stage to the satisfaction of all parties. If, however, no action, or inadequate action, is taken by the factory the local authority can take proceedings under the Act. Should the authority decide that the complaint is not justified then the complainant can initiate proceedings.

3.17 Noise rating (NR) curves

A series of curves, similar to the noise criterion (NC), were devised by Kosten and Van Os [30]. In order to assign a number to a noise on the *noise rating* (NR) scale the band pressure levels of the noise in octave bands has first to be measured. The octave band spectrum is superposed on the noise rating (NR) curves in exactly the same manner as with the NC curves. The NR curves are shown in Figure 3.11.

Kosten and Van Os developed a noise rating procedure for assessing whether the noise from factories is acceptable or not in houses or apartments. The basic rating or criterion is NR 30 for living rooms and NR 25 for bedrooms. Corrections are made to the basic rating for the type of noise, see Table 3.4. A noise with a pure tone (from, for example, a fan) or an impulsive noise (from, say, a pile–driver) is more disturbing than a steady, broad band noise; the basic rating is thus reduced by 5. In general people can put up with noise, if it is during the day and not at night; thus, the rating is increased by 5 for noise during the normal working day. The rating is also increased if the noise only occurs for part of the day.

Economic tie is a curiosity. If amongst the complainants there is one member who works in the factory which is causing the noise, it is implied that more noise can be tolerated from the factory — hence the basic rating is increased by 5.

The type of neighbourhood also influences the criterion. According to the noise rating procedure, people who live in a quiet, expensive locality have the right to expect less noise than those who live in a poor neighbourhood surrounded by heavy industry. The former have their rating reduced whilst the latter have theirs increased considerably. In an area surrounded by industry the background noise levels are, of course, greater.

Noise rating criteria have also been proposed for different types of room, see Table 3.5 [30]. To carry out the rating procedures the octave band pressure levels of the sound in, for example, a private office are measured and a noise rating assigned to the noise. The noise rating is

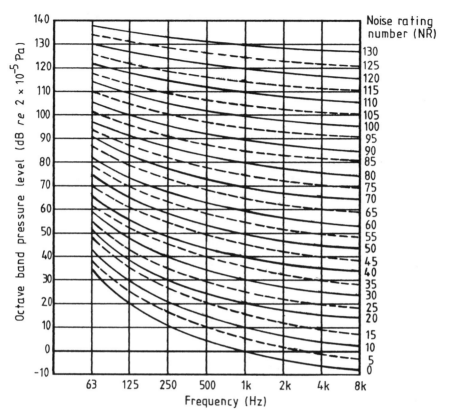

Figure 3.11 Noise rating (NR) curves.

compared with the criterion of 40 and, if the noise rating is less than 40, the noise in the office is acceptable.

Example 3.3

The distant, continuous drone of a fan is causing a disturbance during the night in the bedroom of a house which is located in a suburban area. The octave band pressure levels measured in the room are 22, 32, 38, 24, 18 and 17 dB at 63, 125, 250, 500, 1k and 2k Hz, respectively.

What is the noise rating (NR) number for the noise in the bedroom and what is the noise rating criterion for the bedroom?

Solution Superposition of the octave band spectrum on the noise rating curves of Figure 3.10 will show that the spectrum

TABLE 3.4 Noise rating numbers (NR) and corrections for dwellings

Sleeping room		NR 25
Living room		NR 30
Corrections		
(a)	Pure tone easily perceptible	−5
(b)	Impulsive and/or intermittent	−5
(c)	Noise only during working hours	+5
(d)	Noise during 25% of time	+5
	6%	+10
	1.5%	+15
	0.5%	+20
	0.1%	+25
	0.02%	+30
(e)	Economic tie	+5
(f)	Very quiet suburban	−5
	suburban	
	residential urban	+5
	urban near some industry	+10
	area of heavy industry	+15

falls just below the NR 30 curve. The noise rating number for the noise is thus 30.

The basic criterion for a bedroom is NR 25. The correction for a noise source which is a pure tone is −5. The correction for a suburban locality is 0. The corrected criterion is $25 - 5 + 0 = 20$.

As the noise rating of 30 for the noise exceeds the noise rating criterion of 20 for the particular environment, the noise is likely to result in complaints or dissatisfaction.

The criteria given in Table 3.5 in terms of the noise rating (NR) are similar to those in Table 3.2 for the noise criteria (NC). Smith has reported that in the range NR or NC between 20 and 50 there is little difference between the two [24].

TABLE 3.5 Noise rating (NR) for different environments

Broadcasting studio	15
Concert hall, legitimate theatre 500 seats	20
Class room, music room, TV studio, conference room 50 seats	25
Conference room 20 seats or with public address system, cinema, hospital, church, courtroom, library	30
Private office	40
Restaurant	45
Gymnasium	50
Office (typewriters)	55

3.18 BS 4142:1990. Method of rating industrial noise [31]

This standard is applicable to neighbourhoods where industrial and residential areas are juxtaposed. It is to be applied when the noise source is from some fixed industrial installation in a factory rather than from aircraft or traffic.

According to the standard, the *specific noise source*, i.e., the noise source likely to cause complaints, is first established. Measurements of the noise from the specific noise source are made, and subsequently corrected for tonal or impulsive character to give a rating level. The rating level is compared with the background noise level. Complaints can be expected when the rating level is 10 dB greater than the background.

3.18.1 Measurement of noise level according to BS 4142

Measurements of the noise from the specific noise source should be made with an integrating sound level meter which satisfies type 1 of BS 6698 (see reference [10], Chapter 2) in order to obtain the equivalent continuous A–weighted sound level which in this case is called the *specific noise level*, $L_{\text{Aeq},T}$. The integrating time for the measurement should be between 5 and 60 minutes. The microphone position is outside the building where complaints are expected, 1.2 to 1.5 m above the ground and 3.5 m away from other reflecting surfaces.

For a steady noise a conventional sound level meter which satisfies BS 5969, type 1, can be used (see reference [1], Chapter 2). In this case the A weighting network and the slow time weighting should be used.

3.18.2 Determination of the rating level

The specific noise level is modified according to the tonal or impulsive character of the noise. If the noise has a definite, distinguishable, continuous note (whine, hiss, screech, squeal, hum, etc.), or, if there are significant impulsive regularities in the noise (bangs, clinks, clatters or thumps) 5 dB(A) is added to the measured specific noise level. For a noise which is both tonal and impulsive the correction is still +5 dB(A). The specific noise level adjusted in this way is referred to as the *rating level*, $L_{Ar,T}$.

3.18.3 Measurement of the background noise level

The background noise level is expressed in terms of the A–weighted 90 percentile sound level, $L_{A90,T}$. See Section 3.9 for a description of percentile levels. For this measurement a special sound level meter or environmental noise analyser is required. If such equipment is not available, an ordinary sound level meter — one which satisfies BS 5969 — may be used and the typical low values of sound level recorded.

The standard categorises three different situations that may arise in connection with background noise.

(a) Complaints are expected from a new source or from a modified, existing source. In this case the background noise can be measured before the new source is installed or the existing source modified.

(b) The specific noise source is existing, but is not always operating; in this case the background noise level can be measured quite easily.

(c) The specific noise source is operating continuously. The only possibility is to measure the background at another position deemed to have an equivalent level of background noise.

3.18.4 Assessing the noise for complaint purposes

The likelihood of a noise to provoke complaints depends upon the amount by which the rating level exceeds the background noise. Complaints may be expected if the rating level $L_{Ar,T}$ is 10 dB(A) or more than the background noise level $L_{A90,T}$.

The standard does not involve a correction for the type of locality in the way that the noise rating (NR) procedure does.

The international standard on community noise measurement provides extensive information on terminology and measurement techniques, but does not give guidance on the likelihood of complaints [32].

3.18.5 Effect of atmospheric conditions on sound measurement

It is recommended in the standard that measurements are taken under weather conditions similar to when a complaint was made. Complaints are most likely during conditions of optimum sound propagation which occur when either the measurement position is downwind of the specific noise source or there is a temperature inversion. *Atmospheric or temperature inversion* is a condition in which the temperature of the atmosphere increases with height, reaches a maximum and then decreases as the altitude increases further. It is caused by cooling of the surface of the ground through radiation to the sky. Temperature inversion occurs either one hour before sunset or one hour after sunrise when the cloud cover is less than about half and the wind speed less than 1.5 m/s.

3.19 Closure

This chapter has introduced the reader to several scales and indices used in the measurement and assessment of noise; there are many more which have not been discussed. With the profusion of noise scales it is possible to quote the results of noise measurements — as, for example, with the Concorde supersonic transport — in different scales with different values, thus causing confusion amongst the public. Robinson [33] refers to 'an acoustic Tower of Babel, with a different language of noise rating measures and descriptors to suit every occasion'.

In many ways the situation is made worse by the existence of clever sound level meters and computer–based equipment which can easily measure sound in such scales as EPNdB, phons, etc. Measurements expressed in different scales can often be related by approximate formulae; the book by Schultz [11] is particularly useful in providing such relationships. Of the many noise descriptors in use the equivalent continuous sound level (L_{eq}) is the most universal; it is a useful descriptor for most kinds of environmental noise, including railway noise. As we shall appreciate in Chapter 7, the L_{eq} is also a good measure for assessing hearing damage. The L_{eq} can be measured by equipment which is relatively easy to use and not too expensive. Many environmental noise scales are A–weighted, implying that the loudness of low frequency sound is less significant than the mid and high frequency sound. Other measures, such as the noise criterion (NC) and noise rating (NR) numbers have a similar weighting which also puts less emphasis on the low frequency sound.

The accurate measurement of noise, according to a suitable noise scale, is important in noise rating procedures, if the response of people is to be predicted. There are cases where sound insulation is provided to householders who are exposed to uncomfortable levels of noise, particularly from road or air traffic. For example, in Great Britain, sound insulation is provided for homes subjected to increased levels of 1 dB(A) on the L_{10} (18 hour) traffic index (Land Compensation Act 1973). In the view of the representatives of a government laboratory in Britain [34]:

> Insulation costs can represent a significant component to the total cost of a road and it is important, therefore, that the method should be as accurate as possible so that resources are effectively targeted to eligible dwellings.

In other words, the government should not pay up unless it has to! In other countries there may be a more generous attitude.

The converse of a noise rating procedure is a design procedure for buildings, where it is equally important that the correct criterion is set for the noise in a room in the building and that the criterion is in terms of a scale which accurately reflects the response of people to specific activities in the room. The designer responsible for the noise sources, such as fans and other plant, attempts to ensure that noise levels do not exceed the specification. Setting a criterion too low on the scale gives the designer a difficult task which will involve expensive measures. Smith [24] has stated that reduction of the criterion by 5 units, i.e., 5 on the NR or NC scale or 5 dB(A), is likely to lead to a doubling of the cost associated with the noise reduction measures.

References

[1] The Noise Advisory Council (1975). Noise units. HMSO, London.

[2] J. B. Cronin (1968–69). Noise and the law. *Philosophical Transactions of the Royal Society*, **263A**, pp 325–346.

[3] D. W. Robinson and R. S. Dadson (1956). A re–determination of the equal loudness relations for pure tones. *British Journal of Applied Physics*, **7**, pp 166–181.

[4] L. S. Whittle, S. J. Collins and D. W. Robinson (1972). The audibility of low frequency sound. *Journal of Sound and Vibration*, **21**(4), pp 431–448.

[5] International Organization for Standardization (1987). Acoustics — Normal equal–loudness level contours. ISO 226–1987. (Also British Standards Institution BS 3383:1988.)

[6] Committee on the Problem of Noise (Chairman, Sir A. Wilson) (1963) *Noise* — final report — Cmnd 2056. HMSO, London.

[7] International Organization for Standardization (1975) Method for cal-

culating loudness level. ISO 532 – 1975. (Also British Standards Institution BS 4198:1967.)

[8] K. D. Kryter (1970). *The Effects of Noise on Man*, first edition, p 270. Academic Press, New York.

[9] Reference 6, p 201.

[10] C. J. Rice and J. G. Walker (1982). Subjective acoustics, in *Noise and Vibration* (eds R. G. White and J. G. Walker). Ellis Horwood, Chichester.

[11] T. J. Schultz (1982). *Community Noise Ratings*, second edition. Applied Science Publishers, London.

[12] K. D. Kryter, reference 8, pp 269–331.

[13] International Organization for Standardization (1978). Procedure for describing aircraft noise heard on the ground. ISO 3891–1978. (Also British Standards Institution BS 5727:1979.)

[14] Federal Aviation Administration (1969). Noise Standards, Aircraft Type Certification. Federal Aviation Regulations part 36, 1 December 1969.

[15] US Department of Transportation (Federal Aviation Administration) (1978). Civil Supersonic Aeroplanes (Noise and Sonic Boom Requirements and Decision on EPA Proposals). *Federal Register*, **43**, no 126, Thursday, June 29, 1978.

[16] J. O. Powers (1986). Aircraft noise generation and control: noise around airports, in *Noise Pollution* (eds A. Lara Sáenz and R. W. B. Stephens), Scope 24. John Wiley, Chichester.

[17] M. J. T. Smith (1989). *Aircraft Noise*. Cambridge University Press, Cambridge.

[18] Department of Transport (1988). Calculation of road traffic noise. HMSO, London. (See also P. G. Abbott and P. M. Nelson (1989). The revision of calculation of road traffic noise 1988, *Acoustics Bulletin* (Institute of Acoustics), **14** (1), pp 4–9.)

[19] US Senate (1972). Report to the President and Congress on Noise, Report of the Administrator of the Environmental Protection Agency, Document no 92–63, February 1972.

[20] D. W. Robinson (1971). Towards a unified system of noise assessment. *Journal of Sound and Vibration*, **14** (3), pp 279–298.

[21] I. D. Griffiths and F. J. Langdon (1968). Subjective response to road traffic noise. *Journal of Sound and Vibration*, **8** (1), pp 16–32.

[22] Department of the Environment (1973). Planning and Noise. Circular 10/73, Welsh Office Circular 16/73. HMSO, London. (A new version of Planning and Noise is likely to be issued in 1993.)

[23] Acoustical Society of America (1977). American National Standard for rating noise with respect to speech interference. ANSI S3.14–1977.

[24] T. J. B. Smith (1977). Design targets for background noise levels in buildings. *Noise Control Vibration and Insulation*, March, pp 88–91.

[25] D. Morrison (1978). A review of worldwide automotive noise legislation. *Noise Control Vibration Isolation*, Nov/Dec, pp 377–380.

[26] B. L. Clarkson (1972). Social consequences of noise. *Proceedings of the Institution of Mechanical Engineers*, **186**, no 8, pp 97–107.

[27] C. S. Kerse (1973). *The Law Relating to Noise*. Oyez Press, London.

[28] R. Penn (1979). *Noise Control*. Shaw Press, London.

[29] B. J. Smith, R. J. Peters and S. Owen (1977). *Acoustics*. Longman, London.

[30] C. W. Kosten and G. J. van Os (1962). Community reaction criteria for external noise,in the Control of Noise, NPL Symposium no 12, pp 373–382. HMSO, London.

[31] British Standards Institution (1990). Method of rating industrial noise affecting mixed residential and industrial areas. BS 4142:1990.

[32] International Organization for Standardization (1987). Acoustics — Description and measurement of environmental noise — Part 1: Basic quantities and procedures; Part 2: Acquisition of data pertinent to land use; Part 3: Application to noise limits. ISO 1996 1,2 and 3.

[33] D. W. Robinson (1977). Practice and principle in environmental noise rating. National Physical Laboratory, Teddington. Acoustics Report Ac 81.

[34] Anon. (1989). Calculation of road traffic noise; background to the Technical Memorandum. *Acoustics Bulletin* (Institute of Acoustics) **14** (1), p 4.

4 Sound Insulation and Sound Absorption

Whence is that knocking?
How is't with me, when every noise appals me?
William Shakespeare, *Macbeth*

4.1 Introduction

Sound absorption and sound insulation are two of the most common techniques used in noise control. Sound insulation is used to control the transmission path of the sound and can be achieved by inserting a wall or panel between the source and the listeners. It is a technique used for controlling air–borne sound, and is particularly applied to problems of sound propagation from one room to another in a building. For effective insulation heavy walls of brick or concrete are required. The aim of sound absorption, on the other hand, is to reduce the level of reflected sound in a room in which both the noise source and the listener, whose acoustic conditions it is intended to improve, are situated. Some of the theoretical aspects connected with sound absorption and sound insulation are similar, but the materials which are applied, and the ways in which both techniques are used, are different. Although it is convenient to treat the methods of absorption and insulation as two separate techniques, they are quite often applied together, as in the case of a sound enclosure which is used to reduce the noise from a machine in a factory.

4.2 Basic definitions of sound absorption and sound insulation

Let us assume that between the sound source and the listener there is a wall for the purpose of sound insulation made of a heavy material, such as brick or concrete. Additionally, plane sound waves move normally towards the wall, as shown in Figure 4.1. Only a small fraction of the sound energy is transmitted through the wall, a larger fraction is reflected back towards the noise source. Let α_t be the *sound transmission coefficient* for the wall, i.e., the ratio of the transmitted sound power to the incident sound power, and let α_r be the fraction of the incident sound power which is reflected from the wall.

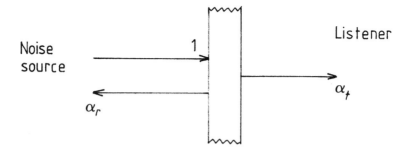

Figure 4.1 Sound insulation.

As there is no loss of energy,

$$\alpha_t + \alpha_r = 1. \tag{4.1}$$

For successful sound insulation the sound transmission coefficient α_t should be as small as possible. Any listener on the same side of the wall as the noise source will suffer an increase in sound. The sound insulation qualities of a wall are very often characterised by a *sound reduction index* instead of the sound transmission coefficient α_t. The sound reduction index at normal incidence R_0 is defined as

$$R_0 = 10 \log \left(\frac{1}{\alpha_t} \right) \text{ dB}. \tag{4.2}$$

The term *transmission loss* (TL) is sometimes used, particularly in the USA, instead of sound reduction index (SRI). A sound reduction index of 40 dB represents good sound insulation; the corresponding value of α_t is

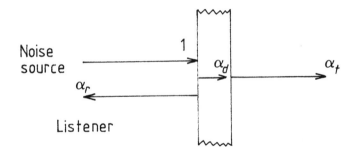

Figure 4.2 Sound absorption.

10^{-4}. The sound reduction index, as defined, is a property of the wall and is independent of the sound fields in the rooms which the wall separates.

In Figure 4.2 some basic concepts related to sound absorption are illustrated. For the purpose of good absorption the wall or barrier should be made from a sound absorbing material, such as mineral wool or fibre glass. As shown in Figure 4.2, α_d represents the ratio of the absorbed sound power to the incident sound power. In other words it is the fraction of the incident sound energy that is converted to another form of energy, mostly heat. For this case:

$$\alpha_r + \alpha_t + \alpha_d = 1. \tag{4.3}$$

To achieve effective sound absorption α_r should be as small as possible and the sum $\alpha_t + \alpha_d$ as large as possible. The sum $\alpha_t + \alpha_d$ is often called the *sound absorption coefficient* α_A. From the point of view of the listener on the same side of the barrier as the noise source it does not matter whether α_A comprises a large value of α_t and a small value of α_d, or a small α_t and large α_d; both conditions provide good sound absorption. A value of α_d close to unity is achieved by a very thick layer of a sound absorbing material. In this case α_t is almost zero and good sound insulation is attained as well as good sound absorption. On the other hand, if the value of α_t is one and the value of α_d zero, perfect sound absorption is obtained, but, of course, there is no sound insulation. It is sometimes said that an open window is a 'perfect absorber of sound'. If a window is opened some sound energy leaves the room and is lost from the room for ever. A similar effect would be obtained within the room if the window were blocked with a thick layer of a good sound absorbing material, such as mineral wool.

So far we have considered some of the general concepts relating to sound insulation and absorption. The approach adopted has been elementary and has neglected certain factors. For example, normal incidence of plane sound waves on the wall has been assumed, whereas in a real room sound would be incident at many angles. Frequency is also a parameter which has not as yet been considered. In the next section sound absorption is dealt with in more detail.

4.3 Normal reflection of a plane sound wave from a surface

The absorption of sound by a wall depends upon the properties of the material of the wall and takes place both at the surfaces and within the material. However, we can study absorption by considering only the reflection of sound wave at a surface of the material.

The surface in question which coincides with the axis Oy of the coordinate system in the plane xy, as shown in Figure 4.3, faces a sound field on the left side of the surface. A plane sound wave, represented by the phasor of sound pressure $\tilde{p}_i$, is normally incident on the surface. As only normal reflection of plane sound waves is considered the problem is one-dimensional. The whole sound field is defined in terms of the co–ordinate x. Note that the origin of the co–ordinate system lies on the surface of the wall.

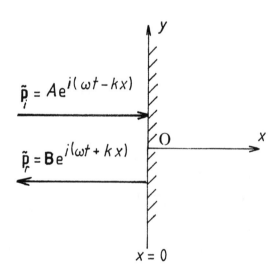

Figure 4.3 Normal reflection of sound from a surface.

To analyse sound reflection it is convenient to apply the phasor formalism, see Section 1.12.2. Thus the phasor for the sound pressure of the incident wave is given by:

$$\widetilde{p}_i = A e^{i(\omega t - kx)}, \qquad (4.4)$$

where A is the amplitude of the sound pressure, ω is the circular frequency and k is the wave number. A similar expression for the sound pressure of the reflected wave is

$$\widetilde{p}_r = \mathbf{B} e^{i(\omega t + kx)}, \qquad (4.5)$$

where $\mathbf{B}$ is the amplitude of the reflected sound pressure wave and in general is complex (hence, the letter B is shown in bold type), because there is usually a phase difference between the reflected and incident waves. The phasors for the particle velocities of the incident and reflected plane sound waves are determined by equations (1.104) and (1.105), respectively.

The incident and reflected waves together form a standing wave (see Section 4.4 and also Section 1.15). To determine the sound field in front of the wall, i.e., to determine the reflected sound wave, the boundary condition must be applied at the surface of the wall ($x = 0$, see Figure 4.3). The boundary condition is formulated in the form of an equality between the specific acoustic impedance of the standing wave in the first medium, e.g. air, and the specific acoustic impedance of the wall at the surface, see Sections 1.18 and 6.3. However, when sound reflection is considered, as well as sound transmission at the interface between two elastic media, the boundary conditions imposed at the interface are that the pressure and normal component of the particle velocity are continuous, see also Section 6.3. These conditions are consequences of the mass and momentum conservation laws.

In the case when the wall is not rigid and is penetrable by the sound wave the concept of the specific acoustic impedance of the wall surface is especially useful in formulating the boundary conditions.

It is expected that in the sound field of a monochromatic wave at any point on the wall surface a linear relationship exists between the sound pressure $\widetilde{p}$ and the component of the particle velocity normal to the surface $\widetilde{u}_n = \vec{\widetilde{u}} \cdot \vec{n}$, where $\vec{n}$ denotes the unit vector normal to the wall surface. Thus,

$$\frac{\widetilde{p}}{\widetilde{u}_n} = \mathbf{z}_n.$$

The ratio $\mathbf{z}_n$ is called the specific acoustic impedance of the wall. The specific acoustic impedance $\mathbf{z}_n$ of the surface depends upon the frequency of the sound wave, properties of the wall material and in general upon the sound field inside the material, since the motion of the surface may be coupled with the wave motion inside the wall.

In certain cases, when the surface is one of *local reaction*, the specific acoustic impedance at any point on the surface does not depend on the spatial distribution of the sound wave inside the material. In our theoretical considerations we shall assume that the wall surface is locally reacting. For a homogeneous material the impedance of the surface is that of the material. The specific acoustic impedance is in general complex, since the phasor of the normal component of the particle velocity can be out of phase with the sound pressure phasor. The specific acoustic impedance of the wall is denoted by the symbol z_n and may be considered to have resistive component r_n and a reactive component x_n. Thus,

$$z_n = r_n + ix_n. \tag{4.6}$$

The specific acoustic impedance z_n represents the mechanical properties of the wall surface, or in this case mechanical properties of the wall material [1]. The sound field in front of the wall is also characterised by a specific acoustic impedance which depends upon x. At the wall surface, at $x = 0$, the specific acoustic impedance of the standing wave must be equal to the specific acoustic impedance of the surface. Thus the complex amplitude **B** can be determined from the boundary condition which expresses the continuity of impedance.

To the left of the surface, for negative values of x, the specific acoustic impedance **z** of the sound field of the standing wave is given by:

$$\mathbf{z} = \frac{\widetilde{\mathbf{p}}_i + \widetilde{\mathbf{p}}_r}{\widetilde{\mathbf{u}}_i + \widetilde{\mathbf{u}}_r}, \tag{4.7}$$

where $\widetilde{\mathbf{u}}_i$ and $\widetilde{\mathbf{u}}_r$ are the phasors of the particle velocities of the incident and reflected waves, respectively. The individual incident and reflected waves are plane and hence the previously obtained relationships, equations (1.104) and (1.105), may be used to relate particle velocities and sound pressures. Thus,

$$\widetilde{\mathbf{u}}_i = \frac{\widetilde{\mathbf{p}}_i}{\rho_0 c_0} \text{ and } \widetilde{\mathbf{u}}_r = -\frac{\widetilde{\mathbf{p}}_r}{\rho_0 c_0}, \tag{4.8}$$

where ρ_0 is the density of the air and c_0 the sound wave velocity. Hence,

$$\mathbf{z} = \rho_0 c_0 \frac{\widetilde{\mathbf{p}}_i + \widetilde{\mathbf{p}}_r}{\widetilde{\mathbf{p}}_i - \widetilde{\mathbf{p}}_r}. \tag{4.9}$$

Using equations (4.4), (4.5) and (4.9), we may write:

$$\mathbf{z} = \rho_0 c_0 \frac{A e^{i(\omega t - kx)} + \mathbf{B}^{i(\omega t + kx)}}{A e^{i(\omega t - kx)} - \mathbf{B}^{i(\omega t + kx)}}. \tag{4.10}$$

At $x = 0$, $\mathbf{z} = \mathbf{z}_n$, the specific acoustic impedance of the wave is equal to that of the surface. Thus,

$$\mathbf{z}_n = \rho_0 c_0 \frac{A + \mathbf{B}}{A - \mathbf{B}} \tag{4.11}$$

and, consequently,

$$\frac{\mathbf{B}}{A} = \frac{\mathbf{z}_n - \rho_0 c_0}{\mathbf{z}_n + \rho_0 c_0}. \tag{4.12}$$

The ratio $\mathbf{B}/A$ is called the *complex reflection ratio* and is denoted by the symbol $\mathbf{r}$. We may write $\mathbf{r}$ in terms of its modulus r and argument θ. Thus,

$$\mathbf{r} = r e^{i\theta}. \tag{4.13}$$

Substituting x_n, specified by the equation (4.6), into equation (4.12), we obtain

$$\mathbf{r} = \frac{\mathbf{B}}{A} = \frac{(r_n - \rho_0 c_0) + i x_n}{(r_n + \rho_0 c_0) + i x_n}. \tag{4.14}$$

The modulus r is given by:

$$r = \sqrt{\frac{(r_n - \rho_0 c_0)^2 + x_n^2}{(r_n + \rho_0 c_0)^2 + x_n^2}} \tag{4.15}$$

and the argument θ is determined by:

$$\tan \theta = \frac{2\rho_0 c_0 x_n}{r_n^2 + x_n^2 - (\rho_0 c_0)^2} \tag{4.16}$$

and

$$\sin \theta = \left(\frac{A}{\mathbf{B}}\right) \frac{2\rho_0 c_0 x_n}{(r_n + \rho_0 c_0)^2 + x_n^2}. \tag{4.17}$$

The sound reflection coefficient α_r, introduced in Section 4.2, is defined as the ratio of the reflected to the incident time–averaged sound energy fluxes (or in other words the ratio of the reflected to the incident sound powers, see equation (1.129)), namely,

$$\alpha_r = \frac{W_r}{W_i} = \frac{\int\limits_S \overline{\vec{I}_r} \cdot \vec{n} dS}{\int\limits_S \overline{\vec{I}_i} \cdot \vec{n} dS}.$$

In the considered case of the plane sound wave at normal incidence to the surface the sound reflection coefficient α_r can be expressed as the ratio of the time–averaged sound intensity $\overline{I}_r$ of the reflected wave to the time–averaged sound intensity $\overline{I}_i$ of the incident wave. Hence,

$$\alpha_r = \frac{\overline{I}_r}{\overline{I}_i},$$

where $\overline{I}_r = \overline{|\vec{I}_r|}$ and $\overline{I}_i = \overline{|\vec{I}_i|}$. Consequently, see equation (1.139),

$$\alpha_r = \frac{|\mathbf{B}|^2}{A^2} = r^2. \tag{4.18}$$

Thus from equation (4.15)

$$\alpha_r = \frac{(r_n - \rho_0 c_0)^2 + x_n^2}{(r_n + \rho_0 c_0)^2 + x_n^2}. \tag{4.19}$$

The sound absorption coefficient at normal incidence α_A is that fraction of the incident time–averaged sound energy flux that is not reflected. It is the fraction of sound that is transmitted to the right of the surface at $x = 0$, see Figure 4.3. Hence,

$$\alpha_A = 1 - \alpha_r = \frac{4 r_n \rho_0 c_0}{(r_n + \rho_0 c_0)^2 + x_n^2}. \tag{4.20}$$

From equation (4.20) we can deduce that when the resistive part of the impedance r_n is zero the sound absorption coefficient α_A is also zero. The coefficient α_A is similarly zero when the reactive part of the specific acoustic impedance x_n tends to infinity. In these cases there is neither sound absorption by the material nor transmission of sound. The reactive part is large in a heavy wall made of, say, concrete.

4.4 Standing waves formed by normal reflection at a surface

For the general case in which the reflection is not perfect the reflected wave has an amplitude less than the amplitude of the incident sound wave and the two waves are not in phase at $x = 0$. In the region of negative x in front of the surface (see Figure 4.3) the incident and reflected waves lose their separate identities and together form a *standing wave*. In order to determine the amplitude of the standing wave we must consider first the component waves separately. Standing waves have already been introduced in Section 1.15 for the case of perfect reflection when the amplitudes of the incident and reflected waves are equal. We consider now a more general case in which the amplitudes are not equal.

In the phasor formalism the sound pressure field of the one–dimensional standing wave is represented by the phasor

$$\tilde{p}(x,t) = Ae^{i(\omega t - kx)} + Be^{i(\omega t + kx + \theta)}, \tag{4.21}$$

which is the sum of the phasors $\tilde{p}_i$ and $\tilde{p}_r$, representing the incident and the reflected sound pressure waves, respectively, see equations (4.4) and (4.5). The amplitudes (moduli) A and B are real and θ is the phase angle. The expression (4.21) for the sound pressure phasor may be transformed into other forms:

$$\begin{aligned}
\tilde{p} &= Ae^{i[\omega t - (kx + \theta/2) + \theta/2]} + Be^{i[\omega t + (kx + \theta/2) + \theta/2]} \\
&= e^{i(\omega t + \theta/2)}\left[(A + B)\cos(kx + \theta/2) + i(B - A)\sin(kx + \theta/2)\right] \\
&= C(x)e^{i[\omega t + \theta/2 + \varphi(x)]} \\
&= C(x)e^{i(\omega t + \Phi_0)},
\end{aligned} \tag{4.22}$$

where $\Phi_0 = \varphi(x) + \theta/2$ is the initial phase and $\varphi(x)$ is defined by the equations

$$\tan\varphi = \frac{B - A}{A + B}\tan(kx + \theta/2) \tag{4.23}$$

$$\sin\varphi = \frac{(B - A)\sin(kx + \theta/2)}{\sqrt{(A + B)^2\cos^2(kx + \theta/2) + (B - A)^2\sin^2(kx + \theta/2)}}. \tag{4.24}$$

The sound pressure amplitude for the standing wave $C(x)$ is expressed by:

$$C(x) = \sqrt{(A + B)^2\cos^2(kx + \theta/2) + (B - A)^2\sin^2(kx + \theta/2)}. \tag{4.25}$$

Similarly, the particle velocity phasor for the one–dimensional standing wave is

$$\tilde{u} = \left[A e^{i(\omega t - kx)} - B e^{i(\omega t + kx + \theta)} \right] / (\rho_0 c_0)$$

$$= \frac{C_1(x)}{\rho_0 c_0} e^{i[\omega t + \theta/2 - \varphi_1(x)]}, \qquad (4.26)$$

where $\varphi_1(x)$ and $C_1(x)$ are defined by:

$$\tan \varphi_1 = \frac{A + B}{B - A} \tan(kx + \theta/2) \qquad (4.27)$$

$$\sin \varphi_1 = -\frac{(A + B) \sin(kx + \theta/2)}{\sqrt{(A - B)^2 \cos^2(kx + \theta/2) + (A + B)^2 \sin^2(kx + \theta/2)}} \qquad (4.28)$$

$$C_1(x) = \sqrt{(A - B)^2 \cos^2(kx + \theta/2) + (A + B)^2 \sin^2(kx + \theta/2)}. \qquad (4.29)$$

Consequently,

$$\tan(\varphi_1 - \varphi) = \frac{(A + B)^2}{(A^2 - B^2)} \tan(kx + \theta/2) \qquad (4.30)$$

and

$$\sin(\varphi_1 - \varphi) = -\frac{2AB \sin(2kx + \theta)}{C_1(x) \cdot C(x)}. \qquad (4.31)$$

The amplitude of the sound pressure $C(x)$ reaches a maximum value equal to $A + B$ when

$$\cos^2(kx + \theta/2) = 1. \qquad (4.32)$$

Similarly, the amplitude $C(x)$ has a minimum value of $A - B$ when

$$\sin^2(kx + \theta/2) = 1. \qquad (4.33)$$

The points in the standing sound wave, where the amplitude is maximal are called *anti–nodes* and are separated by a distance of π/k or half a wavelength. Similarly, the points where the amplitude is minimal are *nodes* and are situated half a wavelength apart.

Note that at points in the standing sound wave, where the sound pressure amplitude is maximal, see equation (4.32), the amplitude of the particle velocity is minimal and equal to $(A - B)/(\rho_0 c_0)$. At the nodes the particle velocity amplitude is maximal and equal to $(A + B)/(\rho_0 c_0)$.

The conditions (4.32) and (4.33) are mathematically equivalent to:

$$kx + \theta/2 = 0, \pm\pi, \pm 2\pi, \text{etc.,} \qquad (4.34)$$
$$kx + \theta/2 = \pm\pi/2, \pm 3\pi/2, \text{etc.,} \qquad (4.35)$$

respectively.

4.4.1 Positions of the anti–nodes and nodes

In the situation illustrated in Figure 4.3 the standing sound wave occupies space from $-\infty$ to 0. Hence, the physically realised nodes and anti–nodes are those for which $x \leq 0$. The above restriction leads to the reduction of conditions (4.34) and (4.35) to the equations

$$x_{\max} = -\left(\frac{\theta}{2k} - \frac{n\pi}{k}\right) \qquad (4.36)$$

and

$$x_{\min} = -\left[\frac{\theta}{2k} - \frac{(2n+\beta)}{2k}\pi\right], \qquad (4.37)$$

respectively, where $n = 0, -1, -2, -3, \text{etc.,}$ and

$$\beta = -1 \text{ for } 0 \leq \theta < \pi$$
$$\beta = 1 \text{ for } \pi \leq \theta < 2\pi.$$

The equation (4.36) describes the positions of the anti–nodes in the standing sound wave. Similarly, equation (4.37) refers to the positions of the nodes. Since β is equal to 1 or -1, the equation

$$\beta = \frac{2k}{\pi}(x_{\min} - x_{\max})$$

or

$$\beta = \frac{4}{\lambda}(x_{\min} - x_{\max}), \qquad (4.38)$$

which is the results from combining the equation (4.36) with the equation (4.37), can be used to check the accuracy in the determination of the positions of the pair: anti–node and node. The pair formed by the anti–node and node (node and anti–node) is numerated by n in the sequence of pairs.

The way in which the amplitude of the standing wave varies with distance is shown in Figure 4.4. In this case x_n, the imaginary part of the

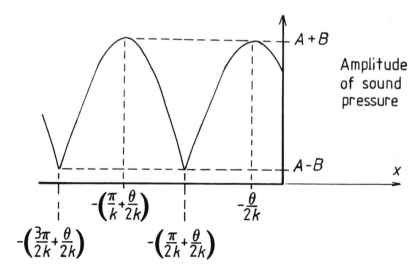

Figure 4.4 Standing wave pattern; x_n and $\sin\theta$ positive.

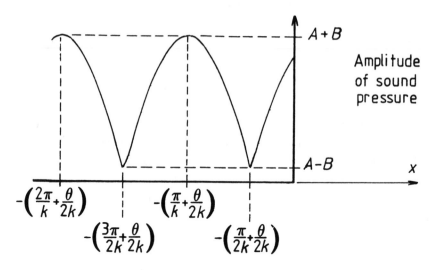

Figure 4.5 Standing wave pattern; x_n and $\sin\theta$ negative.

specific acoustic impedance of the surface, is positive, hence $\sin\theta$ is also positive, see formula (4.17), and $0 \le \theta < \pi$. The amplitude varies from a maximal value $A + B$ to a minimum of $A - B$.

The standing wave pattern for negative x_n (and negative $\sin\theta$) is shown in Figure 4.5. In this case $\pi \le \theta < 2\pi$.

The value of θ may be determined either from equation (4.36) or from equation (4.37), if the values of x_{max} and x_{min}, respectively, are known. It is sufficient to know either the position of an arbitrarily chosen single node or anti–node in the standing wave to deduce the value of β, to find n and, consequently, to be able to determine θ.

Let us assume that x_{max}, the position of a certain anti–node in the standing wave, is known. Let s_1 be the maximal integer part of $|x_{max}|/(\lambda/2)$, which is denoted as

$$s_1 = [2|x_{max}|/\lambda], \tag{4.39}$$

where $0 \leq s_1 \leq 2|x_{max}|/\lambda$, λ is the wavelength and $|x_{max}|$ is the absolute value of x_{max}. The condition

$$0 \leq |x_{max}| - s_1(\lambda/2) < \lambda/4 \tag{4.40}$$

implies that $\beta = -1$ $(0 \leq \theta < \pi)$ and that the pattern of the anti–nodes and nodes in the standing wave is as shown in Figure 4.4 with the first anti–node $(n = 0)$ situated in the closest neighbourhood to the wall. Additionally, for the discussed anti–node $n = -s_1$ and $N_{max} = -n + 1 = s_1 + 1$ denotes the position of the considered anti–node (of x_{max}) in the anti–nodes sequence. Note that $N_{max} = 1, 2, 3 \ldots$ numerates all anti–nodes in their spatial distribution (starting from the wall).

If the condition

$$\lambda/4 \leq |x_{max}| - s_1(\lambda/2) < \lambda/2 \tag{4.41}$$

is fulfilled for x_{max}, $\beta = 1$ or $\pi \leq \theta < 2\pi$. Additionally, for this anti–node $n = -s_1$, $N_{max} = -n + 1$ and the pattern of nodes and anti–nodes is, as shown in Figure 4.5, with the node closest to the wall.

Next, let us assume that instead of x_{max} the position of a certain node x_{min} is known in the standing wave. Let s_2 be the maximal integer part of $|x_{min}|/(\lambda/2)$, namely,

$$s_2 = [2|x_{min}|/\lambda]. \tag{4.42}$$

If

$$\lambda/4 \leq |x_{min}| - s_2(\lambda/2) < \lambda/2 \tag{4.43}$$

is fulfilled for x_{min} then it implies that $\beta = -1$ and in the pattern of anti–nodes and nodes the closest to the wall is the first anti–node, as is illustrated in Figure 4.4. If, however,

$$0 \leq |x_{min}| - s_2(\lambda/2) < \lambda/4 \tag{4.44}$$

then $\beta = 1$ and $\pi \leq \theta < 2\pi$ and the closest to the wall is the first node. In both cases for the considered node $n = -s_2$ and $N_{min} = -n + 1$. Note that $n = 0, -1, -2 \ldots$ numerates the pairs of either anti–nodes and nodes or nodes and anti–nodes. N_{min} describes the position of the considered node (x_{min} position) in the sequence of nodes only.

For the special case of a perfect reflection ($\alpha_r = 1$), the amplitudes of the incident and reflected waves are equal ($B = A$) and the waves are in phase at $x = 0$, since $\theta = 0$. In this case the amplitude is

$$\left[4A^2 \cos^2 kx \right]^{1/2}.$$

The wave which results from a perfect reflection is sometimes called a *stationary wave*.

4.4.2 Sound intensity in a standing wave

The instantaneous sound intensity $\vec{I}$, see appendix B, is the sum of the active and reactive sound intensities. For the standing wave it is given by:

$$\vec{I} = \vec{I}_{ac} + \vec{I}_{rc}$$

$$= [1/(\rho_0 c_0)] \left\{ (A^2 - B^2) \cos^2(\omega t + \varphi(x) + \theta/2) \right.$$

$$\left. + AB \sin[2(kx + \theta/2)] \sin[2(\omega t + \varphi(x) + \theta/2)] \right\} \vec{e}_x,$$

where θ and $\varphi(x)$ are defined by the equations (4.21) and (4.25), respectively.

The time–averaged sound intensity is expressed by the vector:

$$\overline{\vec{I}} = \frac{(A^2 - B^2)}{2\rho_0 c_0} \vec{e}_x.$$

Hence, for the stationary sound wave (standing sound wave with $A = B$) the time–averaged energy flux density vector disappears: $\overline{\vec{I}} = 0$.

4.5 The standing wave tube

The above theory on standing waves forms the basis for the working principles of the standing wave tube, an apparatus which may be used for measuring either the sound absorption coefficient or the specific acoustic impedance of a surface in the case of normal incidence of the sound waves. The construction of a typical apparatus is shown in Figure 4.6. In the circular duct shown the specimen under test is at the right side and the loudspeaker on the left. As illustrated in Figure 4.6 the sound pressure in the duct is measured by a probe from a microphone. The probe is screwed onto a condenser microphone from which the protective grid has been removed. (See Figure 2.4.) The right side of the probe must be supported by a wheel of small dimensions. The microphone, situated on the left side of the apparatus, is held in a carriage which moves over a scale.

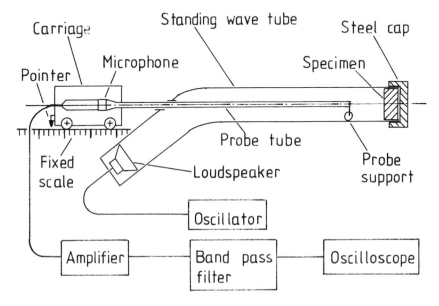

Figure 4.6 The standing wave tube.

There are a number of different realisations of the construction of the standing wave tube. Instead of a duct with a bend some forms of the apparatus have a straight duct with the probe microphone passing through the centre of the loudspeaker. With such a variation in the construction it is necessary to have the loudspeakers specially made, because it would be normally difficult to drill a hole in the magnetic core. With the probe microphone, see Section 2.3.3, the sound pressure is measured at the tip of the probe. It is possible to have the microphone within the duct at a position corresponding to that of the tip of the probe. If this procedure is adopted, the preamplifier and cable from the microphone should be confined within a tube with as small a diameter as possible. The disadvantage of putting the microphone in the duct is that the microphone, with its relatively large size, blocks and distorts the sound field.

4.5.1 Measurement of sound absorption coefficient

In order to measure the sound absorption coefficient the specimen should have behind it a thick, heavy terminating cap. This cap will generally be made of steel and ensures that the intensity transmitted axially out of the duct to the right is as small as possible. Hence, the apparatus permits the measurement of the coefficient α_d, see Section 4.1 and Figure 4.2, since in this case $\alpha_d \simeq \alpha_A$.

In carrying out the measurements the probe or microphone is traversed along the duct. The loudspeaker is at a fixed position and is supplied by an oscillator which generates a sine wave at a particular frequency. The sound pressure signals at different locations of the microphone in the duct are observed on a measuring amplifier or oscilloscope to which the microphone is connected via a preamplifier. Observation of, for example, the oscilloscope enables the amplitude values of the sound pressure signal to be obtained. The amplitude values in the standing wave change from minimal $(A - B)$ to maximal $(A + B)$. The ratio of the signal amplitude at the anti-nodes to the signal amplitude at the nodes is called the *standing wave ratio* SWR, which is a useful parameter in determining the sound (power) absorption coefficient.

From the definition of the standing wave ratio

$$\text{SWR} = \frac{A + B}{A - B} \tag{4.45}$$

and, consequently,

$$\frac{B}{A} = \frac{\text{SWR} - 1}{\text{SWR} + 1}. \tag{4.46}$$

Combining equations (4.18) and (4.46) leads to the relation between the sound absorption coefficient α_A and the standing wave ratio SWR,

$$\alpha_A = 1 - \alpha_r = 1 - \left(\frac{\text{SWR} - 1}{\text{SWR} + 1} \right)^2. \tag{4.47}$$

A typical set of sound absorption coefficients, measured for normal incidence by a standing wave tube, is shown in Figure 4.7. The specimen used comprised a circular sample of polyurethane foam which fitted tightly into the duct without compression and touched the steel terminating cap. The foam was tested with and without a facing of perforated steel sheets of different open areas. The open area is the ratio of the area of the holes to the total plate area, and is generally expressed as a percentage. In the

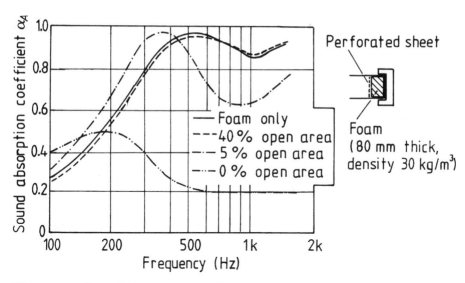

Figure 4.7 Sound absorption coefficients of polyurethane foam faced with perforated steel, obtained in a standing wave tube by A. W. Spencer, City University.

case of zero open area there is a solid sheet in front of the foam. Perforated sheets are often used with sound absorbing materials to provide protection for the material.

4.5.2 Measurement of specific acoustic impedance

The phenomenon of the standing wave may be used to determine the specific acoustic impedance of a surface. In order to measure the specific acoustic impedance at the surface of the material the positions of the nodes, see equation (4.37), and anti–nodes, see equation (4.36), in the standing wave should be first specified. To achieve proper results from the measurements the apparatus must be scaled in such a way that the position of the pointer should correspond to $x = 0$ when the end of the probe is just touching the surface of the specimen. From the position of the first anti–node, given by equation (4.33), the value of the phase angle θ can be obtained easily. In general the value of the phase angle θ can be found from the position of any arbitrarily chosen node or anti–node, as described in Section 4.4.1.

Next, the standing wave ratio should be determined using the procedure described in Section 4.5.1. Taking into account equation (4.46) for the modulus of the complex reflection ratio and equation (4.13), we are able to determine the complex reflection ratio $\mathbf{r}$. From equation (4.14) we consequently find the real r_n and imaginary x_n parts of the specific acoustic impedance of the surface z_n.

It may be possible, depending upon the frequency, to obtain the positions of a series of nodes and anti–nodes and hence obtain more accurate values of the specific acoustic impedance by taking average values of B/A and θ.

Example 4.1

A sample is tested in a standing wave tube at a frequency of 500 Hz. At a certain node, which occurs 80 mm from the surface of the sample, the value of the amplitude of the sound pressure signal is 0.001 Pa. The amplitude of the sound pressure at the first antinode from the surface is 0.016 Pa. Estimate the sound absorption coefficient and the specific acoustic impedance of the surface of the material. Take c_0 to be 344 m/s and ρ_0 to be 1.21 kg/m³.

Solution From the formula (4.42) we get $s_2 = 0$. Consequently, $n = 0$ ($N_{min} = 1$). Finally, $|x_{min}| = 0.08$ fulfils the condition (4.44), $0 \le 0.08 < \lambda/4 = c_0/(4f) = 0.172$ [m], which implies that $\beta = 1$ ($\pi \le 0 < 2\pi$).

The standing wave ratio is 16. Hence from equation (4.46)

$$B/A = 15/17 = 0.8824$$

and from equation (4.47)

$$\alpha_A = 1 - (15/17)^2 = 0.2215.$$

The sound absorption coefficient at 500 Hz is 0.22.

The value of θ is determined either from the position of any anti–node, using formula (4.36), or from the position of any node using equation (4.37). In this example it was found that the first node is at a position of 80 mm from the surface of the sample and that $\beta = 1$. The case is illustrated in Figure 4.5 and refers to $x_n < 0$, i.e., the imaginary part of the specific acoustic impedance is negative. Consequently, from equation (4.37)

$$-0.08 = -[\theta/(2k) - \pi(2k)],$$

where k is $2\pi \times 500/344 = 9.13$ rad/m. Hence,

$$\theta = 2k \times 0.08 + \pi = 4.6 \,\text{rad (or } 263.7°).$$

From equation (4.13) for the complex reflection ratio **r** we obtain

$$\mathbf{r} = 0.8824e^{i4.60} = -0.0964 - i0.8771.$$

From equation (4.14)

$$-0.0964 - i0.8771 = \frac{r_n - 416.24 + ix_n}{r_n + 416.24 + ix_n}$$

or

$$(-0.0964 - i0.8771)(r_n + 416.24 + ix_n) = r_n - 416.24 + ix_n,$$

which is equivalent to the system of two equations

$$-1.0964r_n + 0.8771x_n + 376.13 = 0$$
$$0.8771r_n + 1.0964x_n + 365.07 = 0.$$

Hence,

$$r_n = 46.8 \text{ and } x_n = -370.4. \text{ The units are kgm}^{-2}\text{s}^{-1}.$$

The standing wave tube is a useful apparatus, but it does have some inadequacies. It can only be used in the measurement of the sound absorption coefficient and the specific acoustic impedance for normal incidence of sound waves. Hence, plane waves must exist in the duct. At higher frequencies, or when the wavelength of sound becomes small compared with the diameter of the duct, it is possible for three–dimensional modes of oscillation to occur. In this case not only longitudinal but also transverse modes occur in the tube. The standing wave tube will give false results when the waves are no longer plane. The condition, that only the longitudinal mode is propagated without attenuation in the circular tube, is:

$$f < c_0/(1.7D),$$

where f is the frequency in Hz of the sinusoidal sound wave in the duct, c_0 is the wave velocity of the sound in m/s and D is the duct diameter in mm. Thus, for example, if the duct diameter is 100 mm, the standing wave tube should only be used below a frequency of 2030 Hz.

Additionally, measurements can be performed only to a certain low frequency limit defined by the length of the duct. To find in the standing wave of known frequency the positions of the anti–node and node, see equations (4.36) and (4.37), as well as the standing wave ratio, the probe does not need to be traversed more than a quarter of a wavelength. However, if the frequency of the standing wave is not exactly known, the determination of the position of the anti–node, node (node, anti–node) pair is essential. Hence, in this case the probe must be traversed at least half a wavelength. Consequently, the condition which permits the determination of the sound absorption coefficient and the specific acoustic impedance may be formulated in the form of the inequality

$$L > c_0/(2f),$$

where L is the length of the probe traverse, expressed in m. Thus, if the probe traverse is 1 m the frequency indicated by the above inequality is 172 Hz. This is the lowest sound wave frequency above which results can be guaranteed.

The diameter of the probe in the duct must be small to avoid errors arising from distortion of the sound field. In an apparatus described by Scott [2] the cross–section of the probe was less than $\frac{1}{2}\%$ of that of the main tube.

Manufacturers of sound absorbing materials often quote values for the sound absorption coefficient in their sales catalogues. These absorption coefficients are generally not obtained by using a standing wave tube. International and national standards require that the coefficients be measured by a method that makes use of a reverberation chamber. This method has the advantage of permitting the measurement of the absorption of sound at most angles of incidence. The sound absorption coefficients measured by the reverberation chamber method and the standing wave tube cannot be related. The former method will be described in the next chapter.

4.6 Transmission of sound through a heavy wall

Sound insulation is achieved by installing massive walls, constructed from heavy materials, such as concrete or brick. In materials of this type only a small amount of sound energy is converted into heat by dissipative processes. In general the sound energy transmitted through the wall can be determined by considering the transmission and reflection of normally incident plane sound waves at the two wall surfaces which separate the wall from the surrounding medium (air) [1,3]. In the particular case of sound transmission through a thin, but heavy, solid wall it is possible to adopt a simple approach which is sufficient to derive an expression for the sound power transmission coefficient. If the solid wall is sufficiently thin compared with the wavelength of sound in the solid material, i.e., $h\omega/c_{0w} \ll 1$, where h denotes the thickness of the material and c_{0w} is the sound velocity in the wall material, it is possible to assume that the wall moves like a rigid body under the action of the sound waves.

A schematic illustration of this model is presented in Figure 4.8, where the wall (assumed to be of infinite extent) as well as the directions of propagation of the incident, reflected and transmitted sound waves are shown. The incident, reflected and transmitted sound waves are represented by the sound pressure phasors $\widetilde{p}_i$, $\widetilde{p}_r$ and $\widetilde{p}_t$, respectively.

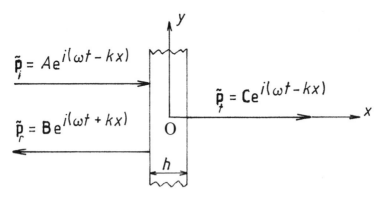

Figure 4.8 Transmission of sound through a heavy wall.

The wall material has a volume density ρ (with units of kgm^{-3}). However, the quantity which characterises the inertia of the wall is the surface density, i.e., the mass per unit area of the wall. The surface density σ (with units of kgm^{-2}) is related to the volume density ρ by the expression,

$$\sigma = \rho h, \tag{4.48}$$

where h is the thickness of the wall in m.

It is assumed in this model that the wall, freely suspended in air, vibrates as a rigid body which generates sound on the outer wall surface. Hence, the thin wall equation of motion is:

$$\widetilde{p}_i + \widetilde{p}_r - \widetilde{p}_t = \rho h \frac{d\widetilde{u}_w}{dt}$$

or

$$\widetilde{p}_i + \widetilde{p}_r - \widetilde{p}_t = \sigma \frac{d\widetilde{u}_w}{dt}, \tag{4.49}$$

where $\widetilde{\mathbf{p}}_i + \widetilde{\mathbf{p}}_r - \widetilde{\mathbf{p}}_t$ denotes the net force acting on unit area of the thin wall and $\widetilde{\mathbf{u}}_w$ is the velocity of the wall. Since the surrounding air remains in contact with both surfaces of the wall, the boundary conditions formulated at the inner and outer wall surfaces are, respectively,

$$\widetilde{\mathbf{u}}_i + \widetilde{\mathbf{u}}_r = \widetilde{\mathbf{u}}_w \qquad (4.50a)$$

$$\widetilde{\mathbf{u}}_t = \widetilde{\mathbf{u}}_w, \qquad (4.50b)$$

where $\widetilde{\mathbf{u}}_i$, $\widetilde{\mathbf{u}}_r$ and $\widetilde{\mathbf{u}}_t$ denote the particle velocity phasors for the incident, reflected and transmitted sound waves, respectively.

Taking into account the following relations (see equation (4.8)):

$$\widetilde{\mathbf{u}}_i = \widetilde{\mathbf{p}}_i/(\rho_0 c_0), \quad \widetilde{\mathbf{u}}_r = -\widetilde{\mathbf{p}}_r/(\rho_0 c_0), \quad \widetilde{\mathbf{u}}_t = \widetilde{\mathbf{p}}_t/(\rho_0 c_0),$$

where

$$\widetilde{\mathbf{p}}_i = \mathbf{A} e^{i(\omega t - kx)}, \quad \widetilde{\mathbf{p}}_r = \mathbf{B} e^{i(\omega t + kx)}, \quad \text{and} \quad \widetilde{\mathbf{p}}_t = \mathbf{C} e^{i(\omega t - kx)}$$

we obtain from equations (4.49) and (4.50), respectively,

$$\mathbf{A} + \mathbf{B} - \mathbf{C} = \frac{i\omega\sigma}{\rho_0 c_0} \mathbf{C} \qquad (4.51)$$

and

$$\mathbf{A} - \mathbf{B} = \mathbf{C}. \qquad (4.52)$$

From equations (4.51) and (4.52) we find that:

$$\frac{\mathbf{C}}{\mathbf{A}} = \frac{1}{1 + i\omega\sigma/(2\rho_0 c_0)}. \qquad (4.53)$$

The sound transmission coefficient α_t is given by:

$$\alpha_t = \frac{|\mathbf{C}|^2}{|\mathbf{A}|^2} = \frac{1}{1 + \omega^2\sigma^2/(4\rho_0^2 c_0^2)} \qquad (4.54)$$

and, as defined by equation (4.2), the sound reduction index R_0 at normal incidence is

$$R_0 = 10 \log \left[1 + \omega^2\sigma^2/(4\rho_0^2 c_0^2)\right]. \qquad (4.55)$$

Except at values of σf below about 400 kgm^{-2}s^{-1} it is the second term in the argument of the logarithm that is most significant. Hence, for the massive wall when $\sigma w \gg 2\rho_0 c_0$ the sound reduction index can be represented by:

$$R_0 = 20\log[w\sigma/(2\rho_0 c_0)] \tag{4.56}$$
$$= 20\log f\sigma - 42.5 \text{ dB}, \tag{4.57}$$

where $\rho_0 c_0 = 420$ kgm^{-2}s^{-1} and $\omega = 2\pi f$.

Equation (4.57), which relates the sound reduction index R_0 to the frequency f and surface density σ, represents the *mass law*, a fundamental principle of sound insulation. The mass law, derived here from a theoretical model which only takes into account normal incidence of the sound waves on the wall surface, is illustrated graphically in Figure 4.9. When the surface density is doubled the sound reduction index increases by 6 dB. Thus, if a concrete wall is increased in thickness from 100 mm to 200 mm, the theory presented in this section indicates an increase in R_0 of 6 dB throughout the frequency range. Similarly, for a given surface density doubling the frequency also increases the sound reduction index by 6 dB.

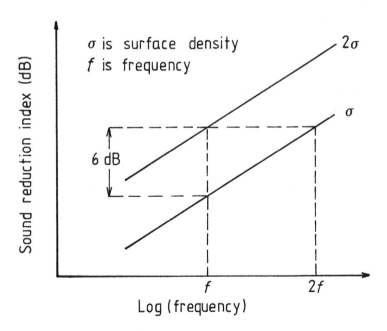

Figure 4.9 Illustration of mass law in terms of frequency f and surface density σ.

It is interesting to note that the specific acoustic impedance of the inner wall surface z_{in} (as seen from the source side) is different from the specific acoustic impedance of the outer wall surface z_{out}. Namely,

$$z_{in} = \frac{(A + B)}{(A - B)} \rho_0 c_0 \qquad (4.58)$$

and

$$z_{out} = \rho_0 c_0.$$

Combining equations (4.51) and (4.52) leads to the expression

$$\frac{(A + B)}{(A - B)} = 1 + \frac{i\omega\sigma}{\rho_0 c_0}. \qquad (4.59)$$

After inserting the expression (4.59) into the equation (4.58), we obtain for the specific acoustic impedance z_{in}:

$$z_{in} = \rho_0 c_0 + i\omega\sigma. \qquad (4.60)$$

4.7 Sound reduction indices at other angles of incidence

As has been mentioned in Section 4.6, the equation (4.54), which specifies the sound transmission coefficient for the transmission of sound through a thin but massive wall, has been derived for plane waves normally incident on the wall surface. The sound transmission coefficient $\alpha_{t,\theta}$ for plane sound waves which have an angle of incidence θ is given by [4]:

$$\alpha_{t,\theta} = \frac{1}{1 + \omega^2\sigma^2 \cos^2\theta/(4\rho_0^2 c_0^2)}. \qquad (4.61)$$

In order to obtain a sound transmission coefficient $\alpha_{t[0,\Theta]}$, which takes into account all angles of incidence from 0 to Θ, a formula devised by Paris [5] is used:

$$\alpha_{t[0,\Theta]} = \frac{\int_0^\Theta \alpha_{t,\theta} \cos\theta \sin\theta d\theta}{\int_0^\Theta \cos\theta \sin\theta d\theta}. \qquad (4.62)$$

Combining the equations (4.61) and (4.62) leads to the expressions for $\alpha_{t[0,\Theta]}$:

$$\alpha_{t[0,\Theta]} = \frac{1}{s^2(1 - \cos^2\Theta)} \ln\left(\frac{1 + s^2}{1 + s^2 \cos^2\Theta}\right) \qquad (4.63)$$

and for the corresponding sound reduction index, see also equation (4.2):

$$R_{[0,\Theta]} = 10 \log \left[\frac{s^2(1 - \cos^2 \Theta)}{\ln(1 + s^2) - \ln(1 + s^2 \cos^2 \Theta)} \right] \text{ dB}, \qquad (4.64)$$

where $s = \omega\sigma/(2\rho_0 c_0)$.

The expression for $R_{[0,\Theta]}$, see equation (4.64), evaluated for $\Theta = 90°$, gives the formula for the *random incidence sound reduction index* $R_{[0°,90°]}$, namely,

$$R_{[0°,90°]} = 10 \log \left[\frac{s^2}{\ln(1 + s^2)} \right] \text{ dB} \qquad (4.65)$$

$$= 10 \log s^2 - 10 \log \left[\ln(1 + s^2) \right] \text{ dB}.$$

For a massive wall and for a sufficiently high frequency ($\sigma\omega \gg 2\rho_0 c_0$) equations (4.2) and (4.54) lead to:

$$R_0 = 10 \log s^2 \text{ dB}.$$

Hence, for values of σf above about 400 $\text{kgm}^{-2}\text{s}^{-1}$ we have

$$R_{[0°,90°]} = R_0 - 10 \log R_0 + 6.4 \text{ dB}. \qquad (4.66)$$

The random incidence sound reduction index (equation (4.66)) takes into account all angles of incidence of sound on the wall. However, the values obtained in this way are too low and a more realistic quantity is the *field incidence sound reduction index* R_{field} which is defined as

$$R_{\text{field}} = R_0 - 5 \text{ dB}. \qquad (4.67)$$

The approximate value of the field incidence sound reduction index is obtained from equation (4.64), if $\Theta = 78°$ [6]. Thus,

$$R_{\text{field}} \simeq R_{[0°,78°]}. \qquad (4.68)$$

In Figure 4.10 the three quantities R_0, R_{field} and $R_{[0°,90°]}$, derived from the equations (4.55), (4.67) and (4.65), respectively, are shown plotted against the product σf.

4.8 Wave coincidence

The sound reduction index may be measured by means of a method in which sound is transmitted through the wall under test from one reverberation chamber to a second reverberation chamber. See Section 5.11.2.

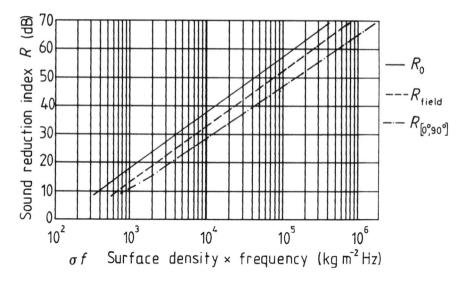

Figure 4.10 Normal incidence R_0, field incidence R_{field} and random incidence $R_{[0°,90°]}$ sound reduction indices as functions of the product of surface density and frequency.

Experimental results, obtained in one third octave bands, for a glass window in a wood and steel frame are shown in Figure 4.11 [4]. The continuous line represents theoretical results based on the expression for the mass law for the window in the case of field incidence. Agreement between theory and experiment is quite reasonable at low frequencies, but there is a discrepancy at the middle frequencies. At 1600 Hz the experimental value is about 18 dB less than the predicted.

The sound reduction index R_{mean}, mentioned in the caption to Figure 4.11, is the arithmetic mean of the indices in dB in one third octave bands from 100 Hz to 3150 Hz.

It may be appreciated from Figure 4.11 that the mass law does not describe completely the sound insulating performance of walls. The phenomenon responsible for the departure from the mass law is *wave coincidence* also called the *coincidence effect*.

Wave coincidence may be explained in terms of Figure 4.12, in which a plane sound wave is shown as having an angle of incidence on wall. The sound wave impinging on the wall is capable of setting the wall into vibration but, since the sound wave front is not parallel to the wall, different elements of the wall vibrate with a different phase. However, the elements of the wall which are separated by a distance equal to the trace wavelength

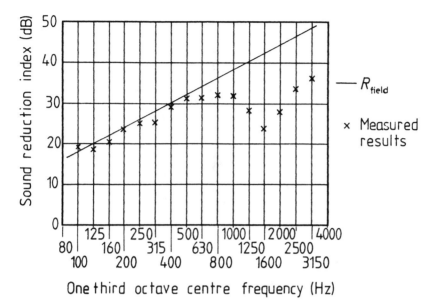

Figure 4.11 Sound reduction indices for a glass window in a wood and steel frame (glass 7.94 mm (5/16 in) thick, panel size 2.7 m×2.15 m); $\sigma = 19.6$ kg/m²; $R_{\mathrm{mean}} = 27$ dB: (a) theoretical results R_{field}, (b) measured results.

are in phase. The *trace wavelength* is the projection of the wavelength λ of the sound wave onto the wall and is equal to $\lambda / \sin \theta$. The vibrations of the wall which are forced in this way are bending or flexural vibrations. Flexural vibrations occur when a bending wave motion is set up in the wall. The bending waves propagate along the wall with the sound waves, they have a wavelength $\lambda_b (= \lambda / \sin \theta)$ and have a velocity of propagation of $c_0 / \sin \theta$.

In the absence of sound waves free bending waves can exist in the wall. Such bending waves are dispersive, i.e., their phase velocity depends upon frequency f. The phase velocity c_b of bending waves in a wall of thickness h is obtained from [7]:

$$c_b = (1.81 h f c_L)^{1/2}. \tag{4.69}$$

The propagation velocity of the longitudinal wave in the wall c_L is given by:

$$c_L = \sqrt{\frac{E}{\rho(1 - \nu^2)}}, \tag{4.70}$$

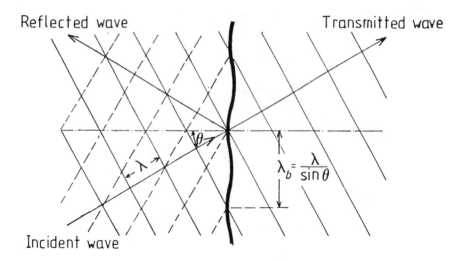

Figure 4.12 Illustration of wave coincidence; sound incident obliquely on a wall in bending.

where E is the Young's modulus of the material of the wall, ρ is the density of the wall and ν is Poisson's ratio.

The amplitude of the wall vibrations are greatest when the velocity of propagation of the forced bending waves, $c_0/\sin\theta$, coincides with the phase velocity of the free bending wave c_b. This condition is referred to as the coincident effect or wave coincidence. When the effect occurs, sound is radiated strongly from the wall in the direction of the transmitted wave, as shown in Figure 4.12. The sound transmission coefficient α_t has a value approaching unity and the sound reduction indices are low, as shown in Figure 4.11.

Thus the condition for the coincident effect is

$$c_0/\sin\theta = (1.81 h f c_L)^{1/2} \tag{4.71}$$

and the frequencies at which coincidence occurs are given by:

$$f = \frac{c_0^2}{1.81 h c_L \sin^2\theta}. \tag{4.72}$$

The lowest frequency for wave coincidence occurs when θ is 90°, i.e., for grazing incidence. This lowest frequency is referred to as the *critical frequency of wave coincidence* f_c and is given by:

$$f_c = \frac{c_0^2}{1.81 h c_L}. \tag{4.73}$$

For the glass window, whose field incidence sound reduction index is shown in Figure 4.11 as a function of frequency, the critical frequency occurs at 1600 Hz.

It is useful to consider the product σf_c. From equations (4.48), (4.70) and (4.73) we have

$$\sigma f_c = \frac{c_0^2}{1.81} \sqrt{\frac{\rho^3 (1 - \nu^2)}{E}}. \tag{4.74}$$

Thus the product σf_c is a function of the properties of the material of the wall. For a given material σf_c is a constant. If the thickness of the wall is doubled the critical frequency is halved. Values of the product σf_c are given in Table 4.1 for various common materials.

TABLE 4.1 Product of surface density σ in kg/m^2 and critical frequency of wave coincidence f_c in Hz for various materials

Material	σf_c
lead	600,000
steel	105,000
concrete	98,000
brick	42,000
glass	38,000
perspex	35,500
hardboard	31,000
plywood	13,000

Obviously, there are limitations for the validity and application of the theoretical model roughly described in this section. The limitations concern especially the relation between the wall thickness and the wavelength of the bending waves. The wavelength of the bending waves should be greater than six times the wall thickness.

4.9 Sound reduction index for an ideal wall

To provide good sound insulation a wall or panel should ideally have:

(a) large mass,
(b) low stiffness, and
(c) large vibration damping.

A low stiffness is required so that the critical frequency f_c occurs at a high frequency. In this way the range over which the mass law is obeyed is as large as possible. Low stiffness is achieved by a low Young's modulus and hence low value of c_L; as a result f_c is large (equations (4.70) and (4.73)).

Vibration damping is inherent in some materials, such as plastics, and in the case of steel the damping of bending vibrations can be achieved by coating the steel with a visco–elastic layer. These techniques are discussed in Chapter 8. An increase in vibration damping increases the sound reduction index in the region above the critical frequency of wave coincidence where the wall or panel is said to be *coincidence–controlled*. The effect of damping is shown schematically in Figure 4.13.

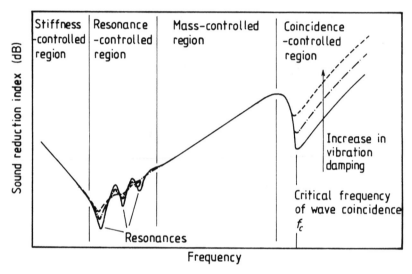

Figure 4.13 Idealised sound reduction index for a wall; effect of vibration damping.

Also shown in Figure 4.13 are regions at low frequencies where the sound reduction index is *resonance–controlled* and *stiffness–controlled*. Normally these regions occur at frequencies which are too low to be of great interest. A real wall or panel is not infinite, consequently there are resonances at the natural frequencies of vibration which depend upon the size of the panel. At the resonance frequencies the values of the sound reduction index are minimum.

A material which in many ways is ideal for sound insulation is lead. A panel made from lead is heavy, has low stiffness and good damping. Lead has no structural use but it can be useful where a flexible wrapping is required.

4.10 Enclosures

The principles of sound insulation are frequently utilised to reduce the noise from a machine by surrounding the machine by a box or enclosure. Enclosures should also make use of sound absorbing material on the inside. They are used for the reduction of air–borne sound and do not reduce structure–borne sound. Enclosures are ineffective at low frequencies, particularly if the machine is causing the floor to vibrate. For the best results it is essential that the enclosure should be total, without any apertures. In many applications this will not be possible to achieve.

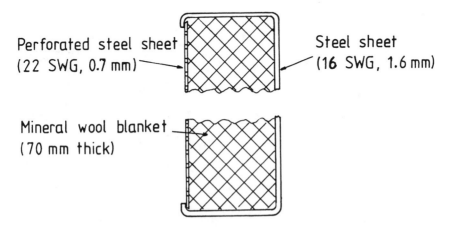

Perforated steel sheet (22 SWG, 0.7 mm)

Steel sheet (16 SWG, 1.6 mm)

Mineral wool blanket (70 mm thick)

Figure 4.14 Cross–section of a typical panel used for sound enclosure.

Enclosures are frequently constructed from panels which have a length greater than their width [8]. A section of a typical panel is shown in Figure 4.14 to illustrate the construction. Sound insulation is achieved by the outer sheet of steel with a typical thickness of 1.6 mm. Sound absorption is provided by a blanket of fibre glass or mineral wool with a thickness from 70 to 100 mm. A perforated steel sheet (typical thickness 0.7 mm) provides some protection for the sound–absorbing material. Perforated sheet with a 30% open area is typically used and still allows sound absorption (see Figure 4.7). Some designs of panel have a vibration damping material on the inside of the outer sheet of steel. Alternatively, the construction of the panel may enable the sound absorbing material to be compressed (usually by about 10%) by the perforated sheet and, because of the compression, provide damping of the flexural vibrations of the outer skin.

Enclosures can either be fabricated by the manufacturer or constructed on site by bolting or locking together standard panels and components. An example of this modular type of construction is shown in Figure 4.15. The figure shows the components of the enclosure, such as doors, windows and silencers. A larger enclosure would probably require a supporting frame.

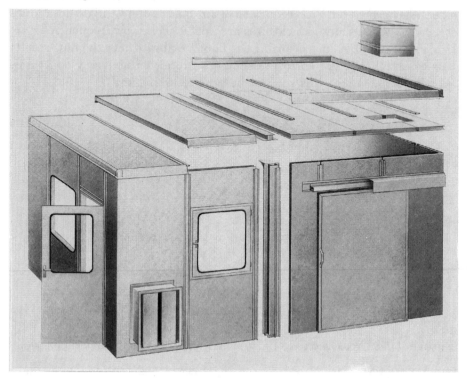

Figure 4.15 Schematic diagram of an enclosure of modular construction (Courtesy of G and H Montage GmbH).

Some enclosures are equipped with doors so that personnel can gain access, or the enclosure may fit tightly round the source [8]. With tight–fitting enclosures a panel is removed if maintenance of the machine is required.

The acoustic performance of enclosures may be specified in various ways. For a component panel the sound reduction indices in different octave bands are the most useful characteristics. However, the sound reduction index is a property of the panel alone and, although a guide to, it does not give information on the attenuation achieved by a complete enclosure in a given situation. The most useful way of specifying the sound attenuation is by determining the *insertion loss*. The sound pressure insertion loss is the difference between the sound pressure levels at a particular point without an enclosure round the machine and with the machine enclosed. Useful as it is, the sound pressure insertion loss is different for each position around the enclosure. It will also depend to some extent on the acoustics of the room in which the measurements are taken. The most effective approach is via the determination of the sound power insertion loss. In this method the sound power level of the enclosed machine (measured preferably in an anechoic chamber or reverberation chamber) is compared with the sound power level of the machine alone. However, these measurements are difficult to perform, because of the extensive facilities which are required.

Some of the results obtained by different methods of specifying enclosure performance are shown in Figure 4.16, as applied to a small enclosure [10]. The noise reduction is the difference between the levels inside and outside the enclosure; it is of limited value. The more useful sound pressure insertion loss gives much lower values than either the sound reduction index for the panels or the noise reduction.

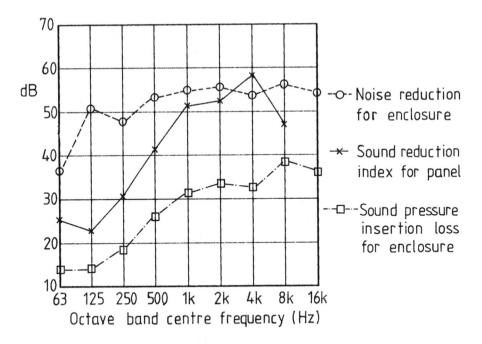

Figure 4.16 Comparison of the different ways of specifying the performance of enclosures.

4.11 Data on sound reduction indices

Typical sound reduction indices are shown in Table 4.2 for various materials and constructions used in building and in noise control. The data in the table have been extracted from references [4] and [11] and from manufacturers' publicity leaflets. The method used for the measurement of the sound reduction indices involves two reverberation chambers, as outlined in Section 5.11.2. The method is described in international and national standards. The measured data correspond approximately to the field incidence sound reduction indices, as defined by equation (4.48). An extensive collection of data is provided by Reynolds [12].

It is sometimes useful to quote a single number to represent the sound reduction indices of a material or structure. The mean sound reduction index is sometimes used [4]. This the mean of the values of the indices which are the available for the quoted one third octave filters. Thus for the results of the glass window shown in Figure 4.11 the mean sound reduction index is 27 dB and is the arithmetic average of the measured sound reduction indices in dB at the one third octaves from 100 to 3150 Hz.

4.12 Closure

Sound insulating properties of sound absorbing materials The difference between sound insulation and sound absorption has been emphasised in this chapter. Sound absorbing materials can, if they are sufficiently thick, provide good sound insulation [13]. As long as the fraction of the incident sound energy flux which is absorbed by the material is very nearly unity ($\alpha_d \to 1$, Figure 4.2), the transmitted sound energy flux is almost zero ($\alpha_t \to 0$). However, porous materials would not normally be used for a wall, as they provide no structural function.

Flanking paths Whilst sound absorbing materials can provide good sound insulation, it is not always the case. There are some materials which provide good sound absorption, but poor sound insulation; an acoustic tile is an example. Acoustic tiles are frequently used in lecture rooms and lobbies for the purpose of reducing the reflected sound. They are often made of fibre glass and have holes on the inside surface. In Figure 4.17 acoustic tiles are shown forming a false ceiling in an office; this is a common application of acoustic tiles and indeed they give better sound absorption with an air space between the tile and a solid surface. In the example presented in Figure 4.17 the office was divided into smaller units after the installation of the false ceiling. Although the dividing partitions may provide adequate sound insulation, sound will be transmitted from one room to the next through the acoustic tiles which provide only poor sound

TABLE 4.2 Sound reduction indices for different materials

Material description (see notes below table)	Sound reduction index (dB) at third octave centre frequency (Hz)																			
	100	125	160	200	250	315	400	500	630	800	1000	1250	1600	2000	2500	3150				
1. Brick, single row	34	36	36	37	38	37	40	40	40	41	46	47	51	54	56	57				
2. Brick, 2 rows	39	43	38	41	45	52	54	55	51	53	55	60	70	79	82	87				
3. Concrete, reinforced (thickness 51 mm)	32	38	37	37	40	41	42	40	37	34	36	40	43	46	50	54				
4. Concrete, reinforced (thickness 102 mm)	36	38	39	40	37	35	38	41	43	47	48	51	55	57	61	65				
5. Noisemaster acoustic blocks;	29	25	23	23	22	19	22	26	29	33	35	36	40	43	46	47				
6. Chipboard, 19 mm thick	15	17	16	16	18	20	22	25	26	29	30	30	29	26	27	31				
7. Plywood, 6 mm thick	8	9	9	10	13	12	15	16	18	20	21	22	25	27	27	28				
8. Wood joist floor	17	11	21	22	29	31	30	33	34	34	38	39	42	42	38	34				
9. Wood joist floor	22	21	22	32	39	36	38	41	42	42	46	47	51	52	51	54				
10. Gyproc metal stud partition	18	16	29	28	26	30	32	35	38	42	46	48	49	47	44	36				
11. Plasterboard (12.7 mm)	–	22	26	26	27	27	27	27	28	29	30	31	32	33	32	32				

Notes:

1. Brick, 114 mm thick, $\sigma = 184$ kg/m^2
2. Brick, 2 rows of 114 mm thick separated by air cavity of 58 mm, total thickness 286 mm, $\sigma = 369$ kg/m^2
3. Concrete, reinforced, thickness 51 mm, area 2.42×12 m, $\sigma = 117$ kg/m^2
4. Concrete, reinforced, thickness 102 mm, area 2.42×12 m, $\sigma = 234$ kg/m^2
5. Noisemaster acoustic blocks; unpainted pelletised concrete blocks with internal Helmholtz resonator, partially filled with acoustic foam; each block with surface area 390×190 mm, thickness 100 mm, $\sigma = 122$ kg/m^2
6. Chipboard, 19 mm thick sheets on wooden frame, $\sigma = 15.5$ kg/m^2
7. Plywood, 6 mm thick sheets on wooden frame, $\sigma = 3.4$ kg/m^2
8. Wood joist floor; 22 mm thick wooden floor supported on wooden joists with 12.7 mm plasterboard attached to lower side of joists, total thickness 210 mm, $\sigma = 35$ kg/m^2
9. Wood joist floor similar to above but with mineral wool pugging, total thickness 260 mm, $\sigma = 50$ kg/m^2
10. Gyproc metal stud partition; two 12.7 mm plasterboards separated by 48 mm steel studs, total thickness 75 mm, $\sigma = 24$ kg/m^2
11. Plasterboard (12.7 mm) covered by 0.4 mm lead sheet, $\sigma = 15$ kg/m^2

Continued on next page

TABLE 4.2 Sound reduction indices for different materials (cont.)

Material description (see notes below table)	Sound reduction index (dB) at third octave centre frequency (Hz)															
	100	125	160	200	250	315	400	500	630	800	1000	1250	1600	2000	2500	3150
12. Glass window	13	15	15	15	17	18	21	23	25	27	29	30	31	30	28	22
13. Double glass window	17	23	25	26	29	30	33	34	36	38	39	42	45	45	45	36
14. Barrier mat	15	17	17	19	20	22	23	24	25	28	30	31	31	32	33	34
15. PVC curtain	6	8	9	10	13	11	13	12	12	11	12	12	12	10	10	11
16. PVC sheet sandwiching fibre glass	5	13	13	12	13	17	14	16	18	21	27	31	34	36	39	40
17. Pipe lagging material	12	16	16	18	18	21	23	20	21	27	31	36	39	42	45	45

Notes:

12. Glass window; 4 mm thick, well–sealed in 3.6×1.8 m wood frame, $\sigma = 12$ kg/m^2
13. Double glass window; two 4 mm thick sheets with 100 mm cavity, well–sealed in 3.6×1.8 m wood frame, total thickness 108 mm, $\sigma = 23$ kg/m^2
14. Barrier mat; free hanging curtain of loaded PVC (4.5 mm thick) with jute backing, total thickness 5 mm, $\sigma = 7.5$ kg/m^2
15. PVC curtain; transparent, flexible, 4 mm thick in strips 380 mm wide with 50% overlap, $\sigma = 6$ kg/m^2
16. PVC sheet sandwiching fibreglass (0.6 mm reinforced PVC, 20 mm fibreglass, 1.2 mm reinforced PVC), total thickness 22 mm, $\sigma = 5$ kg/m^2
17. Pipe lagging material (4 mm lead with 25 mm fibreglass), total thickness 29 mm $\sigma = 27$ kg/m^2

insulation, particularly at low frequencies.

The partition between rooms shown in Figure 4.17 provides sound insulation poorer than expected because of flanking paths through the acoustic tiles and around the partition. Care must always be taken to prevent flanking paths.

The full benefit of a panel with a high sound reduction index will not be achieved if there are weak links in the path of the air–borne sound. Such weak links may take the form of a window or, worse still, a hole. If diffraction effects are neglected, sound is transmitted through the hole with a sound power transmission coefficient of one for the area of the hole. For example, let us assume that a partition has a surface area of 10 m^2 and a sound reduction index of 40 dB. There is a hole in the partition of area 10^{-2} m^2. The equivalent sound reduction index for the partition and

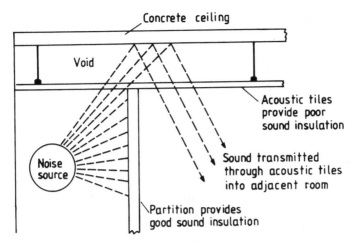

Figure 4.17 Illustration of how sound can form a flanking path around a sound insulating partition through acoustic tiles.

hole is 30 dB — a loss of 10 dB.

The flanking paths can have a route which is either air–borne, as shown in Figure 4.17, or structure–borne. Floors, ceilings and walls can provide the routes for flexural waves, if they are in contact with vibrating machines. Sound is re–radiated from those parts of the building which are vibrating. Sound from these flanking paths can be reduced by the techniques of vibration isolation, described in Chapter 8. It is important to apply vibration isolation if the full benefits of sound insulation are to be achieved.

Double panels   Sound insulation appears to be limited by the mass law. Is it possible to achieve better insulation than the mass law indicates? Yes, if double panel structures are used. The use of independent panels separated by an air space can give sound reduction indices that increase at the rate of 12 dB for every doubling of frequency [6]. However, flanking paths have to be totally eliminated to achieve the full potential of double panels. Two totally independent enclosures, one inside the other, would form a double panel and give very high attenuation of sound.

The problems associated with the interaction of sound waves with structures are discussed in detail in textbooks by Cremer and Heckl [14], Fahy [15], Junger and Feit [16] and Lyon [17].

It was intended in this chapter to introduce the reader to the concepts of sound absorption and sound insulation. It is hoped that the understanding achieved is better than that of Thomas Carlyle and Marcel Proust who wrote their great works in cork–lined studies in the belief that they were eliminating the noise from the outside world. Proust was probably more successful at avoiding external noise from his neighbours when he spent his holidays at the Grand Hotel, Cabourg, where for the sake of quiet he took five adjoining rooms on the top floor; unfortunately he may not have succeeded in escaping from structure–borne noise.

References

[1] S.Temkin (1981). *Elements of Acoustics.* John Wiley, New York.

[2] R. A. Scott (1946). An apparatus for the accurate measurement of the acoustic impedance of sound absorbing materials. *Proceedings of the Physical Society,* **58**, pp 253–264.

[3] L. E. Kinsler and A. R. Frey (1962). *Fundamentals of Acoustics,* second edition. John Wiley, New York.

[4] E. N. Bazley (1966). *The Airborne Sound Insulation of Partitions.* HMSO, London.

[5] E. T. Paris (1928). On the coefficient of sound absorption measured by the reverberation method. *Philosophical Magazine,* **5**, pp 489–497.

[6] I. L. Ver and C. I. Holmer (1971). Interaction of Sound Waves with Solid Structures, in *Noise and Vibration Control* (ed. L. L. Beranek). McGraw-Hill, New York.

[7] E. Volterra and E. C. Zachmanoglou (1965). *Dynamics of Vibrations.* Charles E. Merrill Books, Columbus, Ohio.

[8] I. L. Ver (1973). Reduction of Noise by Acoustic Enclosures, in *Isolation of Mechanical Vibration, Impact and Noise* (eds J. C. Snowdon and E. E. Ungar). A Colloquium presented at the American Society of Mechanical Engineers Design Engineering Technical Conference, Cincinnati, Ohio, Sept 1973. AMD vol 1.

[9] L. W. Tweed and D. R. Tree (1977). A model of close-fitting acoustical enclosures. *Proceedings of NOISE-CON 1977,* pp 319–330.

[10] J. S. Anderson and D. G. Bull (1979). User's Guide to Acoustic Enclosures. *Noise Control Vibration Isolation,* February, pp 51–54.

[11] T. Smith, P. E. O'Sullivan, B. Oakes and R. B. Conn (Eds) (1971). *Building Acoustics.* British Acoustical Society Special Volume No 2. Oriel Press, Newcastle–upon–Tyne.

[12] D. D. Reynolds (1981). *Engineering Principles of Acoustics.* Allyn and Bacon, Boston.

[13] A. J. King (1965). *The Measurement and Suppression of Noise.* Chapman and Hall, London.

[14] L. Cremer and M. Heckl (1973). *Structure–Borne Sound.* Springer–

Verlag, Berlin.

[15] F. Fahy (1985). *Sound and Structural Vibration*. Academic Press, London.

[16] M. C. Junger and D. Feit (1986). *Sound, Structures and their Interaction*, second edition. The MIT Press, Cambridge, Massachusetts.

[17] R. H. Lyon (1975). *Statistical Energy Analysis of Dynamical Systems*. The MIT Press, Cambridge, Massachusetts.

5 Room Acoustics

All those rather imprecise terms used to describe admired acoustics can be applied; warm but not mushy, bright but not harsh, resonant but not soggy, distinct but not dry or brittle, blended but not coagulated, lively but not hard.

Andrew Porter, music critic of the *Financial Times*, on the Avery Fisher Hall, New York, 1976.

5.1 Introduction

The sound that is heard in a room includes direct sound and sound reflected from the walls, ceiling, floor and objects in the room. The reflected sound depends upon the nature of the room. If the same machine is taken from one factory to another, the sound pressure levels of the noise will be different in the two factories, even though the measurements may have been carried out in an identical manner.

It has already been emphasised that sound is a wave motion. Sound waves in enclosed spaces have to satisfy the boundary conditions at the walls and ceiling, etc. Consequently, the general solution of the sound wave equation which fulfils the boundary conditions is a series of discrete functions of frequency, called *eigenfunctions* or *normal modes of vibration*. Every eigenfunction is characterised by a unique frequency of oscillations which is known as the *natural frequency*, eigenfrequency or normal mode frequency. The normal modes, and the natural frequencies which correspond to them, depend on the shape and size of the enclosure.

226

Thus wave motion in an enclosed space can occur only at a certain frequency of vibration, the natural frequency. For a large room the normal modes are so numerous in the audible range of frequencies that it is impossible in practice to use an approach based on wave motion in order to make predictions about the sound field. Also, for enclosures with a complicated geometry the wave equation cannot be solved analytically. Therefore, instead of using the wave equation the concept adopted in *geometrical room acoustics* is that of *sound rays*.

In geometrical ray acoustics sound, like light, propagates along rays which are lines that are always perpendicular to surfaces of constant phase. The rays are shown emanating from a source in Figure 5.1 for rays in a plane. The sound ray model is valid when (a) the amplitude of the wave and (b) the speed of sound do not change significantly with distances comparable to a wavelength. Both conditions (a) and (b) should be satisfied in the ray model. Each ray of sound is independent of the others; there is no interchange of energy between rays or beams of sound, hence phenomena such as wave diffraction cannot be taken into account.

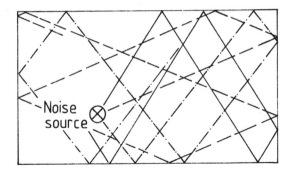

Figure 5.1 Reflection of rays of sound in a room.

The sound ray approach is quite reasonable at high frequencies. High frequency sound from sources is often very directional and forms straight beams like rays of light. For the case of the reflection of sound the proper application of the sound ray model is restricted to specular reflection, as in optics. The wavelength should be small compared with the dimensions of the reflecting surface, but large with respect to the size of the surface roughness. Thus surfaces should be large and smooth for the method of geometrical ray acoustics to give good results. The method is often used, even at low frequencies, because there is no easily applicable alternative.

5.2 The diffuse field

Close to the source of sound the direct field of sound dominates, but further away from the source it is the reflected sound which may be stronger. The reflected sound is said to form a *diffuse field* in the room when the reflections are so numerous that rays of sound are arriving at a point in the room from all directions and all phase shifts of the rays are equally probable. There is no preferred direction for the reflected sound at any point in the room; any direction is equally probable. To achieve a diffuse field there should be no surface areas of the room boundaries where the sound absorption is very much greater or less than the average. In a diffuse field the averaged sound energy density, and hence the sound pressure level, are constant throughout the volume of the room. It should be noted that the idea of a diffuse field does not involve individual standing or travelling waves. The influence of the shape of the room on the sound field does not occur in this concept.

The expression reverberant sound is often used to describe the reflected sound. A reverberant sound field may, or may not, be diffuse. A reverberant field is not diffuse if one surface of the room has reflective properties which are very much different from the rest of the room. A diffuse field may not exist in the transient stage just after the source has been switched on, but the field may become diffuse once steady state conditions have been achieved.

The mean–squared pressure $\overline{p_r^2}$ of the reflected sound is constant in space for a diffuse field and depends upon the acoustic power of the source, the properties of the medium in which the waves propagate and the properties of the surface of the room, as follows:

$$\overline{p_r^2} = \frac{4W\rho_0 c_0}{R_c}, \tag{5.1}$$

where W is the sound power of the source,

$\rho_0 c_0$ is the characteristic acoustic resistance and

R_c is called the *room constant.*

See references [1,2] for a derivation of equation (5.1).

The room constant is defined in terms of the *average sound absorption coefficient* $\overline{\alpha}$ of the surfaces of the room for the room as a whole:

$$R_c = \frac{S\overline{\alpha}}{1 - \overline{\alpha}}, \tag{5.2}$$

where S is the total surface area of the room.

5.2.1 Average sound absorption coefficient

The average sound absorption coefficient $\overline{\alpha}$ takes into account the absorption of all the room surfaces. In many rooms the ceiling, for example, is made from one material and its entire surface S_1 is associated with a particular sound absorption coefficient α_1. Similarly, the walls of area S_2 may be made of a uniform material and characterised by a sound absorption coefficient α_2, and so on. Thus $\overline{\alpha}$ the average sound absorption coefficient is related to α_1, α_2, etc., in the following way:

$$S\overline{\alpha} = S_1\alpha_1 + S_2\alpha_2 + \ldots \tag{5.3}$$

The products $S_1\alpha_1, S_2\alpha_2$, etc., are sometimes called the *absorption areas* and the product $S\overline{\alpha}$ the *total equivalent absorption area*. For small values of $\overline{\alpha}$ the product $S\overline{\alpha}$ will be approximately equal to the room constant.

5.2.2 Mean (or statistical) sound absorption coefficient

It is important to understand the significance of the individual absorption coefficients used in equation (5.3), α_1, α_2, etc., which are associated with particular surfaces of the room. The term sound absorption coefficient has already been introduced in Section 4.3 for the case of sound at normal incidence on a surface. Sound rays at a surface in a room are arriving randomly at all angles of incidence. The sound absorption coefficient is different for each angle of incidence. There are two ways in which this random incidence can be accounted for in the absorption coefficient.

Firstly, the absorption coefficient, taking into account all angles of incidence, can be calculated from the specific acoustic impedance of the surface (which may have been measured in the standing wave tube, Section 4.5.2). The calculation procedure is described by Morse and Ingard [3] and by Kuttruff [4]. What is achieved by this calculation is an absorption coefficient which is defined as the ratio of the sound power absorbed by a surface to the sound power incident upon the surface when the sound field is diffuse. This absorption coefficient has been called the *mean sound absorption coefficient* by many authors, including Morse and Ingard [3] and Cremer and Müller [5]. Embleton [6] prefers the term *statistical sound absorption coefficient*.

The second possibility is to make use of absorption coefficients that are based entirely upon measurement. The measurement method is based upon the work of Sabine on reverberation time and is described later in Sections 5.5 and 5.11.1. Absorption coefficients based upon the measurements are sometimes called the *Sabine absorption coefficients* [6]. In general the values of the Sabine absorption coefficient are greater than the mean absorption coefficients.

5.3 The direct field

The space where the boundaries of a room have no effect on the propagation of sound is called the *direct field* (because the sound is arriving at an observation point directly from the source and not indirectly from the boundaries) or the *free field* (because the field is free of obstructions which could reflect the sound). Imagine a point source of sound suspended in the middle of the room. A point source produces the same sound field as a simple source of the same strength, see Section 1.17. The sound waves propagate from the point source of sound as spherical, concentric waves. The values of the sound intensity of such a sound source are the same on the spherical surface of the wave front. Let the sound power of the point source be W and let the time–averaged sound intensity at a distance r from the point source be $\overline{I}$. Hence, see equation (1.193),

$$W = 4\pi r^2 \overline{I}. \tag{5.4}$$

The mean–squared sound pressure at the distance r is $\overline{p_d^2}$. Thus from equations (1.192) and (5.4),

$$W = 4\pi r^2 \frac{\overline{p_d^2}}{\rho_0 c_0}$$

and

$$\overline{p_d^2} = \frac{W \rho_0 c_0}{4\pi r^2}. \tag{5.5}$$

Alternatively, the same point source of sound (with the same sound power W) may be placed on the floor in the room. If the surface of the floor is hard and perfectly reflecting, the wave front becomes hemispherical (see Section 1.17.1) and the relationship between the time–averaged sound intensity $\overline{I_1}$ at a distance r from the sound source and the sound power W becomes

$$W = 2\pi r^2 \overline{I_1}, \tag{5.6}$$

which indicates that

$$\overline{I_1} = 2\overline{I}. \tag{5.7}$$

Thus, by moving the source from the middle of the room space to the floor, values of the time–averaged intensity and the mean square pressure have been doubled. This is equivalent to an increase of the sound pressure level of 3 dB.

Similarly, let the same sound source be placed at the intersection of two surfaces, such as a wall and the floor. In this case the source is able to radiate only into a quarter of a sphere. So, if the time–averaged sound intensity at a distance r from the sound source is denoted by $\overline{I_2}$, then

$$W = \pi r^2 \overline{I_2} \tag{5.8}$$

and hence

$$\overline{I_2} = 4\overline{I}. \tag{5.9}$$

The increase in sound pressure level is 6 dB.

Finally, the sound source may be placed in the corner of the room at the intersection of three surfaces. The source now radiates into a space of an eighth of a sphere and $\overline{I_3}$, the time–averaged sound intensity at a distance r from the point source, is related to the sound power W by the expression,

$$W = \frac{\pi r^2}{2}\overline{I_3}. \tag{5.10}$$

Hence,

$$\overline{I_3} = 8\overline{I}. \tag{5.11}$$

In this case the sound pressure level is increased by 9 dB.

The results of the discussion concerning the relation between the positions of the sound source and the resultant sound intensities are summarised in Figure 5.2.

Equation (5.5) may be generalised to account for the different situations illustrated in Figure 5.2 by introducing a *directivity factor* Q. Modification of equation (5.5) leads to:

$$\overline{p_d^2} = \frac{W\rho_0 c_0 Q}{4\pi r^2}. \tag{5.12}$$

The directivity factor Q is 1, 2, 4 or 8 for the point source located in free space in the middle of the room, on the floor, at the intersection of two surfaces and in a corner, respectively. In the majority of cases the directivity factor is 2 and hence equation (5.12) becomes

$$\overline{p_d^2} = \frac{W\rho_0 c_0}{2\pi r^2}. \tag{5.13}$$

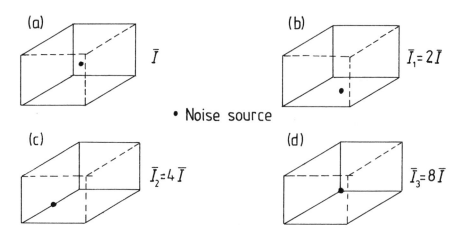

Figure 5.2 Effect of position of a point source of sound on the time–averaged sound intensity; (a) in the middle of the room , (b) on the floor, (c) at the intersection of two surfaces and (d) in a corner.

Equation (5.13) applies when, for example, a machine generating sound is located on the floor of a factory.

It should be noted that so far it has been assumed that the source is a point source. Such a source radiates sound uniformly in all directions. Most sources of sound, however, are directional, particularly at high frequencies.

The directionality of sound sources is also accounted for by a directivity factor Q_{axis} which for a particular axis, is defined as the ratio of the time–averaged sound intensity at a point located on a designated axis of the directional sound source at an arbitrarily chosen distance from the centre of the source to the time–averaged sound intensity at the same point from a point source of the same sound power [7]. From the definition, we have (for the far field)

$$Q_{axis} = \frac{4\pi r^2 \overline{p^2}_{d,axis}}{W \rho_0 c_0}, \tag{5.14}$$

where $\overline{p^2}_{d,axis}$ is the mean–squared sound pressure at a distance r from the source on the designated axis.

In the majority of cases the designated axis is the axis of maximum radiation. The directivity factor differs from unity because of the position of the source and the directional characteristics of the source. For a given position of a sound generator it indicates how effective the source is in radiating the available sound energy in a chosen direction.

5.4 The combined diffuse and direct fields

At an arbitrarily chosen point in a room both the direct and the diffuse fields exist and contribute to a value of the mean–squared sound pressure at the point. The mean–squared pressure at any point in the considered enclosure is the sum of the sound pressure contributions $\overline{p_d^2}$ and $\overline{p_r^2}$ from the direct and diffuse fields, respectively. Thus,

$$\overline{p^2} = \overline{p_d^2} + \overline{p_r^2}$$
$$= \frac{W \rho_0 c_0 Q}{4\pi r^2} + \frac{4W \rho_0 c_0}{R_c}. \tag{5.15}$$

With the aid of the definitions of sound pressure level, equation (1.231), and sound power level L_W, equation (1.239), equation (5.15) may be written in its logarithmic form as

$$L = L_W + 10 \log\left(\frac{Q}{4\pi r^2} + \frac{4}{R_c}\right) + 10 \log\left(\frac{\rho_0 c_0}{400}\right) \text{ dB}, \tag{5.16}$$

where L is the sound pressure level at a distance r from the source.

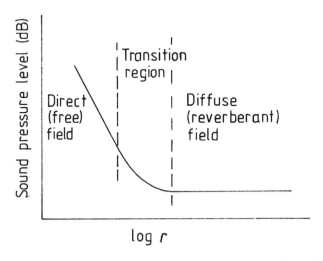

Figure 5.3 Idealised dependence of sound pressure level on distance r from the source.

The value of the specific acoustic resistance $\rho_0 c_0$ for air in a room with a typical temperature and normal atmospheric pressure is 415 kgm^{-2}s^{-1}, hence equation (5.16) becomes,

$$L = L_W + 10\log\left(\frac{Q}{4\pi r^2} + \frac{4}{R_c}\right) + 0.2 \text{ dB.} \qquad (5.17)$$

The general way in which the sound pressure level varies with the logarithm of the distance r from the source for a given sound power is shown in Figure 5.3. Close to the source, but in the far field, the sound pressure level decays at the rate of 6 dB for every doubling of the distance r; in other words the *inverse square law* can be applied and hence in this case the observation point is also in the direct or free field. At larger distances from the source the sound pressure is independent of distance and the measurement point is in the diffuse field. Close to the wall of the room the diffuse field no longer exists and it is difficult to predict the sound pressure level. Between the free field and the diffuse field is a transition region where both fields contribute to the sound pressure level.

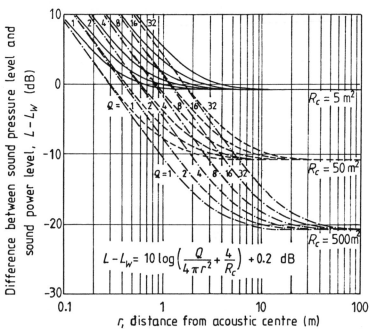

Figure 5.4 Difference between sound pressure level and sound power level as a function of distance r for values of the directivity factor Q greater than 1.

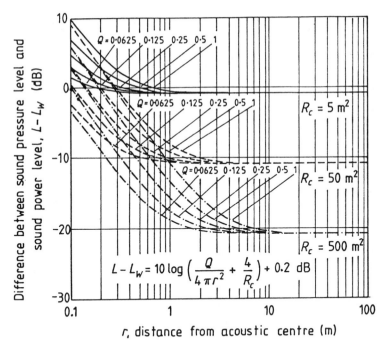

Figure 5.5 Caption as for Figure 5.4, but Q less than 1.

In Figures 5.4 and 5.5 the difference between the sound pressure level and sound power level is shown plotted against the logarithm of the distance from the source for different values of the directivity factor Q and the room constant R_c. The curves in Figure 5.4 apply for values of Q greater than unity and the curves in Figure 5.5 for values of Q less than unity.

Equation (5.15) is an idealised expression in which it is assumed that a diffuse field is formed in the room; in many situations a diffuse field will not be achieved. It is also assumed, equation (5.14), that the sound originates from a source which has dimensions small compared with the wavelength; with larger sources the sound pressure may be constant close to the source and then decays with distance according to the inverse square law.

Equation (5.17) is sometimes known, particularly in relation to audio–engineering, as the Hopkins–Stryker equation [8].

5.5 Reverberation time

The acoustic properties of a room are most commonly characterised by the reverberation time. This is the time required for a sound to decay in intensity to almost zero. In a great cathedral, if we clap our hands, it may take a few seconds for the sound to fade away. Such an interior space has a large reverberation time and is said to be 'live'. A room with a small reverberation time is described as 'dead'.

The first study of reverberation was made by W.C. Sabine [9]. Without the aid of modern equipment he measured with a stop watch the time for the sound from an organ pipe to fade away to silence. He observed that in equal time intervals the sound energy of the diffuse field in a room decayed by the same fraction of its initial value. In other words the averaged (in space) sound energy density $E (E = \langle E_a \rangle)$ decreased exponentially with time t, as shown below:

$$E = E_0 e^{-t/\tau}, \tag{5.18}$$

where E_0 is the averaged initial sound energy density (at $t = 0$) and τ is a time constant characterising the properties of the room. Equation (5.18) describes the temporal decay of the sound energy density after switching off a sound source at time $t = 0$.

For a diffuse sound field not only equation (5.18) holds, but also

$$\overline{p^2}(t) = \overline{p^2}(0)e^{-t/\tau}, \tag{5.19}$$

where $\overline{p^2}(t) = (1/\Delta t) \int_{t-\Delta t}^{t} p^2(t_1)dt_1$ is the mean–squared sound pressure over time Δt and where $2\pi/\omega \ll \Delta t \ll \tau$. Sabine introduced the term reverberation time as a quantity which determines the rate of decay of sound. The reverberation time is defined as the time required for the mean–squared sound pressure (or averaged sound energy density) to decay to one millionth of its initial value. In the measurement method outlined in Section 5.9 the square of the sound pressure is averaged over a short period of time (much shorter than the averaging time used in conventional sound level meters described in Section 2.1). The logarithm of the ratio of the mean–squared sound pressure to the square of the reference sound pressure is the sound pressure level, see equation (1.231). Thus the reverberation time may also be defined as the time required for the sound pressure level to decrease by 60 dB.

Taking into account the definition of the reverberation time and equation (5.18), one obtains the relation between the reverberation time T and the time constant τ.

$$T = 6\tau \ln 10$$
$$T = 13.8\tau. \tag{5.20}$$

The meaning of the reverberation time is illustrated in Figure 5.6. If the decay of mean–squared sound pressure is exponential as implied by equation (5.19), the sound pressure level against time is a straight line, and the reverberation time may be obtained from the slope of this line.

In practical situations a decrease of 60 dB may not be attainable. Often the estimation of the reverberation time has to be based on a decay of 30 dB or less. As long as the decay of sound pressure level is characterised by a single straight line, the reverberation time should be the same.

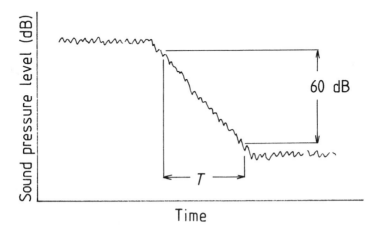

Figure 5.6 Illustration of reverberation time.

Sabine [10] discovered experimentally that the reverberation time depends only on the properties of the room, such as the dimensions of the room and its total equivalent absorption area which is defined as the product of the surface area of the room and the average sound absorption coefficient. The reverberation time does not depend on how the absorptive material is distributed in the room, as long as an exponential decay is achieved. It follows from Sabine's experimental work that the reverberation time relates to the room volume and the total equivalent absorption area $S\overline{\alpha}$ as:

$$T = \frac{55.3V}{c_0 S\overline{\alpha}} \text{ s,} \qquad (5.21)$$

where V is the volume of the room in m^3,
         S is the surface area of the room in m^2,
         c_0 is the sound velocity in m/s and

$\overline{\alpha}$ is the average sound absorption coefficient
of the surfaces in the room.

The evaluation of the product $S\overline{\alpha}$, the total equivalent absorption area of the room, has already been discussed in Section 5.2.1. Equation (5.21) can be used to calculate the reverberation time T, using measured values of the absorption coefficients. As we shall see later, the measured values of α (listed in Table 5.1) are themselves based upon measurements of reverberation time. As mentioned in Section 5.2.2 the term Sabine absorption coefficient is sometimes used for values of the coefficient based on measurements of the reverberation time.

For a typical room temperature of 20°C the sound velocity c_0 is 343.5 m/s (see Table 1, Chapter 1, for values at other temperatures). The incorporation of the value of c_0 into equation (5.21) leads to:

$$T = \frac{0.161V}{S\overline{\alpha}}. \qquad (5.22)$$

Equations (5.21) or (5.22) are known as the *Sabine formulae* for reverberation time.

The laws of reverberation, equations (5.18) and (5.21) are only valid for diffuse fields. In a diffuse field they can be applied not only to steady state conditions in a room when the sound energy is absorbed as quickly as it is supplied, but also to the case of sound decaying with time.

The work of Sabine, from which equation (5.20) can be deduced, was empirical [10]. However, it is possible to obtain the same expression analytically by considering the balance of sound energy in a room which has a diffuse field [11,12]. In the analytical derivation an expression for the absorbed sound power P_W is required and is

$$P_W = \frac{1}{4}Ec_0S\overline{\alpha}, \qquad (5.23)$$

where E is the averaged energy density of the sound in the room,
c_0 is the velocity of sound in the room and
$S\overline{\alpha}$ is the total equivalent absorption area.
Equation (5.23) was first derived by Franklin [13].

If $\overline{\alpha} = 0$, according to equation (5.22) the reverberation time T is infinite. This result is acceptable, because with no absorption of sound there is no decay of sound energy. On the other hand, if $\overline{\alpha} = 1$, i.e., perfect absorption, a finite value is obtained for the reverberation time from equation (5.22). However, the expected value is zero, because for perfect absorption — as in the open air — there is no reflected sound.

So the Sabine formula (5.22) fails to satisfy the asymptotic condition $\overline{\alpha} = 1$. For this reason the Sabine formula is generally restricted to rooms for which the values of the average sound absorption coefficient are small. A formula which does satisfy both asymptotic conditions, $\overline{\alpha} = 0$ and $\overline{\alpha} = 1$, was provided by Eyring [14].

5.6 Eyring formula for reverberation time

The Eyring formula may be derived quite simply, provided that use is made of the concept of *mean free path*. In the kinetic theory of gases the average distance travelled by a molecule in time between two subsequent collisions with other molecules is called the mean free path. In statistical room acoustics the term is used for the average distance travelled by a ray of sound between successive reflections at the surfaces. For a room of volume V and total surface area S the mean free path of a sound ray d is given by [15]:

$$d = 4V/S. \tag{5.24}$$

Consider a ray of sound incident on a surface which has a mean sound absorption coefficient α, see Section 5.2.2.

The sound intensity of the incident sound ray is I and the sound intensity of the reflected sound at the wall is $I(1 - \alpha)$.

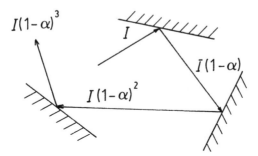

Figure 5.7 Reflection of sound at different surfaces (Eyring formula derivation).

The situation is diagrammatically illustrated in Figure 5.7. The reflected sound ray is incident on a second surface characterised by the same sound absorption coefficient. The sound intensity of the reflected sound ray is now $I(1 - \alpha)^2$. After n reflections the reflected sound has an intensity of $I(1 - \alpha)^n$. The reverberation time is the time required for the sound intensity to decay to 10^{-6} of its original value. Thus the number of reflections n_T, occurring during the reverberation time T, can be obtained from the equation:

$$I(1 - \alpha)^{n_T} = 10^{-6}I.$$

Hence,

$$n_T = -\frac{6}{\log(1 - \alpha)}. \tag{5.25}$$

The number of sound reflections which occur during the reverberation time depends only on the mean sound absorption coefficient. For example, if $\alpha = 0.1$, $n_T = 131$; if $\alpha = 0.5$, $n_T = 20$. On the other hand, the rays of sound travel with the sound velocity c_0 and hence, see equation (5.24), the average time between reflections t_r is given by:

$$t_r = \frac{4V}{Sc_0}. \tag{5.26}$$

The reverberation time T corresponds to n_T reflections, thus

$$T = \frac{4V}{Sc_0}n_T. \tag{5.27}$$

Equations (5.25) and (5.27) lead to an expression for the reverberation time:

$$T = -\frac{24V}{c_0 S \log(1 - \alpha)}.$$

Since $2.303 \log(1 - \alpha) = \ln(1 - \alpha)$,

$$T = -\frac{55.3V}{c_0 S \ln(1 - \alpha)}. \tag{5.28}$$

Finally, if a value of c_0 of 343.5 m/s is used:

$$T = \frac{0.161V}{S[-\ln(1 - \alpha)]}. \tag{5.29}$$

In the derivation of equation (5.29) it has been assumed that the mean sound absorption coefficient α is the same at each reflection. In most rooms the mean sound absorption coefficient will vary from surface to surface. However, provided that values of the individual mean sound absorption coefficients at each reflection do not depart much from the value of the average sound absorption coefficient $\bar{\alpha}$, it is possible to replace α

in equation (5.29) by $\overline{\alpha}$. Adoption of this assumption allows (5.29) to be expressed as:

$$T = \frac{0.161V}{S[-\ln(1-\overline{\alpha})]} \text{ s.} \tag{5.30}$$

Equation (5.30) is the *Eyring formula* for reverberation time. A similar expression was also derived by Norris [16].

In addition to the concept of a mean free path for sound rays, the Eyring model combined the idea of image sources, originally used in optics. The Norris approach is based on the main assumption that E the averaged (in space) energy density of the sound in a room decays, after switching off the sound source, by $(1 - \overline{\alpha})$ of its initial value during the time which the sound takes to travel the mean free path of a sound ray. Hence,

$$E = E_0(1-\overline{\alpha})e^{-(Sc_0/4V)t}, \tag{5.31}$$

where E_0 is the averaged initial sound energy density. Equation (5.31) leads to the same expression for reverberation time as in (5.30). Both the Eyring and Norris models are valid for a diffuse sound field and can be applied to highly reverberant rooms, so-called live rooms, as well as for dead rooms which are characterised by high values of the mean sound absorption coefficients.

The term $\ln(1-\overline{\alpha})$ in the Eyring formula (5.30) may be expanded as a power series:

$$\ln(1-\overline{\alpha}) = -\overline{\alpha} - \frac{\overline{\alpha}^2}{2} - \frac{\overline{\alpha}^3}{3} - \cdots$$

If the value of $\overline{\alpha}$ is small only the first term in the power series is important and the Eyring formula (5.30) is reduced to Sabine's formula (5.22).

5.7 Sound absorption coefficients

In Table 5.1 values are listed of the sound absorption coefficients for different materials at frequencies from 125 Hz to 4 kHz. The values in the table are based on the measurement of the reverberation time in a reverberation chamber. Reverberation chambers are described in Section 5.11 and the method of measuring the absorption coefficient in Section 5.11.1. As mentioned already in Section 5.2.2 the coefficients in the table are sometimes referred to as Sabine sound absorption coefficients, because their evaluation depends upon use of the Sabine formula for reverberation time (equation (5.22)).

TABLE 5.1 Sound absorption coefficients for different materials. Thickness h is in mm, surface density σ is in kg/m^2. (These are Sabine coefficients based on the measurements of reverberation time)

| | | h | σ | Sound absorption coefficients at octave centre frequency (Hz) | | | | | |
				125	250	500	1k	2k	4k
1.	Mineral wool slab	51	4.4	0.25	0.65	0.80	0.85	0.90	0.90
2.	Mineral wool slab	102	8.8	0.55	0.90	0.90	0.85	0.90	0.95
3.	Fibre glass slab	51	1.0	0.20	0.45	0.65	0.75	0.80	0.80
4.	Fibre glass slab	102	2.0	0.45	0.75	0.80	0.85	0.90	0.85
5.	Polyurethane foam slab	6	0.18	0.02	0.04	0.10	0.20	0.50	0.90
6.	Polyurethane foam slab	12	0.36	0.04	0.10	0.20	0.55	0.85	1.00
7.	Polyurethane foam slab	25	0.75	0.08	0.26	0.58	0.92	1.00	0.95
8.	Polyurethane foam slab	50	1.50	0.12	0.48	0.88	0.95	0.85	0.85
9.	Mineral wool acoustic tiles attached to solid surface	16	5.9	0.10	0.20	0.70	0.90	0.75	0.60
10.	Mineral wool acoustic tiles with 51 mm cavity behind (width 67 mm with cavity)	16	5.9	0.30	0.60	0.75	0.80	0.65	0.75
11.	Mineral wool acoustic tiles with 305 mm cavity behind (width 321 mm with cavity)	16	5.9	0.70	0.45	0.65	0.85	0.70	0.75
12.	Fibre glass acoustic tiles with 400 mm void	16	–	0.50	0.40	0.60	0.75	0.60	0.35
13.	Wooden panel; polished walnut veneer on plywood	5	2.4	0.35	0.25	0.20	0.15	0.05	0.05
14.	Solid, polished wood panel	51	–	0.10	0.07	0.05	0.04	0.04	0.04
15.	Wooden panel; hardwood veneer on chipboard with longitudinal grooves opening into cylindrical recesses to form Helmholtz resonators	25	–	0.29	0.25	0.45	0.62	0.97	0.38
16.	Brick wall, untreated	114	184	0.02	0.02	0.03	0.04	0.05	0.05
17.	Brick wall, plastered	130	180	0.02	0.02	0.02	0.03	0.03	0.04
18.	Special acoustic plaster	13	–	0.10	0.15	0.25	0.45	0.55	0.60
19.	Glass windows; glazed	4	12	0.2	0.12	0.1	0.07	0.05	0.02
20.	Noisemaster acoustic blocks; unpainted pelletised concrete blocks with internal Helmholtz resonator, partially filled with acoustical foam	100	122	0.20	0.72	0.90	0.55	0.52	0.50
21.	Concrete floor	–	–	0.01	0.01	0.02	0.02	0.02	0.02
22.	Wood block floor	–	–	0.05	0.04	0.05	0.08	0.1	0.1

continued on next page

TABLE 5.1 Sound absorption coefficients for different materials. Thickness h is in mm, surface density σ is in kg/m^2. (These are Sabine coefficients based on the measurements of reverberation time) (continuation)

		h	σ	Sound absorption coefficients at octave centre frequency (Hz)					
				125	250	500	1k	2k	4k
23.	Linoleum on floor	–	–	0.03	0.05	0.05	0.08	0.1	0.1
24.	Pile carpet and underfelt on concrete floor	–	–	0.1	0.2	0.3	0.4	0.5	0.6
25.	Pile carpet and underfelt on wood joist floor	–	–	0.2	0.3	0.3	0.4	0.5	0.6
26.	Curtains, medium, hung straight and close to wall	–	–	0.05	0.08	0.25	0.30	0.30	0.40
27.	Curtains, medium, double folds spaced away from the wall	–	–	0.10	0.35	0.4	0.5	0.5	0.6
28.	PVC sheets sandwiching fibreglass (0.6 mm reinforced PVC, 20 mm fibreglass, 1.2 mm reinforced PVC)	22	5	0.06	0.20	0.55	0.92	1.0	1.0

Various references and sources have been used to compile this table [17–20].

The presence of people can change very much the absorptive properties of a room, because people are good absorbers of sound. With people it is not possible to identify a particular surface area and corresponding sound absorption coefficient. The absorption caused by the presence of people is defined in terms of an equivalent absorption area in m^2, i.e., the product $S\alpha$. In Table 5.2 are tabulated the values of the equivalent absorption for people with different clothing and in different situations [21–23].

TABLE 5.2 Equivalent absorption area in m^2 for persons and seats

	Absorption area (m^2) at octave centre frequencies (Hz)					
	125	250	500	1k	2k	4k
One seat, upholstered in fabric	0.20	0.30	0.35	0.40	0.45	0.40
One seat, upholstered in leather	0.15	0.25	0.35	0.35	0.25	0.20
One person sitting in seat upholstered in fabric	0.25	0.40	0.45	0.55	0.55	0.50
One person standing, lightly dressed	0.15	0.30	0.55	0.80	0.90	0.80
Musician sitting with instruments	0.45	0.85	1.05	1.25	1.25	1.15

Often people are packed into a room in such a way that they cover the floor completely. An audience at a concert hall is a good example. In this case it is better to regard the people as being equivalent to a surface layer with a certain sound absorption coefficient; thus S and α should be known independently. Values of the absorption coefficient for large groups of people in different situations are shown in Table 5.3 [24].

TABLE 5.3 Sound absorption coefficients of audience forming a continuous covering in an auditorium

	Sound absorption coefficients at octave centre frequencies (Hz)					
	125	250	500	1k	2k	4k
Occupied; audience, orchestra and chorus area	0.60	0.74	0.88	0.96	0.93	0.85
Unoccupied; 'average' cloth-covered, well-upholstered seating areas (seats with perforated bottoms)	0.49	0.66	0.80	0.88	0.82	0.70
Unoccupied; leather–covered seating areas	0.44	0.54	0.60	0.62	0.58	0.50

5.8 Absorption of sound by the atmosphere

Dissipation of sound energy occurs not only at the boundaries of the room, but also during the propagation of sound in the air. The absorption of sound by the atmosphere of the room at a given frequency is a function of the temperature and the relative humidity. The decay of the sound pressure amplitude P depends upon distance x travelled by the sound and has an exponential form:

$$P = P_0 e^{-mx},$$

where P_0 is the initial sound pressure amplitude and m is a sound attenuation coefficient, often expressed in units of Nepers per metre (Np/m); $1\,\text{Np} \equiv 8.69$ dB. If air attenuation is included, the Sabine formula equation (5.22), is modified and becomes:

$$T = \frac{0.161V}{S\bar{\alpha} + 8mV}\ \text{s.} \tag{5.32}$$

Values of the sound attenuation coefficient m are given in tables in an American National Standard [25]. The Standard also gives the formulae for the calculation of m. The sound attenuation coefficient depends upon temperature as well as humidity, but, as rooms are mostly kept at a

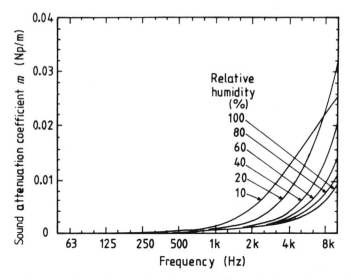

Figure 5.8 Attenuation of sound in air at 20° C; variation of sound attenuation coefficient m with frequency for different values of relative humidity.

temperature of about 20° C, values for this temperature only will be presented. The coefficient m as a function of relative humidity for different frequencies is shown in Figure 5.8.

5.9 Measurement of reverberation time

A typical arrangement for measuring the reverberation time is shown schematically in Figure 5.9. A suitable source of sound is required which can be switched off quickly. Traditionally an impulsive source, such as a starting pistol, was used. A starting pistol is still useful, if one requires for the reverberation time a single number which is characteristic of the whole frequency range. A better source of sound is one that provides a continuous band–limited noise which can be very quickly discontinued by a switch. The Brüel and Kjaer type 4205, for example, is such a source. It comprises two loudspeakers in a single box which are fed by a battery–operated unit containing a noise generator (producing pink noise), a power amplifier and octave filters. The loudspeaker is placed preferably in a corner of the room so that as many modes of oscillation as possible are excited in the room. However, on some occasions it may be more representative to place the loudspeaker where, for example, a human speaker is normally situated.

To receive the signal a sound level meter with an octave filter set is

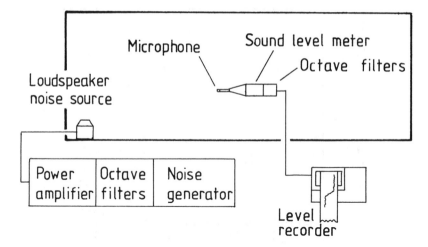

Figure 5.9 Schematic diagram of the apparatus used for the measurement of reverberation time.

required. The centre frequency of the octave filter set must be adjusted to be the same centre frequency as that of the signal sent to the noise source. In this way the ratio of sound signal to background noise can be maximised. The microphone should not be placed too close to the noise source. For a source close to the front of the room, a position about two thirds of the way back is typical for the receiver position.

The trace of intensity level against time, as shown for an idealised case in Figure 5.6, is obtained on a level recorder which receives the signal from the ac output of the sound level meter. The level recorder is a chart recorder with a logarithmic potentiometer and the capacity to carry out averaging by means of the writing speed of the pen. The averaging time increases as the writing speed of the pen decreases.

5.9.1 Typical reverberation times

Some typical values of the reverberation time at 1000 Hz are given in Table 5.4 for different rooms. Further details of reverberation times of typical rooms are given by Furrer [26] for many kinds of rooms, Parkin and Humphreys [27] for studios, for modern churches [28], and Lewers and Anderson [29] for large churches. Table 5.4 should only be regarded as a guide to the order of magnitude of reverberation times.

5.10 Anechoic chambers

Anechoic chambers are special rooms for acoustical measurements in which

TABLE 5.4 Typical reverberation times at 1000 Hz for different rooms

Type of room	Reverberation time (s)
Anechoic chamber	0.1
TV studios (1000 m^3)	0.4
Sound studio (1000 m^3)	0.9
Department stores, restaurants	0.5
Living rooms	0.5
Offices, small classrooms	1.0
School assembly hall, church hall (c 1700 m^3)	2.0
Large lecture room (2600 m^3) empty	3.0
full	1.1
Modern churches	1.8 – 2.0
Gymnasiums (no acoustic treatment)	5.0 – 6.0
(with sound absorbing walls)	2.3 – 2.7
Factories (no ceiling treatment)	2.5 – 3.5
(with ceiling treatment)	1.2 – 1.8
Large churches and cathedrals	5.0 – 10.0

the walls of the room are perfectly absorbing, i.e., the surfaces of the room have a sound absorption coefficient which, ideally, is unity. The name anechoic derives from the fact that no echoes can exist in the room. In an anechoic chamber a field is simulated, called a direct or free field, which is without reflections of sound and similar to the field which exists out of doors. In Figure 5.10 a photograph of an anechoic chamber is shown with a ship's whistle as the noise source. Anechoic chambers are used whenever a free field of sound is required; in particular they are used for measuring the sound power of sources and for determining the directivity pattern of sources.

The free field conditions within an anechoic chamber are generally achieved by lining the walls with wedges of sound absorbing material, as can be seen in Figure 5.10. The space between the wedges forms a reversed horn and allows a continuous and gradual transition from the characteristic specific resistance of the air to that of the material. Anechoic chambers are designed to give good sound absorbing properties above a

Figure 5.10 Ship's whistle located in an anechoic chamber.

certain frequency, called the *cutoff frequency*. At frequencies above the cutoff frequency the sound absorption coefficient of the wedges is 0.99 or more. Below the cutoff frequency the sound absorption coefficient is less than 0.99 and the chamber is not so effective. The dimensions of the wedges used in Figure 5.10 are shown in Figure 5.11. The wedges are 600 mm long on a base which is a cube of 200 mm side; the cutoff frequency is about 180 Hz. The length l_w of the wedge is the most parameter which controls the cutoff frequency f_{ac}. It is possible to deduce from the article of Duda [30] that

$$l_w = -1.44 \log f_{ac} + 3.85 \text{ m.} \tag{5.33}$$

For a cutoff frequency of 50 Hz to be achieved a wedge length of 1400 mm is required.

Although the cutoff frequency is a useful guide to the performance of an anechoic chamber, it is the inverse square law test which is the real test of the effectiveness of the room. In this test a small, omni–directional sound source is placed close to the centre of the room and the sound pressure level is measured at increasing distances from the source.

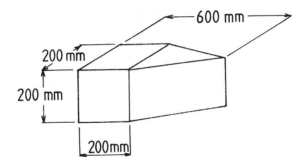

Figure 5.11 Sound absorbing wedge used in the anechoic chamber shown in Figure 5.10.

Departures from the inverse square law arise because of reflections from the wall, and increase as the wall is approached [31]. For this reason Duda [30] suggested that the microphone should be located $\lambda/4$ or 600 mm, whichever is the greater, from the wall. In order to satisfy the International Standard [32] on anechoic chambers for sound power measurements the fluctuations from the inverse square law should be within ± 1 dB in the frequency range from 800 Hz to 5000 Hz and ± 1.5 dB above and below this frequency range. In the Health and Safety Executive document [33] it is considered that deviations of ± 2 dB are acceptable in anechoic chambers which are used for industrial purposes.

The sound absorbing wedges in anechoic chambers are generally made from mineral wool, fibre glass or polyurethane foam. All these materials can have good sound absorbing properties. Polyurethane foam is a dangerous material in that it can burn easily and during combustion gives off dense black fumes. It is relatively cheap and can be cut or drilled easily. It is in use in the chamber shown in Figure 5.10.

There are different ways of fixing wedges to the walls and ceiling of an anechoic chamber, depending on the materials used. One method is illustrated in Figure 5.12. In this case, as the wedges were made from foam, a hole could easily be drilled in the base. Six wedges were threaded onto a wooden dowel. The dowel was attached to battens on the walls of the chamber. The wedges in the anechoic chamber illustrated in Figure 5.10 were fixed in this way [34].

Sound insulation It is important to prevent external air–borne noise penetrating the anechoic chamber. Good sound insulation must be provided by the walls and door. Typically, for a small or moderately sized chamber the walls are made from concrete of thickness 150 mm. In order

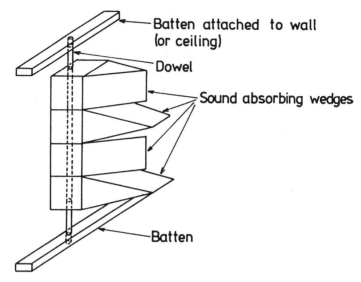

Figure 5.12 Method of fixing wedges to the wall by means of a wooden dowel.

to achieve a sound reduction index similar to that of the concrete the door should be made from two plates of steel with a sound absorbing material sandwiched between the plates. The door should have a bevelled edge so that a good seal on the bearing surfaces of the door can be ensured by strips of foam which are compressed on closure. Constructional details of such a door are given by Sharland [35].

Structure–borne sound Structure–borne sound must be prevented from radiating into the chamber. Normally the chamber is supported by pads of rubber or cork, or a material that acts as a spring. In this way the chamber on the springs forms a single degree of freedom system. The springs have to be selected to provide vibration isolation of the chamber. Guidance on this topic can be found in handbooks and textbooks [36,37]. In general, the natural frequency of the system should be at least half that of the lowest frequency it is intended to use in the chamber. Thus, if the inverse square law is obeyed down to a frequency of 100 Hz and it is intended to make measurements at that frequency in the chamber, the natural frequency should be 50 Hz or less. It may be difficult to find springs to satisfy this aim. In some cases air springs have been used.

Semi–anechoic chamber The acoustic performance of an anechoic chamber is of necessity altered when machinery, whose sound is to be measured, is introduced. Often the sound power and directional charac-

teristics of a sound source need to be measured in the anechoic chamber
in a manner which represents its situation in a factory or workshop. To
achieve this there should be no wedges on the floor so that the machine is
placed on the hard, reflecting surface of the chamber floor (often of concrete). A chamber with a hard, reflecting floor is called a semi–anechoic
chamber and is widely used for the measurement of the sound power from
machines, and even from vehicles [38]. The acoustic performance, as characterised by adherence to the inverse square law, is not as good as that of
the fully anechoic room.

Floor   In the case of a fully anechoic chamber the floor has to be
considered carefully. The floor is often a mesh of tensioned wires [30].
With this type of floor it is easy to enter and leave, but it is not so suitable
for supporting machinery, as any heavy objects have to be mounted on
columns which are screwed or bolted to the floor. There is always the
possibility that vibrating machinery mounted directly onto the floor of
the chamber will cause vibration of the chamber and the radiation of
structure–borne sound. In one case this problem has been overcome by
mounting the machine directly onto the raft of the main building through
holes in the floor of the chamber [34].

Far field requirements   In most cases measurements will be required
in the far field of the sound source. See Section 1.19 for a discussion of the
far field and also [39–41]. In the far field the sound pressure and particle
velocity are in phase and the inverse square law is obeyed. In this region
accurate measurements of the sound power and the directional properties
can be made. The following conditions are required for the measurement
point, at a distance r from the source, to be in the far field:

$$r \gg \lambda \text{ and } r \gg 2\pi D^2/\lambda, \tag{5.34}$$

where λ is the wavelength of the generated sound and D is the maximum
dimension of the source. It follows from the above conditions that $r \gg D$.

Closer to the sound source is the near field. The near field may be
divided into a hydrodynamic near field and a geometric near field. In
the hydrodynamic near field the pressure and particle velocity are out of
phase and no reliable estimations of sound power and directionality can
be made in this region from measurements of sound pressure level. The
geometric near field lies between the hydrodynamic near field and the far
field; in this region the pressure and particle velocity may be almost in
phase. The relationship between the time–averaged sound intensity $\bar{I}$ and

the mean–squared sound pressure p_{rms}^2 (or $\overline{p^2}$),

$$\overline{I} = \frac{p_{rms}^2}{\rho_0 c_0},$$ (5.35)

applies; thus estimates of sound power are possible [40]. However, the decay of time–averaged sound intensity with distance may not follow the inverse square law and maxima or minima may be superimposed upon the decay. Consequently, this region should not be used to obtain accurate results for the directionality of sound. Note that in the case of a simple source the relationship (5.35) always applies.

For sound power measurements the distance r should be greater than or equal to $2D$, but never less than 1 m [32]. Thus for a source with a maximum dimension of 0.5 m a distance r of 1 m should be satisfactory for measurements of sound power. At distances so close to the source the measurement point will be in the geometric near field.

For many machines it may be difficult to determine the acoustic centre from which the distance r should be measured. For a machine in which all parts appear to radiate sound equally in all directions the geometric centre is the acoustic centre. In some machines one component may be the major source of sound. For example, in Figure 5.10 the sound radiates from the mouth of the horn of the whistle; the centre of the mouth is in this case the point from which r is measured.

5.10.1 Measurement of sound power in anechoic chambers

For a directional sound source the time–averaged sound intensity, and hence mean–squared pressure, varies with position of the measurement point on a sphere of radius r and centre at the source. The sound power W is obtained by integrating the time–averaged sound intensity $\overline{I}$ over the surface of the sphere S:

$$W = \int_S \overline{I} dS$$ (5.36)

and from equation (5.35)

$$W = \int_S \frac{p_{rms}^2}{\rho_0 c_0} dS$$ (5.37)

for measurements of the mean–squared sound pressure p_{rms}^2 in the far field (or the geometric near field). Note that equation (5.4) is a special case of

(5.36) for a point source of sound. A common procedure for determining the sound power by measurement is to divide the area S of the sphere into n equal areas S_n. A microphone is placed at or near the geometric centre of each curved surface and the sound pressure levels, and hence the mean–squared sound pressures $\overline{p_1^2}$, $\overline{p_2^2}$, etc., associated with each of the n areas, are measured. A free field microphone should be used which is pointed at the geometric centre of the noise source. Thus

$$W = \left(\overline{p_1^2} S_n + \overline{p_2^2} S_n + \ldots + \overline{p_n^2} S_n \right) / (\rho_0 c_0) \tag{5.38}$$

and as

$$S_n = \frac{4\pi r^2}{n}$$

$$W = \frac{4\pi r^2}{\rho_0 c_0 n} \sum_{i=1}^{n} \overline{p_n^2}. \tag{5.39}$$

Or,

$$W = \frac{4\pi r^2}{\rho_0 c_0} \langle \overline{p^2} \rangle, \tag{5.40}$$

where

$$\langle \overline{p^2} \rangle = \sum_{i=1}^{n} \overline{p_n^2} \Big/ n \tag{5.41}$$

and $\langle \overline{p^2} \rangle$ is the spatial average of the n measurements of the mean–squared sound pressure. For radiation of sound into a hemisphere and for measurements on the surface of the hemisphere, we have:

$$W = \frac{2\pi r^2}{\rho_0 c_0} \langle \overline{p^2} \rangle . \tag{5.42}$$

In terms of decibels the equations corresponding to (5.40) and (5.42) are, for radiation into a full sphere,

$$L_W = \langle L \rangle + 10 \log 4\pi r^2 \text{ dB}, \tag{5.43}$$

and for radiation into a hemisphere,

$$L_W = \langle L \rangle + 10 \log 2\pi r^2 \text{ dB}, \tag{5.44}$$

where $\langle L \rangle$ is the spatially–averaged sound pressure level which is obtained from equation (5.41), applied to either a sphere or a hemisphere. Note that equations (5.43) and (5.44) are approximate, as they neglect the correction for $\rho_0 c_0$. See equation (5.15). Normally 0.2 dB should be subtracted from the right sides of equations (5.43) and (5.44).

There are various schemes for the positions of the microphone on the sphere or hemisphere. Twenty positions are recommended by the International Standard for a sphere and ten for a hemisphere [32]. These positions are listed in Table 5.5 in terms of the x, y and z co-ordinates, related to the distance r from the source. In a semi–anechoic chamber the hard reflecting surface, on which the sound source is located, forms the xy plane. Obviously, the z axis is perpendicular to this surface. For hemispherical radiation the first ten points, corresponding to positive values of z, are required. Alternatively, Petersen and Gross list the co–ordinates for 8 or 12 points on a sphere [42].

Errors in sound power measurements are discussed by Wheeler [43] and Yang and Ellison [39]. In general, failure to achieve far field conditions leads to errors which increase as the source becomes more directional. For example, for a highly directional source, such as a quadrupole, the error is 5 dB for $r = 1$ m at a frequency of 100 Hz [43].

Sound power of a source with axial symmetry The elemental area dS on the surface of the sphere is given in terms of spherical co–ordinates (r, θ, ψ) by:

$$dS = r^2 \sin\theta d\theta d\psi. \tag{5.45}$$

See Figure 5.13. The sound power W is written in terms of the mean–squared sound pressure $\overline{[p(r,\theta,\psi,t)]^2}$ as follows:

$$W = \int_S \frac{\overline{[p(r,\theta,\psi,t)]^2}}{\rho_0 c_0} dS \tag{5.46}$$

$$= \frac{r^2}{\rho_0 c_0} \int_0^{2\pi} \int_0^{\pi} \overline{[p(r,\theta,\psi,t)]^2} \sin\theta d\theta d\psi. \tag{5.47}$$

Many sources have axial symmetry, for example, loudspeaker cones, circular air jets, ship's whistles, etc.

If, see Figure 5.13, z is the axis of symmetry for the sound field of the source, the sound pressure is independent of angle ψ. Thus, the sound power of the source is

$$W = \frac{2\pi r^2}{\rho_0 c_0} \int_0^{\pi} \overline{[p(r,\theta,t)]^2} \sin\theta d\theta. \tag{5.48}$$

TABLE 5.5 Microphone positions for a free field (semi–anechoic and anechoic chamber)

Position	x/r	y/r	z/r
1	−0.99	0	0.15
2	0.50	−0.86	0.15
3	0.50	0.86	0.15
4	−0.45	0.77	0.45
5	−0.45	−0.77	0.45
6	0.89	0	0.45
7	0.33	0.57	0.75
8	−0.66	0	0.75
9	0.33	−0.57	0.75
10	0	0	1.0
11	0.99	0	−0.15
12	−0.50	0.86	−0.15
13	−0.50	−0.86	−0.15
14	0.45	−0.77	−0.45
15	0.45	0.77	−0.45
16	−0.89	0	−0.45
17	−0.33	−0.57	−0.75
18	0.66	0	−0.75
19	−0.33	0.57	−0.75
20	0	0	−1.0

The sound radiation from a propeller fan is shown in Figure 5.14. This type of fan has axial symmetry about the axis of rotation.

Example 5.1

The propeller fan, whose sound radiation measured at a distance of 1.2 m is shown in Figure 5.14, has axial symmetry about the z axis. What is the sound power of the fan in the octave band of centre frequency 8 kHz? Take $\rho_0 c_0 = 420 \, \mathrm{kg \, m^{-2} s^{-1}}$.

Solution The band pressure level (BPL) may be obtained from Figure 5.14 for each angle. As symmetry is not perfect the av-

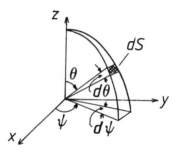

Figure 5.13 Spherical polar co–ordinates.

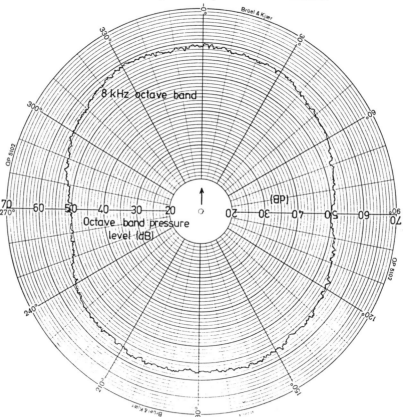

Figure 5.14 Sound radiation from a propeller fan in the octave band of centre frequency 8 kHz.

erage on either side of the axis will be used. Thus the BPLs at 30° and 330° should be the same, but are 58.0 dB and 59.2 dB, respectively. The arithmetic average is 58.6 dB and this is the value which is used in Table 5.6 at $\theta = 30°$.

TABLE 5.6 Radiation from propeller fan, Example 5.1

degrees	BPL at θ and $360° - \theta$ (dB)		Average BPL (dB)	$\overline{[p(\theta,t)]^2}$ (Pa²)	$\overline{[p(\theta,t)]^2}\sin\theta$ (Pa²)
0	59.0		59.0	3.18×10^{-4}	0
30	58.0	59.2	58.6	2.96×10^{-4}	1.48×10^{-4}
60	54.5	54.6	54.6	1.15×10^{-4}	1.00×10^{-4}
90	50.0	50.5	50.3	0.43×10^{-4}	0.43×10^{-4}
120	54.2	54.2	54.2	1.05×10^{-4}	0.91×10^{-4}
150	57.0	56.6	56.8	1.92×10^{-4}	0.96×10^{-4}
180	57.8		57.8	2.41×10^{-4}	0
				$\sum =$	4.78×10^{-4}

$$\int_0^\pi \overline{[p(\theta,t)]^2}\sin\theta d\theta = 4.78 \times 10^{-4} \times \pi/6 \text{ Pa}^2$$

$$= 2.50 \times 10^{-4} \text{ Pa}^2$$
$$W = 2 \times \pi \times 1.2^2 \times 2.50 \times 10^{-4}/420$$
$$W = 5.38 \text{ }\mu\text{W}.$$

The sound power level L_W is obtained from equation (1.239):

$$L_W = 10\log(5.38 \times 10^{-6}/10^{-12})$$
$$= 67.3 \text{ dB re } 10^{-12} \text{ W}.$$

5.10.2 Measurement of directivity factor and directivity index

Anechoic chambers are particularly valuable for measuring the directional properties of sound radiation, particularly when measurements are made in the far field. From the definition of directivity factor Q_{axis}, see equation (5.14), we may write approximately,

$$L_W = L_{axis} + 10\log 4\pi r^2 - 10\log Q_{axis} \text{ dB}, \qquad (5.49)$$

where L_{axis} is the sound pressure level at a distance r on a particular axis:

$$L_{axis} = 10 \log \frac{\overline{p^2_{d,\,axis}}}{(2 \times 10^{-5})^2} \text{ dB.}$$

Subtraction of equation (5.49) from (5.43) leads to:

$$10 \log Q_{axis} = L_{axis} - \langle L \rangle \text{ dB,} \qquad (5.50)$$

where $10 \log Q_{axis}$ is called the *directivity index* DI_{axis}. Thus in general

$$DI = 10 \log Q. \qquad (5.51)$$

Equation (5.50) applies to radiation into a sphere and is exact. For hemi–spherical radiation the equivalent equation is

$$DI_{axis} = L_{axis} - \langle L \rangle + 3 \text{ dB.} \qquad (5.52)$$

Usually, the most important directivity index is the one that corresponds to the axis in which the intensity of sound is a maximum or to the axis of symmetry. For a sound source with axial symmetry, as in Figure 5.14, the directivity index can be obtained directly from the polar diagram.

Example 5.2

Obtain the directivity factor and directivity index for the direction of maximum radiation for the propeller fan for which the polar plot of the sound radiation is presented in Figure 5.14.

Solution Q_{axis} can be obtained from equation (5.14). Using the data from Example 5.1 we can see that in the direction of maximum radiation (0 degrees) $\overline{p^2_{d,\,axis}} = 3.18 \times 10^{-4} \text{ Pa}^2$. Hence, from equation (5.14)

$$Q_{axis} = \frac{4\pi \times 1.2^2 \times 3.18 \times 10^{-4}}{5.38 \times 10^{-6} \times 420}$$
$$= 2.55.$$
$$DI_{axis} = 10 \log 2.55$$
$$= 4.05 \text{ dB.}$$

Q_{axis} could also be obtained from equation (5.49) with $L_W = 67.3 \text{ dB}$ and $L_{axis} = 59.0 \text{ dB}$. Remember, however, that equation (5.49) is approximate.

5.11 Reverberation chambers

Reverberation chambers are specially constructed rooms with highly reflecting walls. The interior surfaces are generally concrete or steel. These rooms are particularly used for the measurement of the sound power of sources and sound absorption coefficients. Two reverberation chambers are needed for the measurement of sound reduction indices.

In a reverberation chamber the diffuse field predominates apart from a small region around the noise source. Thus in equation (5.15) the term representing the diffuse field is greater than that for the direct field:

$$\frac{4}{R_c} > \frac{Q}{4\pi r^2}.$$

If the effect of the direct field is negligible, equation (5.17) becomes

$$L = L_W + 10\log\frac{4}{R_c} + 0.2\,\text{dB}, \tag{5.53}$$

where R_c, the room constant, is defined in equation (5.2). Equation (5.53) may be used to estimate L_W, the sound power level of the source, from one measurement of L, the sound pressure level. The room constant R_c must be known and is estimated from the reverberation time T and the volume of the chamber V. In a reverberation chamber the average sound absorption coefficient $\overline{\alpha}$ is very small and the room constant may be approximated to $S\overline{\alpha}$. Thus from equation (5.22)

$$R_c \simeq S\overline{\alpha} = 0.161\,V/T. \tag{5.54}$$

In reality the procedure for measuring the sound power level is more complicated. Full details of the requirements which should be fulfilled by reverberation chambers are given in various national and international standards [44,45]. To satisfy the standards the chamber should have a volume greater than $200\,\text{m}^3$, if frequencies as low as the one third octave centre frequency of 100 Hz are to be tested. As one measurement point is clearly insufficient to determine L at least three measurement points should be used at half a wavelength apart or, alternatively the microphone should be traversed at constant speed over a distance of 3 m [44]. In the case of sources which produce discrete tones more measurement points are needed [46]. All measurement points need to be at least half a wavelength from a surface in the room. The sound pressure levels measured at the different points should be the same. In fact this is difficult to achieve at low frequencies and even in a good chamber the standard deviation of the sound pressure levels may be as great as 3 dB at 125 Hz [44].

There are many normal modes of vibration for the reverberation chamber. The values of the natural frequencies for the higher modes are very close together. Additionally the modes are distributed in many different directions in space. Thus a diffuse field is easily attained at the high frequencies. At low frequencies, however, the natural frequencies are relatively widely separated in frequency. This can cause a problem in the case of a sound source which radiates at discrete frequencies, because the frequency of the tones from the source may fall between two natural frequencies of the chamber, and the room modes cannot be excited by the source. Certain chamber shapes can be chosen to ensure that the modes are distributed as uniformly as possible at low frequencies. The standards give six sets of recommended ratios for height, width to length of a rectangular chamber; one set is $1 : 2^{1/3} : 4^{1/3}$ [44,45]. Often a non–rectangular chamber is used.

Another way of overcoming the problem of wide modal spacing at low frequencies and improving the diffusivity of the field is to use a rotating diffuser [47,48]. One design for a rotating diffuser is made from 0.8 mm steel sheet in the form of a cone with a maximum diameter of 2.6 m, attached to a tubular frame [48]. It is driven round by a motor at a rate of about 15 to 30 rev/min. The purpose of the diffuser is to shift the modes of the room to different frequencies; it acts as a modulator which transforms the natural frequency of a single mode into several frequencies centred around the mode. The diffusers can also be made more simply from fixed curved plates of perspex which are suspended from the ceiling [49].

An additional way of improving the diffusivity at low frequencies is by introducing sound absorbers into the reverberation chamber. This may at first seem paradoxical. However, without absorption a particular mode can only be excited over a narrow frequency range, but with an absorber the resonance curve is broadened and some sound can be excited even between modes of vibration. It is recommended that the sound absorption coefficient in the chamber should be 0.16 below 200 Hz [44]. The presence of both a rotating diffuser and sound absorbers reduces the standard deviation in the sound pressure level at low frequencies. A design for a low frequency panel absorber to give a maximum sound power absorption of nearly 0.8 at 125 Hz is shown in Figure 5.15 [50]. The introduction of eight absorbers reduces the reverberation time at 125 Hz from 9 s to 3 s. Such absorbers are required in a concrete chamber. There is no need for

absorbers if the walls are made from steel panels with a surface density of
72 kg/m² [51].

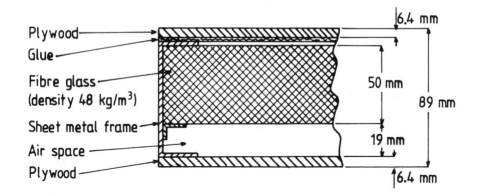

Figure 5.15 Low frequency panel absorber for use in a reverberation
chamber.

5.11.1 Measurement of sound absorption coefficients

The measurement of the sound absorption coefficient at normal incidence
in a standing wave tube has already been discussed. Coefficients have
also been quoted for use in calculations in room acoustics (Table 5.1); it
was mentioned that these coefficients were measured with the aid of a
reverberation chamber. Measurements by this method should conform to
the relevant international standard [52].

The reverberation time T of the reverberation chamber is given by

$$T = \frac{0.161V}{S\overline{\alpha}} \text{ s},\qquad(5.55)$$

where V is the volume of the chamber, S is its surface area and $\overline{\alpha}$ the
average sound absorption coefficient.

The reverberation time T will be reduced to T_1 when a sample of
sound absorbing material is placed on the floor or when a person is in the
room. If the sample of material has a surface area S_1 and α_1 is the sound
absorption coefficient at a particular frequency, then

$$T_1 = \frac{0.161V}{S\overline{\alpha} + S_1\alpha_1}\qquad(5.56)$$

and from equations (5.55) and (5.56)

$$S_1 \alpha_1 = \frac{0.161V}{T_1} - \frac{0.161V}{T}. \tag{5.57}$$

The product $S_1 \alpha_1$ would be required, if the absorption of people or chairs were to be determined.

As mentioned earlier (Section 5.2.2) the sound absorption coefficients obtained by this method are sometimes referred to as Sabine absorption coefficients [6]. Values obtained by this method are listed in Tables 5.1 and 5.2. The Sabine absorption coefficients are greater than the mean (or statistical) absorption coefficients. Indeed the sound absorption co-efficients obtained in a reverberation chamber can often be greater than unity. Because the estimates of absorption coefficient from measurements of the decay of sound (Tables 5.1 and 5.2) are probably too large, Emble-ton [6] has recommended that the expression for the room constant R_c, see equation (5.2), should be modified to:

$$R_c = S\overline{\alpha}, \tag{5.58}$$

where $\overline{\alpha}$ is obtained from, for example, Tables 5.1 and 5.2.

5.11.2 Measurement of sound reduction indices

The measurement of the sound reduction index of a panel, wall or door, etc., is carried out by placing the test specimen between two reverberation chambers so that sound from one chamber, where the source of sound is located, is transmitted into the second chamber only through the test specimen. The arrangement is shown schematically in Figure 5.16 where the left hand chamber is the source room and the right chamber is the receiving room. Recommended dimensions are given in the standard [53]; the source and receiver rooms should be at least 100 m³ in volume and the test panel should have an area of at least 2.5 m². Average band pressure levels are measured in the two rooms, normally in one third octave bands. The sound reduction index R in dB is obtained from the following expression:

$$R = \langle L_1 \rangle - \langle L_2 \rangle + 10 \log S - 10 \log S_2 \overline{\alpha_2} \text{ dB}, \tag{5.59}$$

where $\langle L_1 \rangle$ is the average band pressure level in the source room,
          $\langle L_2 \rangle$ is the average band pressure level in the receiving room,
          S is the surface area of the panel under test in m², and

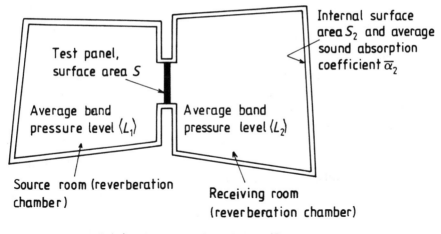

Test panel, surface area S

Average band pressure level $\langle L_1 \rangle$

Internal surface area S_2 and average sound absorption coefficient $\overline{\alpha}_2$

Average band pressure level $\langle L_2 \rangle$

Source room (reverberation chamber)

Receiving room (reverberation chamber)

$$R = \langle L_1 \rangle - \langle L_2 \rangle + 10\log S - 10\log S_2\,\overline{\alpha}_2 \quad dB$$

Figure 5.16 Measurement of sound reduction index.

$S_2\overline{\alpha_2}$ refers to the receiving room of surface area S_2 in m^2 and average sound absorption coefficient $\overline{\alpha_2}$.

The data presented in Table 4.2 were obtained by the method described here.

5.12 Closure

The importance of room acoustics is intrinsically acknowledged in the increasing adoption of the sound power level as a means of specifying the sound from machinery sources. The sound power level is a property of the sound source alone and is independent of the surfaces of the room in which the source is located. According to the directive from the European Commission, designers have to provide information on potentially noisy machines and sound power level is the most suitable means of specification. The European Commission has a further interest in noise from the point of view of trade and competition. The Commission is systematically considering noise sources and issuing regulations which describe how measurements of sound power should be made. So far, measurement methods for lawn mowers, tower cranes, compressor sets and welding gear have been published [54].

In most cases it is not necessary to use an anechoic chamber or a reverberation chamber for the measurement of sound power. For small machines measurements quite close to the source can be used, even in a typical, semi–reverberant room; measurements of sound pressure level at different distances away from the source will indicate whether or not the inverse square is obeyed and, if not, a correction can be applied [39].

Sound intensity probes and meters can also be used to measure sound power and will be valuable in cases where the diffuse field is contributing to the sound pressure level at the measurement point. The time–averaged contribution to the sound intensity from the diffuse field is zero. Sound intensity meters are also useful where measurements have to be made in the presence of strong background noise.

Room acoustics is an important subject for many reasons. In the design of concert halls, the installation of sound reinforcement systems and in many other cases a knowledge of the acoustics of the space is essential. In this chapter we are mostly concerned with sound in the workplace and in factories. Factory halls are often very large enclosed spaces with a height which is relatively small compared with the length. Consequently, the shape of the room prevents the establishment of a diffuse field. Furthermore factory halls are likely to have many machines which lack the large, smooth surfaces necessary for specular reflection. Machines tend to scatter the sound and, although they may appear to have little effect on the variables in the Sabine formula, their presence may cause a considerable reduction in the reverberation time [55]. Thus the Sabine formula is unlikely to give a good prediction of the reverberation time. Additionally, the constant sound pressure level which is attained in a diffuse field will not be achieved; the sound level may continuously decrease as the distance from the source increases. An empirical formula for the estimation of reverberation time T in seconds at 1 kHz in factory halls has been devised by Friberg [56]; this formula takes into account the height h in m of the hall, $\overline{\alpha}$ the average sound absorption coefficient of the ceiling at 1 kHz and the concentration of fittings in the factory. Friberg's expression is:

$$T = 0.15h - 1.8\overline{\alpha} + 1.8 + k_T \text{ s}, \qquad (5.60)$$

where k_T is a constant which depends upon h, $\overline{\alpha}$ and the density of furnishings [56]. For a medium concentration of furnishings k_T is 0. The expression is based on a wide range of measurements in factories. However, in some enclosed industrial spaces, such as the bottling hall of a dairy, a diffuse field — and a constant sound pressure level — may exist in most of the space.

References

[1] L. L. Beranek (1954). *Acoustics*, pp 311–312. McGraw–Hill, New

York.

[2] A. D. Pierce (1981). *Acoustics: an introduction to its physical principles and applications*, chapter 6. McGraw–Hill, New York.

[3] P. M. Morse and K. U. Ingard (1968). *Theoretical Acoustics*, pp 579–580. McGraw–Hill, New York.

[4] H. Kuttruff (1979). *Room Acoustics*, pp 26–29, second edition. Applied Science Publishers, London.

[5] L. Cremer and H. A. Müller (1982). *Principles and Applications of Room Acoustics*, Vol 1, pp 204–206. Applied Science Publishers, London.

[6] T. F. W. Embleton (1971). Sound in large rooms, in *Noise and Vibration Control* (ed. L. L. Beranek), p 221. McGraw–Hill, New York.

[7] Reference 1, p 109.

[8] H. F. Hopkins and N. R. Stryker (1948). A proposed loudness– efficiency rating for loudspeakers and the determination of system power requirements for enclosures. *Proceedings of the Institute of Radio Engineers*, **36**, pp 315–335.

[9] W. C. Sabine (1964). Reverberation, in *Collected Papers on Acoustics*, pp 3–68. Dover Publications, New York.

[10] Reference 9. Section entitled 'Exact Solution' in paper on Reverberation, pp 43–52. (Reproduced in *Architectural Acoustics* (ed. T. D. Northwood), Benchmark Papers in Acoustics/10, pp 133–141, Dowden, Hutchinson and Ross, 1977.)

[11] Reference 5, pp 189–207.

[12] L. E. Kinsler, A. R. Frey, A. B. Coppens and J. V. Sanders (1982). *Fundamentals of Acoustics*, third edition. John Wiley, New York.

[13] W. S. Franklin (1903). Derivation of equation of decaying sound in a room and definition of open window equivalent of absorbing power. *Physics Review*, **16**, pp 372–374.

[14] C. F. Eyring (1930). Reverberation time in dead rooms. *Journal of the Acoustical Society of America*, **1**, pp 217–235. (Reprinted in Architectural Acoustics (ed. T. D. Northwood), Benchmark Papers in Acoustics/10, pp 145–162, Dowden, Hutchinson and Ross, 1977.)

[15] Reference 5, p 218.

[16] R. F. Norris. A Discussion of the true coefficient of sound absorption — A derivation of the reverberation formula. (Reprinted in *Architectural Acoustics* (ed. T. D. Northwood), Benchmark Papers in Acoustics/10, pp 142–143, Dowden, Hutchinson and Ross, 1977.)

[17] E. J. Evans and E. N. Bazley (1960). *Sound Absorbing Materials*. National Physical Laboratory, HMSO, London.

[18] D. D. Reynolds (1981). *Engineering Principles of Acoustics.* Allyn and Bacon, Boston.

[19] Department of Education and Science (1975). *Acoustics in Educational Buildings.* Building Bulletin 51. HMSO, London.

[20] L. L. Beranek (1954). *Acoustics.* McGraw–Hill, New York.

[21] W. Furrer (1964). *Room and Building Acoustics and Noise Abatement.* Butterworths, London.

[22] H. Kuttruff, reference 4, p 156.

[23] U. Kath and W. Kuhl (1965). Messungen zur Schallabsorption von Polsterstuhlen mit und ohne Personen. *Acustica*, **15**, pp 127–131.

[24] L. L. Beranek (1960). Audience and seat absorption in large halls. *Journal of Acoustical Society of America*, **32**, (6) pp 661–670.

[25] American National Standards Institute (1978). ANSI SI.26–1978 (ASA 23–1978). Method for the calculation of the absorption of sound by the atmosphere.

[26] Reference 21, pp 97–158.

[27] P. H. Parkin and H. R. Humphrey (1969). *Acoustics, Noise and Buildings*, chapter 5. Faber and Faber, London.

[28] D. Lubman and E. A. Wetherill (eds) (1985). *Acoustics of Worship Spaces.* American Institute of Physics.

[29] T. H. Lewers and J. S. Anderson (1984). Some acoustical properties of St Paul's Cathedral, London. *Journal of Sound and Vibration*, **92** (2), pp 285–297.

[30] J. Duda (1977). Basic design considerations for anechoic chambers, *Noise Control Engineering*, **9** (2), pp 60–67.

[31] M. E. Delany and E. N. Bazley (1970). Monopole radiation in the presence of an absorbing plane. *Journal of Sound and Vibration*, **13** (3), pp 269–279.

[32] International Organization for Standardization (1977). Acoustics: Determination of sound power level of noise sources – precision methods for anechoic and semi–anechoic rooms. ISO 3745–1977. (Also British Standards Institution BS 4196: part 5:1981.)

[33] Health and Safety Executive (1990). Procedures for noise testing. Noise Guide no 7. HMSO, London.

[34] J. S. Anderson (1977). The design construction and operation of the anechoic chamber in the Department of Mechanical Engineering, The City University, Research Memorandum ML 96.

[35] I. J. Sharland (1972). *Woods Practical Guide to Noise Control*, p 163. Woods of Colchester.

[36] J. S. Anderson and M. Bratos–Anderson (1987). *Solving Problems in*

Vibrations. Longman Scientific, Harlow.

[37] C. M. Harris and C. E. Crede (eds) (1976). *Shock and Vibration Handbook*, second edition. McGraw–Hill, New York.

[38] International Organization for Standardization (1981). Acoustics: Determination of sound power level of noise sources — engineering methods for free field conditions over a reflecting plane. ISO 3744–1981. (Also British Standards Institution BS 4196: part 4:1981.)

[39] S. J. Yang and A. J. Ellison (1985). *Machinery Noise Measurement.* Oxford University Press, Oxford.

[40] D. A. Bies (1976). Uses of anechoic and reverberation rooms for the investigation of noise sources. *Noise Control Engineering*, **7** (3), pp 154–163.

[41] D. A. Bies and C. H. Hansen (1988). *Engineering Noise Control.* Unwin Hyman, London.

[42] A. P. G. Petersen and E. E. Gross, Jr (1974). *Handbook of Noise Measurement*, seventh edition. General Radio.

[43] P. D. Wheeler (1982). Measurement and diagnosis of machinery noise, in *Noise and Vibration*, (eds R. G. White and J. G. Walker). Ellis Horwood, Chichester.

[44] International Organization for Standardization (1975). Acoustics — Determination of sound power levels of noise sources — Precision methods for broad–band sources in reverberation rooms. ISO 3741–1975. (Also British Standards Institution BS 4196: part 1:1981.)

[45] American National Standards Institute (1972). Methods for the determination of sound power levels of small sources in reverberation rooms. ANSI S1.21–1972.

[46] International Organisation for Standardization (1975). Acoustics — Determination of sound power levels of noise sources — Precision methods for discrete–frequency and narrow band sources in reverberation rooms. ISO 3742–1975. (Also British Standards Institution BS 4196: part 2:1981.)

[47] C. E. Ebbing (1971). Experimental evaluation of moving sound diffusers for reverberant rooms. *Journal of Sound and Vibration*, **16** (1), pp 99–118.

[48] C. I. Holmer (1976). Qualification of an acoustic research facility for sound power determination. *Noise Control Engineering*, **7** (2), pp 87–92.

[49] H. Kuttruff (1979), reference 4, page 238.

[50] J. T. Rainey, C. E. Ebbing and R. A. Ryan (1976). Modifications required to permit qualification of a 269 m³ reverberation room. *Noise*

Control Engineering, **7** (2), pp 81–86.

[51] R. P. Harmon (1976). Development of an acoustic facility for determination of sound power emitted by appliances. *Noise Control Engineering*, **7** (2), pp 110–114.

[52] International Organization for Standardization (1985). Methods for measurement of sound absorption coefficients in a reverberation room. ISO 354–1985. (Also British Standards Institution BS 3638:1987.)

[53] International Organization for Standardization (1978). Measurement of sound insulation in buildings and of building elements — part I recommendations for laboratories; part II statement of precision requirements; part III laboratory measurements of airborne sound insulation of building elements. ISO–140 (Also British Standards Institution BS 2750: parts 1, 2 and 3:1980.)

[54] The Council of the European Communities (1984). Council Directives 84/533–537/EEC of 17 September on the approximation of the laws of the Member States relating to the permissible sound power level of compressors, tower cranes, welding generators, power generators, powered hand–held concrete–breakers and picks and lawnmowers. *Official Journal of the European Communities L 300*, **27**, pp 123–178.

[55] M. Hodgson [1983]. Measurements of the influence of fittings and roof pitch on the sound field in panel–roof factories. *Applied Acoustics*, **16**, pp 369–391.

[56] R. Friberg [1975]. Noise reduction in industrial halls obtained by acoustical treatment of ceilings and walls. *Noise Control and Vibration Reduction*, March, pp 75–79.

6 Silencers

There is no such thing as an empty space or an empty time.
There is always something to see, something to hear. In fact, try
as we may to make a silence, we cannot. For certain engineering
purposes, it is desirable to have as silent a situation as possible.
Such a room is called an anechoic chamber, its six walls made
of special material, a room without echoes. I entered one at
Harvard University several years ago and heard two sounds, one
high and one low. When I described them to the engineer in
charge, he informed me that the high one was my nervous system
in operation, the low one my blood in circulation. Until I die
there will be sounds.

John Cage, *Silence*

6.1 Introduction

There are occasions when the transmission path of sound from source to
observer can be confined within a duct or tube. For example, if a fan is
used to ventilate a room, the air and sound are ducted to the room. In
this case it is possible to reduce in the room the noise from the fan by
modifications to the transmission path of the sound. The modification
can be achieved by inserting a silencer in the duct between the fan and
the termination of the duct in the room.

In the present chapter we will consider the transmission of sound
through ducts and the application of silencers to reduce the transmitted
sound. Our discussion will be confined to plane waves.

6.1.1 Types of silencers

The silencers most commonly used may be classified as either dissipative or reactive. In *dissipative silencers* there is a dissipation of sound energy by conversion into other forms of energy, mostly heat. This type of silencer makes use of sound–absorbent materials of the type described in Chapter 4. On the other hand there is no dissipative process in reactive silencers. The working principle of *reactive silencers* is based upon the fact that at some discontinuity in the transmission path sound waves are reflected back to the source and very little energy is transmitted downstream. In other words the sound energy is kept within the duct system and not allowed to escape from the end of the duct. Reactive silencers are extensively used for reducing combustion noise transmitted down the exhaust pipe of internal combustion engines. Reactive silencers are sometimes referred to as mufflers or acoustic filters. Dissipative silencers are frequently used in ventilation and air conditioning systems. Some silencers incorporate the mixed characteristics of both dissipative and reactive types.

Both reactive and dissipative silencers are passive components of the system in the sense that they are placed in a duct and left there; they require no supply of power to function. Active silencers make use of sound energy which is introduced into a duct by a loudspeaker. If there is a phase difference of 180° between the existing sound waves in the duct and the introduced sound waves, a cancellation effect takes place and no sound is transmitted through the duct termination.

6.2 Methods of specifying the attenuation of a silencer

The attenuation achieved by a silencer may be quoted as a transmission loss or an insertion loss. The transmission loss is a property of the silencer itself and is defined in terms of the *sound (power) transmission coefficient* α_t. The sound transmission coefficient is the ratio of the transmitted sound power to the incident sound power (see equation (1.129) and the definition of sound power given in Section 1.13). The propagation directions of the waves entering and leaving the silencer are shown in Figure 6.1. Thus,

$$\alpha_t = \frac{W_t}{W_i} = \frac{\text{transmitted sound power}}{\text{incident sound power}} \qquad (6.1)$$

and the *transmission loss* (L_{TL}) is given by:

$$L_{\mathrm{TL}} = 10\log(1/\alpha_t) \text{ dB}. \qquad (6.2)$$

Figure 6.1 Sound waves in a silencer.

The definition (6.2) is the same as that for the sound reduction index; see Chapter 4, equation (4.2).

The transmission loss is entirely a property of the silencer and does not depend upon the lengths of the pipes which are upstream or downstream.

In most practical situations the transmission loss is not sufficient to characterise the sound reduction resulting from the introduction of a silencer. A silencer is generally installed with upstream and downstream pipes of finite lengths. In order to take into account the effect of the installation of the silencer sound levels are measured with and without the silencer in position. If no silencer is used and only a straight pipe exists between the source and the pipe termination, a relatively high sound pressure level L_1 is measured at a certain point in the free field of the radiated sound. See Figure 6.2a.

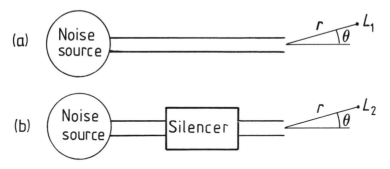

Figure 6.2 Illustration of insertion loss measurements; (a) without silencer and (b) with silencer.

When a silencer is placed in the pipe and the distance from the source to the termination remains the same the sound pressure level L_2 measured at the same position would be expected to be less than L_1, see Figure 6.2b. The difference between the two sound pressure levels defines the *insertion loss.* Thus, insertion loss L_{IL} is given by:

$$L_{\text{IL}} = L_1 - L_2 \text{ dB.} \tag{6.3}$$

The insertion loss, as defined above, is expressed in terms of sound pressure level, but could also be described in terms of band pressure levels or sound power levels. The insertion loss depends upon the position of the silencer in the pipe and is different from the transmission loss.

6.3 Plane waves of sound in ducts

In any system of ducts or silencers a straight duct or pipe is an important component. Usually it is assumed that plane sound waves are propagated along the duct. This assumption is valid for sufficiently low frequencies of the sound waves entering the duct. If the frequency is lower than the *cutoff frequency* for the lowest transverse mode, only the longitudinal mode (plane wave) is propagated without attenuation in the duct [1].

In many applications a system of ducts is considered. In some situations the system of ducts can be treated as a duct with sudden area changes or a duct with side branches. Even for these cases, characterised by quite complicated geometry, the one–dimensional model of sound propagation is adopted, if the sound waves are of sufficiently low frequency. The sudden enlargement of the duct cross–section causes the appearance of a reflected wave. Hence, only part of the energy of the primary incident sound wave is transmitted to the duct with the different duct cross–section. The incident, reflected and transmitted waves are assumed to be plane, although in reality, at least in the close vicinity of the sudden area change, this is not the case.

6.3.1 Transition region in a duct

Boundary conditions for a duct transition region, which is characterised by an enlargement of the duct cross–section, are obtained from the linearised equations which express the conservation laws of mass and momentum, see Sections 1.6 and 1.7. In the case when the transition region is compact, i.e., its width is small in comparison with the wavelength of sound, the boundary conditions can be reduced to matching conditions which express the conservation laws in the form of equality of volume velocities and sound pressures at both sides of the change in cross–section ($x = 0$), see Figure 6.3.

The matching conditions at the transition section ($x = 0$) are:

$$\tilde{\mathbf{u}}_1(0)S_1 = \tilde{\mathbf{u}}_2(0)S_2, \tag{6.4}$$

$$\tilde{\mathbf{p}}_1(0) = \tilde{\mathbf{p}}_2(0), \tag{6.5}$$

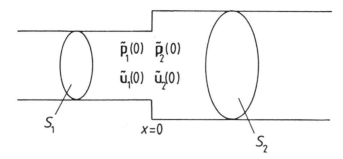

Figure 6.3. Transition region in a duct; enlargement in cross–section from S_1 to S_2.

where $\widetilde{\mathbf{p}}_1$ and $\widetilde{\mathbf{p}}_2$ are the sound pressure phasors of the resulting sound waves in parts 1 and 2 of the duct; $\widetilde{\mathbf{u}}_1$ and $\widetilde{\mathbf{u}}_2$ are the corresponding particle velocity phasors. The system of equations (6.4) and (6.5) is equivalent to:

$$\widetilde{\mathbf{p}}_1(0) = \widetilde{\mathbf{p}}_2(0), \tag{6.6}$$

$$\widetilde{\mathbf{p}}_1(0)/(\widetilde{\mathbf{u}}_1(0)S_1) = \widetilde{\mathbf{p}}_2(0)/(\widetilde{\mathbf{u}}_2(0)S_2). \tag{6.7}$$

Equation (6.7) can be presented in the form

$$\mathbf{Z}_1(0) = \mathbf{Z}_2(0). \tag{6.8}$$

Equations (6.6) and (6.8) express the continuity of pressure and acoustic impedance at the transition cross–section (see also Section 4.3). Note that the reflected wave is taken into account in the definition of the acoustic impedances $\mathbf{Z}_1$ and $\mathbf{Z}_2$.

In the literature [1] the term the *acoustic impedance of the tube* is used in relation to the incident sound wave. It is defined as the ratio of the sound pressure to the volume velocity of the incident sound wave. This definition of the impedance of the tube is useful, e.g., in the determination of the transmission coefficient α_t at the junction of two straight tubes, as illustrated in Figure 6.4, where the second tube of larger diameter is either infinitely long or anechoically terminated.

The acoustic impedance of the first tube, see Figure 6.4, defined in terms of the incident wave, is $\mathbf{Z}_1^+ = \widetilde{\mathbf{p}}_i/(\widetilde{\mathbf{u}}_i S_1)$ and the acoustic impedance of the second tube is $\mathbf{Z}_2^+ = \widetilde{\mathbf{p}}_t/(\widetilde{\mathbf{u}}_t S_2)$, where $\widetilde{\mathbf{p}}_i$ and $\widetilde{\mathbf{p}}_t$ denote the sound pressure phasors for the incident and transmitted sound waves, respectively. Similarly, $\widetilde{\mathbf{u}}_i$ and $\widetilde{\mathbf{u}}_t$ are the phasors of the particle velocities of the

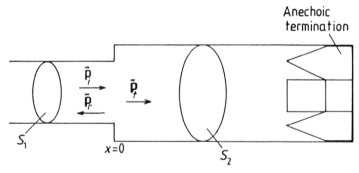

Figure 6.4 System of two straight ducts; second duct anechoically terminated.

incident and transmitted sound waves. Finally, taking into account equations (6.1), (6.4), (6.5) and (1.82), we obtain for the sound transmission coefficient

$$\alpha_t = \frac{4S_1 S_2}{(S_1 + S_2)^2} = \frac{4Z_1^+ Z_2^+}{(Z_1^+ + Z_2^+)^2}. \tag{6.9}$$

From equation (6.9) we can deduce that $\alpha_t = 1$ when $Z_2^+ = Z_1^+$. Hence, if the acoustic impedances of both tubes Z_2^+ and Z_1^+ are equal, the maximal sound power is transmitted from the first to the second tube.

6.3.2 Plane sound waves in a straight duct

Let us consider an incident plane sound wave and a reflected sound wave in a straight duct of length l and of constant cross–sectional area S. As only plane waves are to be considered, one spatial co–ordinate x is required to define the sound pressure or particle velocity in the sound wave as functions of the position in the duct. The origin of the co–ordinate system $(x = 0)$ is assumed to be at the beginning of the duct, as shown in Figure 6.5.

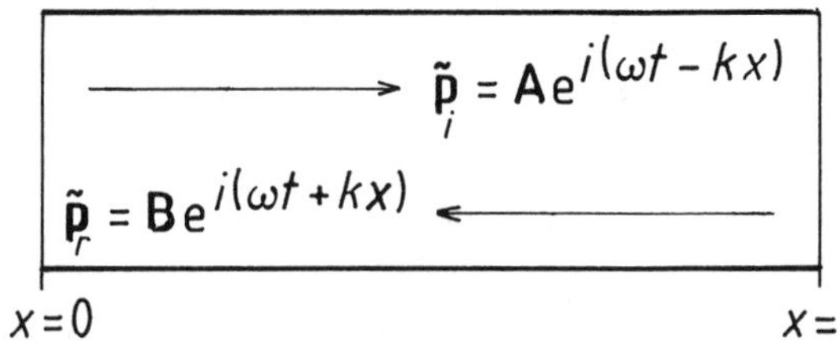

$$\tilde{p}_i = A e^{i(\omega t - kx)}$$

$$\tilde{p}_r = B e^{i(\omega t + kx)}$$

$$x = 0 \qquad\qquad\qquad\qquad x = l$$

Figure 6.5 Sound pressure phasors of incident and reflected waves in a duct.

The sound pressure describing the incident sound wave is expressed by the phasor $\widetilde{\mathbf{p}}_i$, which may be presented in terms of the complex amplitude $\mathbf{A}$ as

$$\widetilde{\mathbf{p}}_i = \mathbf{A}e^{i(\omega t - kx)}, \tag{6.10}$$

where ω is the circular frequency and k the wave number. The phasor of the particle velocity $\widetilde{\mathbf{u}}_i$ for the incident sound wave is related to the phasor of the sound pressure $\widetilde{\mathbf{p}}_i$ by equation (1.104). As was already mentioned, a convenient parameter describing the propagation of a sound wave in a system of pipes of different cross–sections is the volume velocity, which is represented by the phasor $\widetilde{\mathbf{U}}$ and is related to the particle velocity phasor $\widetilde{\mathbf{u}}$ as

$$\widetilde{\mathbf{U}}(x) = \widetilde{\mathbf{u}}(x)S(x), \tag{6.11}$$

where S denotes the cross–sectional area. Thus $\widetilde{\mathbf{U}}_i$, the phasor of the volume velocity of the incident sound wave, is expressed in terms of the sound pressure as,

$$\widetilde{\mathbf{U}}_i = \frac{\widetilde{\mathbf{p}}_i S}{\rho_0 c_0} = \frac{\mathbf{A}}{(\rho_0 c_0 / S)} e^{i(\omega t - kx)}. \tag{6.12}$$

Similarly, the sound pressure phasor $\widetilde{\mathbf{p}}_r$ of the reflected wave may be written in terms of its complex amplitude $\mathbf{B}$ as

$$\widetilde{\mathbf{p}}_r = \mathbf{B}e^{i(\omega t + kx)} \tag{6.13}$$

and the volume velocity phasor $\widetilde{\mathbf{U}}_r$ for the reflected sound wave in a duct is given by:

$$\widetilde{\mathbf{U}}_r = -\frac{\widetilde{\mathbf{p}}_r}{\rho_0 c_0} S = -\frac{\mathbf{B}}{(\rho_0 c_0 / S)} e^{i(\omega t + kx)}. \tag{6.14}$$

The acoustic impedance $\mathbf{Z}_x$ at the cross–section $S(x)$ of the duct is the ratio of the sound pressure $\widetilde{\mathbf{p}}(x)$ to the volume velocity $\widetilde{\mathbf{U}}(x) = \widetilde{\mathbf{u}}S$ (see also equation (1.218) and Section 1.18). Thus,

$$\mathbf{Z}_x = \frac{\widetilde{\mathbf{p}}_i(x) + \widetilde{\mathbf{p}}_r(x)}{\widetilde{\mathbf{U}}_i(x) + \widetilde{\mathbf{U}}_r(x)} \tag{6.15}$$

or,

$$\mathbf{Z}_x = \left(\frac{\rho_0 c_0}{S}\right) \frac{\mathbf{A}e^{-ikx} + \mathbf{B}e^{ikx}}{\mathbf{A}e^{-ikx} - \mathbf{B}e^{ikx}}. \tag{6.16}$$

If there is no reflected wave, $\mathbf{B} = 0$ and the acoustic impedance is that of a freely progressing plane wave in the pipe, namely, $\rho_0 c_0 / S$. When there is no incident wave, $\mathbf{A} = 0$ and the acoustic impedance is $-(\rho_0 c_0 / S)$.

The acoustic impedance $\mathbf{Z}_x$, defined by equation (6.16), is a function of the co-ordinate x. Hence, the acoustic impedance for the plane waves in the duct at $x = 0$ is

$$\mathbf{Z}_0 = \left(\frac{\rho_0 c_0}{S}\right) \frac{\mathbf{A} + \mathbf{B}}{\mathbf{A} - \mathbf{B}}. \tag{6.17}$$

and at $x = l$ the acoustic impedance $\mathbf{Z}_l$ is given by:

$$\mathbf{Z}_l = \left(\frac{\rho_0 c_0}{S}\right) \frac{\mathbf{A}e^{-ikl} + \mathbf{B}e^{ikl}}{\mathbf{A}e^{-ikl} - \mathbf{B}e^{ikl}}. \tag{6.18}$$

By eliminating the complex amplitudes $\mathbf{A}$ and $\mathbf{B}$ from equations (6.17) and (6.18), we obtain the relation between $\mathbf{Z}_l$ and $\mathbf{Z}_0$:

$$\mathbf{Z}_0 = \left(\frac{\rho_0 c_0}{S}\right) \frac{\mathbf{Z}_l + i(\rho_0 c_0 / S)\tan kl}{\rho_0 c_0 / S + i\mathbf{Z}_l \tan kl}. \tag{6.19}$$

Let us define complex amplitudes $\mathbf{p}_0$ and $\mathbf{U}_0$ of the sound pressure $\widetilde{\mathbf{p}}$ and volume velocity $\widetilde{\mathbf{U}}$ at position x in the duct:

$$\mathbf{p}_0(x) = \widetilde{\mathbf{p}}e^{-i\omega t} = (\widetilde{\mathbf{p}}_i + \widetilde{\mathbf{p}}_r)e^{-i\omega t}. \tag{6.20}$$

$$\mathbf{U}_0(x) = \widetilde{\mathbf{U}}e^{-i\omega t} = (\widetilde{\mathbf{U}}_i + \widetilde{\mathbf{U}}_r)e^{-i\omega t}. \tag{6.21}$$

The complex amplitudes $\mathbf{p}_0(x)$ and $\mathbf{U}_0(x)$ characterise harmonic oscillations at position x in the sound field of the duct. The amplitudes are different at different cross-sections of the duct. Taking into account equations (6.10) and (6.13), we obtain for $\mathbf{p}_0$ the expression

$$\mathbf{p}_0 = (\mathbf{A} + \mathbf{B})\cos kx + i(\mathbf{B} - \mathbf{A})\sin kx \tag{6.22}$$

and from equations (6.12) and (6.14)

$$\mathbf{U}_0 = \frac{(\mathbf{A} - \mathbf{B})}{(\rho_0 c_0 / S)}\cos kx - i\frac{(\mathbf{A} + \mathbf{B})}{(\rho_0 c_0 / S)}\sin kx. \tag{6.23}$$

Consequently, at $x = 0$:

$$\mathbf{p}_0(0) = \mathbf{A} + \mathbf{B} \tag{6.24}$$

and

$$\mathbf{U}_0(0) = \frac{\mathbf{A} - \mathbf{B}}{(\rho_0 c_0 / S)}. \tag{6.25}$$

At $x = l$:

$$\mathbf{p}_0(l) = (\mathbf{A} + \mathbf{B}) \cos kl + i(\mathbf{B} - \mathbf{A}) \sin kl \tag{6.26}$$

and

$$\mathbf{U}_0(l) = \frac{(\mathbf{A} - \mathbf{B})}{(\rho_0 c_0 / S)} \cos kl - i\frac{(\mathbf{A} + \mathbf{B})}{(\rho_0 c_0 / S)} \sin kl. \tag{6.27}$$

Finally, elimination of $\mathbf{A}$ and $\mathbf{B}$ from equations (6.26) and (6.27) by taking into account equations (6.24) and (6.25) leads to the expressions for the amplitudes $\mathbf{p}_0(0)$ and $\mathbf{U}_0(0)$ in terms of the amplitudes $\mathbf{p}_0(l)$ and $\mathbf{U}_0(l)$:

$$\mathbf{p}_0(0) = \cos kl \, \mathbf{p}_0(l) + i \left(\frac{\rho_0 c_0}{S} \right) \sin kl \, \mathbf{U}_0(l) \tag{6.28}$$

and

$$\mathbf{U}_0(0) = i \frac{\sin kl}{(\rho_0 c_0 / S)} \mathbf{p}_0(l) + \cos kl \, \mathbf{U}_0(l). \tag{6.29}$$

The equations developed here will be found useful in subsequent sections in this chapter.

6.4 Expansion chamber silencer

A simple type of reactive silencer is formed from three pipe sections, the middle one of which has the largest cross–sectional area S, as shown in Figure 6.6. The largest section pipe forms an expansion chamber.

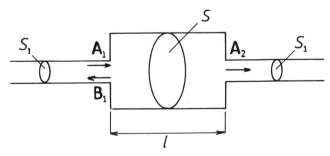

Figure 6.6 Expansion chamber silencer.

A sound wave with an amplitude $\mathbf{A}_1$ is transmitted towards the change in section in the left pipe. Since a discontinuity in the cross–section exists at the interface between the left pipe and the middle pipe (where the area suddenly increases from S_1 to S), a sound wave of amplitude $\mathbf{B}_1$ is reflected from the interface and appears in the left pipe (see also equation (6.13)). The amplitude of the sound pressure for the wave transmitted downstream of the expansion chamber is $\mathbf{A}_2$.

To obtain relations between the complex amplitudes $\mathbf{A}_1$, $\mathbf{B}_1$ and $\mathbf{A}_2$, and hence to be able to determine α_t, the sound transmission coefficient for the expansion chamber silencer (where $\alpha_t = |\mathbf{A}_2|^2/|\mathbf{A}_1|^2$), we must consider the system of equations (6.28) and (6.29), as well as the matching conditions at $x = 0$ and $x = l$, see Figure 6.6. The system of equations (6.28) and (6.29) describes the relation between the complex amplitudes, see equations (6.20) and (6.21), of plane sound waves in a duct of length l at the beginning ($x = 0$) and at the end of the duct ($x = l$). The expansion chamber is also assumed to be a duct of length l. Hence, for the expansion chamber:

$$\begin{aligned}
\mathbf{p}_0(0) &= \cos kl\,\mathbf{p}_0(l) + i(\rho_0 c_0/S)\sin kl\,\mathbf{U}_0(l), \\
\mathbf{U}_0(0) &= \cos kl\,\mathbf{U}_0(l) + i\,[S\sin kl/(\rho_0 c_0)]\,\mathbf{p}_0(l),
\end{aligned} \tag{6.30}$$

where $\mathbf{p}_0$ and $\mathbf{U}_0$ denote the complex amplitudes of the sound pressure and volume velocity in the plane sound wave existing in the expansion chamber. The matching conditions at $x = 0$, see also equations (6.4) and (6.5), are

$$\mathbf{p}_0(0) = \mathbf{A}_1 + \mathbf{B}_1, \tag{6.31}$$

$$\mathbf{U}_0(0) = (\mathbf{A}_1 - \mathbf{B}_1)S_1/(\rho_0 c_0). \tag{6.32}$$

Similarly, the matching conditions at $x = l$ are

$$\mathbf{p}_0(l) = \mathbf{A}_2, \tag{6.33}$$

$$\mathbf{U}_0(l) = \mathbf{A}_2 S_2/(\rho_0 c_0). \tag{6.34}$$

The system of equations (6.28) and (6.29) combined with the pair of matching conditions (6.31), (6.32) and the pair (6.33), (6.34) leads to a further system of equations:

$$\mathbf{A}_1 + \mathbf{B}_1 = \cos kl\,\mathbf{A}_2 + i(S_2/S)\sin kl\,\mathbf{A}_2, \tag{6.35}$$

$$\mathbf{A}_1 - \mathbf{B}_1 = (S_2/S_1)\cos kl\,\mathbf{A}_2 + i(S/S_1)\sin kl\,\mathbf{A}_2. \tag{6.36}$$

Addition of equations (6.35) and (6.36) gives

$$2\mathbf{A}_1 = \mathbf{A}_2\left[(1 + S_2/S_1)\cos kl + i(S_2/S + S/S_1)\sin kl\right].$$

Hence,

$$|\mathbf{A}_1|^2 = (1/4)|\mathbf{A}_2|^2\left[(1 + S_2/S_1)^2\cos^2 kl + (S_2/S + S/S_1)^2\sin^2 kl\right].$$

Finally, the sound transmission coefficient α_t for the expansion chamber silencer is given by:

$$\alpha_t = \frac{|\mathbf{A}_2|^2}{|\mathbf{A}_1|^2} = \frac{4}{(1 + S_2/S_1)^2\cos^2 kl + (S_2/S + S/S_1)^2\sin^2 kl}. \qquad (6.37)$$

For the expansion chamber silencer with a downstream pipe of the same cross–sectional area S_1 as the upstream pipe the sound transmission coefficient α_t is given by:

$$\alpha_t = \frac{1}{\cos^2 kl + (1/4)(S_1/S + S/S_1)^2\sin^2 kl} \qquad (6.38)$$

and hence the transmission loss L_{TL}, see equation (6.2), is

$$L_{\mathrm{TL}} = 10\log\left[\cos^2 kl + (1/4)(S_1/S + S/S_1)^2\sin^2 kl\right]. \qquad (6.39)$$

For $l = 0.5$ m and $S/S_1 = 5, 10, 15$ and 20 the transmission loss, see equation (6.39), is shown plotted against frequency in Figure 6.7.

The variation of transmission loss with frequency forms a repetitive pattern which is maintained as long as plane waves of sound exist in the duct system. At certain circular frequencies $(0, \pi c_0/l, 2\pi c_0/l, 3\pi c_0/l,$ etc.), as is shown in Figure 6.7, the transmission loss is zero and the sound transmission coefficient is equal to 1. It is seen from equation (6.39) that zero transmission loss occurs when

$$kl = n\pi, \qquad (6.40)$$

where $n = 0, 1, 2 \ldots$ etc. In this case the sound is transmitted through the expansion chamber without any reflections at the junctions at $x = 0$ and $x = l$. It is as though the expansion chamber did not exist, but was replaced by a straight pipe.

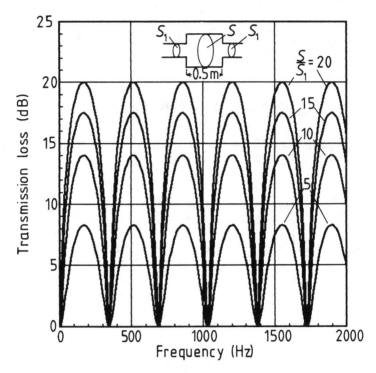

Figure 6.7 Transmission loss of an expansion chamber silencer as a function of frequency f; $l = 0.5$ m, $S/S_1 = 5$, 10, 15 and 20; $c_0 = 345$ m/s.

The transmission loss attains maximum values when

$$kl = (2n - 1)\pi/2, \qquad (6.41)$$

where $n = 1, 2, 3 \ldots$ etc. The maximum values of the transmission loss, see equation (6.39), are determined from the equation

$$L_{\mathrm{TL\,max}} = 20\log\left[(1/2)(S/S_1 + S_1/S)\right] \text{ dB.} \qquad (6.42)$$

The values of $L_{\mathrm{TL\,max}}$ increase with increasing values of the area ratio S/S_1. For example, when $S/S_1 = 20$, $L_{\mathrm{TL\,max}} = 20.0$ dB. To obtain substantial attenuation an expansion chamber of large cross–section is required. Such a silencer is bulky and in addition plane waves may no longer exist in the chamber, thus the theory outlined here would become invalid.

6.5 Acoustic impedance in a duct closed by a rigid cap

Consider a section of duct, such as is shown in Figure 6.5, with a rigid termination at one of its ends, $x = l$. At $x = 0$ there is a source of plane sound waves, e.g., a light and flexibly supported piston which generates sound waves of, say, frequency ω and amplitude $\mathbf{A}$ for the sound pressure wave, see equation (6.10). Because of the rigid termination at $x = l$, the amplitude of the volume velocity at $x = l$ must be equal to zero, i.e., $\mathbf{U}_0(l) = 0$. Thus the boundary condition at $x = l$, see equation (6.27), is:

$$\frac{\mathbf{A} - \mathbf{B}}{(\rho_0 c_0 / S)} \cos kl - i \frac{\mathbf{A} + \mathbf{B}}{(\rho_0 c_0 / S)} \sin kl = 0$$

or

$$\frac{\mathbf{A} + \mathbf{B}}{\mathbf{A} - \mathbf{B}} = -i \cot kl.$$

Combining the above equation with equation (6.17), we obtain an expression for the input acoustic impedance $\mathbf{Z}_0$:

$$\mathbf{Z}_0 = -i \left(\frac{\rho_0 c_0}{S} \right) \cot kl. \tag{6.43}$$

The acoustic impedance $\mathbf{Z}_x$ at the cross–section $S(x)$ of the duct may be expressed in terms of the input acoustic impedance $\mathbf{Z}_0$, see equations (6.16) and (6.17):

$$\mathbf{Z}_x = \left(\frac{\rho_0 c_0}{S} \right) \left[\frac{\left(1 - e^{i2kx}\right) + \mathbf{Z}_0(S/\rho_0 c_0)\left(1 + e^{i2kx}\right)}{\left(1 + e^{i2kx}\right) + \mathbf{Z}_0(S/\rho_0 c_0)\left(1 - e^{i2kx}\right)} \right]. \tag{6.44}$$

Note that the acoustic impedances in the duct may be determined directly from the expressions for the sound pressure and particle velocity in a standing wave, see equations (4.22) and (4.26), provided that the appropriate boundary conditions are taken into account.

6.5.1 Acoustic resonance in a duct closed by a rigid cap

Resonance and anti–resonance occur in a mechanical system when the imaginary part of the input impedance is zero [2,3]. However, an additional condition relating to the input resistance should be formulated to differentiate between resonance and anti–resonance. In general, see Section 6.6 which refers to open ducts, an investigation of extrema of the input resistance leads to the specification of resonance (and anti–resonance).

Namely, minimum of the resistive part of the input impedance refers to resonance and, similarly, maximum of the resistance specifies anti–resonance.

In the case of the tube with one end rigidly terminated, as was considered in Section 6.5, the condition for resonance is reduced to $Z_0 = 0$, since the input impedance consists only of an imaginary part. This condition implies that for a finite, even very small, amplitude of the sound pressure at $x = 0$ the amplitude of the volume velocity $U_0(0)$ is infinite.

Hence, the condition for resonance in a rigidly terminated duct is reduced to:

$$\cot kl = 0. \tag{6.45}$$

Thus,

$$kl = (2n - 1)\pi/2,$$

where $n = 1, 2, 3\ldots$ etc. and, consequently, the frequencies at which resonance occurs are given by the formula

$$f = \frac{(2n - 1)c_0}{4l}. \tag{6.46}$$

Since $c_0 = \lambda f$, the formula (6.46) can be reduced to a simple relation between the length of the duct l and the wavelength λ:

$$l = (2n - 1)\lambda/4.$$

Thus when $n = 1$ the length of the duct is a quarter of the wavelength and when $n = 2$ the length of the duct is three quarters of a wavelength, etc. A duct closed at one end is often called a *quarter wave tube*.

It should be noted that resonance occurs in the system (in this case in the tube with a rigidly terminated wall) when the frequency of the incident sound wave is equal to the one eigenfrequency (natural or characteristic frequency) of the system. Consequently, the corresponding eigenmode is excited.

The modal patterns for the quarter wave tube in terms of displacement (particle velocity, volume velocity) for $n = 1$ and $n = 2$ are shown in Figure 6.8.

6.6 Acoustic impedance in an open duct

In Sections 6.5 we considered a duct with a rigid cap at $x = l$. Let us analyse the case of a duct open at $x = l$. As for a rigidly terminated duct the assumption is made that at $x = 0$ there is a source of plane waves, see Section 6.5.

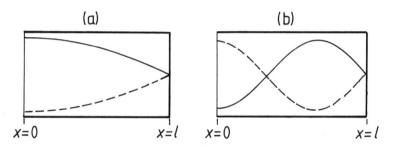

Figure 6.8 Modal shapes in a quarter wave tube; (a) $n = 1$ and (b) $n = 2$.

The input acoustic impedance at $x = 0$ is expressed as a function of the acoustic impedance at $x = l$ by the formula (6.19) and the acoustic impedance $\mathbf{Z}_x$ is given by equation (6.44). For a duct with a rigid cap $\mathbf{Z}_l$, the acoustic impedance at $x = l$, is infinite. However, the acoustic impedance $\mathbf{Z}_l$ for a duct with an open end at $x = l$ depends upon the duct geometry and may be expressed as

$$\mathbf{Z}_l = \frac{\rho_0 c_0}{S} [R_1 + iX_1], \tag{6.47}$$

where $(\rho_0 c_0/S)R_1$ and $(\rho_0 c_0/S)X_1$ are the resistive and reactive parts of the acoustic impedance $\mathbf{Z}_l$, respectively.

If the duct is terminated at $x = l$ in an infinite flange R_1 and X_1 are the same as the *piston impedance functions* [2,4], or *piston resistance* and *reactance functions*. The piston impedance functions specify the radiation impedance of a rigid piston in an infinite flange. The theory relevant to sound radiation from a baffled piston is not discussed in this book. The full expressions for R_1 and X_1, the piston impedance functions, defined for a rigid piston in an infinite baffle can be found in other texts [1,2].

Taking into account the equations (6.19) and (6.47), we obtain for the input acoustic impedance (at $x = 0$) the equation,

$$\mathbf{Z}_0 = \left(\frac{\rho_0 c_0}{S}\right) \frac{R_1 + i(\tan kl + X_1)}{1 - X_1 \tan kl + iR_1 \tan kl} \tag{6.48}$$

which defines the relation between the input acoustic impedance and the radiation impedance of a duct terminated in an infinite flange.

6.6.1 Acoustic resonance in an open duct

Resonance occurs in a duct, see also Section 6.5.1, when the reactive (imaginary) part of the input impedance is zero and the resistive part of the input impedance reaches a minimum. The first condition $(\mathrm{Im}\,(\mathbf{Z}_0) = 0$, see equation (6.48)) leads to:

$$\tan(kl - n\pi)_{\mathrm{r,\,ar}} = \frac{\left(1 - X_1^2 - R_1^2\right) \pm \sqrt{\left(1 - X_1^2 - R_1^2\right) + 4X_1^2}}{2X_1} \qquad (6.49)$$

$(n = 0, 1, 2, 3,\ldots$ etc.) which specifies resonance (subscript r) and anti–resonance (subscript ar) in a duct. Zero values of $\mathrm{Im}\,(\mathbf{Z}_0)$ coincide with the maxima and minima of $\mathrm{Re}\,(\mathbf{Z}_0)$. As the minima of $\mathrm{Re}\,(\mathbf{Z}_0)$ define resonance, the condition for resonance is given by:

$$\tan(kl - n\pi)_{\mathrm{r}} = \frac{1 - \sqrt{1 + 4X_1^2/(1 - X_1^2 - R_1^2)^2}}{2X_1/(1 - X_1^2 - R_1^2)}. \qquad (6.50)$$

For a circular duct with the cross–sectional radius a such that $ka \ll 1$ (low frequency approximation) the squares of the piston impedance functions X_1 and R_1 may be neglected in comparison with unity and equation (6.50) is reduced to:

$$\tan(kl - n\pi)_{\mathrm{r}} = -X_1, \qquad (6.51)$$

where, see [2],

$$X_1 = \frac{8ka}{3\pi}. \qquad (6.52)$$

Consequently, from equations (6.51) and (6.52)

$$kl - n\pi = \arctan\left(-\frac{8ka}{3\pi}\right), \qquad (6.53)$$

where $n = 0, 1, 2, \ldots$ etc.

Since $ka \ll 1$, equation (6.53) may be approximated by:

$$kl - n\pi = -\frac{8ka}{3\pi}. \qquad (6.54)$$

Hence, the frequencies at which resonance occurs are given by:

$$f = \frac{nc_0}{2\left[l + 8a/(3\pi)\right]}. \qquad (6.55)$$

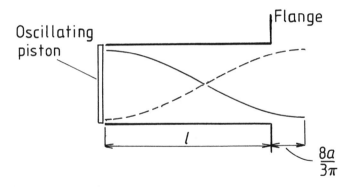

Figure 6.9 Modal shape for sound waves in a duct with an open flange at one end and an oscillating piston at the other (left end).

For the case of $n = 1$ the wavelength λ is related to the duct length l as

$$\lambda = 2[l + 8a/(3\pi)]. \tag{6.56}$$

The quantity $8a/(3\pi)$, added to the physical length of the duct l, is referred to as the *end correction*.

Eigenmodes of oscillation for a duct with an open flange at one end are shown in Figure 6.9.

6.7 Side branch silencer in the form of a quarter wave tube

A common form of silencer is a side branch to the main duct, as shown in Figure 6.10. The side branch is a straight tube of length l and cross–sectional area S_{sb}. The co–ordinate system shown in Figure 6.10 is chosen in such a manner that the position of the entrance of the side branch refers to $x = 0$.

Additionally, it is assumed that the transition region (around $x = 0$) is compact, i.e., the diameter (or width) of the side branch is small compared with the wavelength of the incident sound wave. Hence, the matching conditions at $x = 0$ in the form of the mass and momentum conservation equations, similar to equations (6.4) and (6.5), can be applied.

Let the incident plane sound wave entering the transition region $(x = 0)$ in the main duct of cross–section S be represented by a sound pressure phasor $\widetilde{\mathbf{p}}_i = \mathbf{A}_i e^{i(\omega t - kx)}$ and, consequently, by a volume velocity phasor $\widetilde{\mathbf{U}}_i = \widetilde{\mathbf{p}}_i S/(\rho_0 c_0)$, see Section 6.3 and also Figure 6.10. Let $\widetilde{\mathbf{p}}_r = \mathbf{A}_r e^{i(\omega t + kx)}$ and $\widetilde{\mathbf{p}}_t = \mathbf{A}_t e^{i(\omega t - kx)}$ denote, respectively, the sound pressure phasors for the reflected and transmitted sound waves travelling

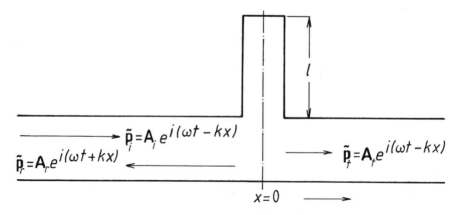

Figure 6.10 Side branch formed by a straight tube.

in the main duct. Note that $\mathbf{A}_i$, $\mathbf{A}_r$ and $\mathbf{A}_t$ are complex amplitudes for the incident, reflected and transmitted waves. Finally, $\widetilde{\mathbf{U}}_r = -\widetilde{\mathbf{p}}_r S/(\rho_0 c_0)$ and $\widetilde{\mathbf{U}}_t = \widetilde{\mathbf{p}}_t S/(\rho_0 c_0)$ are the volume velocity phasors for the reflected and transmitted waves in the main duct. The sound pressure in the standing wave at the entrance to the side branch duct (at $x = 0$) can be represented by the phasor $\widetilde{\mathbf{p}}_{\mathrm{sb}}(0) = \mathbf{A}_{\mathrm{sb}}(0)e^{i\omega t}$, where $\mathbf{A}_{\mathrm{sb}}$ is the complex amplitude of the standing wave in the side branch tube, see Section 4.4. The volume velocity at $x = 0$ in the standing wave in the side branch is denoted by the phasor $\widetilde{\mathbf{U}}_{\mathrm{sb}}(0)$. Consequently, with the above notation the matching conditions which express the momentum and mass conservation laws are

$$\widetilde{\mathbf{p}}_i(0) + \widetilde{\mathbf{p}}_r(0) = \widetilde{\mathbf{p}}_t(0) = \widetilde{\mathbf{p}}_{\mathrm{sb}}(0), \tag{6.57}$$

$$\widetilde{\mathbf{U}}_i(0) + \widetilde{\mathbf{U}}_r(0) = \widetilde{\mathbf{U}}_t(0) + \widetilde{\mathbf{U}}_{\mathrm{sb}}(0) \tag{6.58}$$

or,

$$\mathbf{A}_i + \mathbf{A}_r = \mathbf{A}_t = \mathbf{A}_{\mathrm{sb}}(0), \tag{6.59}$$

$$\left(\frac{S}{\rho_0 c_0}\right)(\mathbf{A}_i - \mathbf{A}_r) = \left(\frac{S}{\rho_0 c_0}\right)\mathbf{A}_t + \mathbf{U}_{\mathrm{sb}}(0), \tag{6.60}$$

where $\mathbf{U}_{\mathrm{sb}}(0) = \widetilde{\mathbf{U}}_{\mathrm{sb}}(0)e^{-i\omega t}$.

The system of equations (6.59) and (6.60) is equivalent to another system which comprises equation (6.59) and a second equation (6.61) formed by dividing equation (6.60) by equation (6.59), namely,

$$\left(\frac{S}{\rho_0 c_0}\right)\left(\frac{\mathbf{A}_i - \mathbf{A}_r}{\mathbf{A}_i + \mathbf{A}_r}\right) = \frac{S}{(\rho_0 c_0)} + \frac{1}{\mathbf{Z}_{sb}}, \qquad (6.61)$$

where $\mathbf{Z}_{sb}$ denotes the input acoustic impedance for the side branch duct.

Equation (6.61) expresses the relation between the acoustic impedances at the transition region where the side branch is joined to the main pipe. The reciprocal of the acoustic impedance at the cross-section of the main duct just before the transition region is equal to the sum of the reciprocal of $\mathbf{Z}_{sb}$ and the reciprocal of the acoustic impedance of the freely progressing wave transmitted downstream of the junction.

The input acoustic impedance $\mathbf{Z}_{sb}$ for the side branch tube is given by the equation (6.43), i.e.,

$$\mathbf{Z}_{sb} = -i\frac{\rho_0 c_0}{S_{sb}}\cot kl. \qquad (6.62)$$

The acoustic performance of a side branch inserted in a duct can be assessed in terms of the sound transmission coefficient α_t or the transmission loss L_{TL} for the system. The aim is to achieve maximal transmission loss in the main duct. The sound transmission coefficient can be found from the system of equations (6.59) and (6.61). From equation (6.61) we have

$$\frac{\mathbf{A}_r}{\mathbf{A}_i} = -\frac{(\rho_0 c_0/S)/\mathbf{Z}_{sb}}{2 + (\rho_0 c_0/S)/\mathbf{Z}_{sb}}. \qquad (6.63)$$

Inserting the reflection ratio $\mathbf{A}_r/\mathbf{A}_i$, determined from equation (6.63) into equation (6.59), we obtain an expression for the ratio $\mathbf{A}_t/\mathbf{A}_i$:

$$\frac{\mathbf{A}_t}{\mathbf{A}_i} = \frac{2}{2 + (\rho_0 c_0/S)/\mathbf{Z}_{sb}}. \qquad (6.64)$$

From equations (6.62) and (6.64) we have:

$$\frac{\mathbf{A}_t}{\mathbf{A}_i} = \frac{1}{1 + i[S_{sb}/(2S)]\tan kl}. \qquad (6.65)$$

The sound transmission coefficient α_t, see equation (6.1) for the definition, is given by:

$$\alpha_t = \frac{|\mathbf{A}_t|^2}{|\mathbf{A}_i|^2} \qquad (6.66)$$

or,

$$\alpha_t = \frac{1}{1 + (S_{sb}^2/4S^2)\tan^2 kl}. \tag{6.67}$$

Finally, for the transmission loss, see equation (6.2), we have:

$$L_{TL} = 10\log\left(1 + \frac{S_{sb}^2}{4S^2}\tan^2 kl\right) \text{ dB.} \tag{6.68}$$

The dependence of the transmission loss on kl (where $kl = 2\pi fl/c_0$) is shown for different values of the area ratio S_{sb}/S in Figure 6.11.

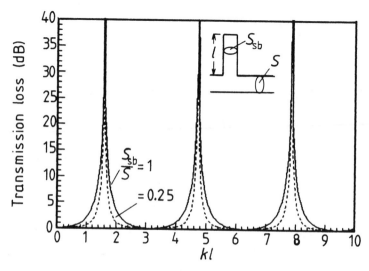

Figure 6.11 Transmission loss of a quarter wave tube as a function of kl.

The transmission loss is infinite when $\tan kl = \infty$ or $\cot kl = 0$, hence the corresponding frequencies are given by equation (6.46). These frequencies are the same as the resonant frequencies of the side branch tube. For the lowest resonant frequency the length of the side branch tube is equal to a quarter of a wavelength, as shown in Figure 6.8a. A quarter wave tube is used as a side branch silencer to reduce the sound at one frequency which is usually the fundamental frequency of the resonant oscillation in the side branch, that is, when $kl = \pi/2$.

The quarter wave tube is particularly useful for reducing the noise from a source producing a tone at a low frequency; at low frequencies the dissipative type of silencer is not effective. This type of silencer has been applied to gas jet burners and compressor inlets. The disadvantage of the quarter wave tube is that the side branch must be long, if it is used to attenuate a low frequency sound. In general the angle at which the side branch is inclined to the main duct is arbitrary. Sometimes it may be convenient for constructional reasons to incline the side branch at an angle of, say, 45° to the main duct.

When constructing a quarter wave tube it is generally advisable to have some means of varying the length; for example, to use a piston on a plunger, so that the system can be tuned to give the maximum attenuation.

6.8 The Helmholtz resonator

A difficulty with the application of the quarter wave tube as a silencer is that a long length of tube is required for the attenuation of low frequencies. The Helmholtz resonator is a compact device which can be used as a side branch in a similar manner to the quarter wave tube.

Let us consider a system of two tubes , as is schematically represented in Figure 6.12.

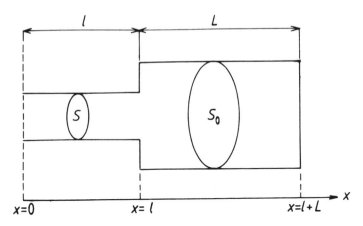

Figure 6.12 System of two connected tubes.

The tube of length l and of cross–sectional area S has a junction at $x = l$ with the second tube, rigidly terminated at $x = l + L$, of length L and cross–sectional area S_0. It is assumed that at the entrance to the first tube, at $x = 0$, there is a source of plane sound waves and, consequently, in both tubes plane standing sound waves exist, since the transition region fulfils the compactness condition.

From the matching conditions for the acoustic impedance at $x = l$, see Figure 6.12, it is possible to determine the input acoustic impedance for the whole system, say, $\mathbf{Z}_0 = R_0 + iX_0$, and hence find the resonant frequencies of the system. This procedure maintains its validity for the particular case when the whole system can be treated as compact, i.e., the dimensions of the system are much smaller than the wavelength of sound; such a system has most of the characteristics of a Helmholtz resonator.

The acoustic impedance at the cross–section $S(l)$ of the tube of length l may be expressed, see equation (6.44), as a function of the input acoustic impedance $\mathbf{Z}_0$:

$$\mathbf{Z}(l) = \frac{\rho_0 c_0}{S} \left\{ \frac{(1 - e^{i2kl}) + \mathbf{Z}_0[S/(\rho_0 c_0)](1 + e^{i2kl})}{(1 + e^{i2kl}) + \mathbf{Z}_0[S/(\rho_0 c_0)](1 + e^{i2kl})} \right\} \tag{6.69}$$

or,

$$\mathbf{Z}(l) = \frac{R_0 + i(\rho_0 c_0/S)\delta}{\cos^2 kl + [S/(\rho_0 c_0)]^2 (R_0^2 + X_0^2)\sin^2 kl}, \tag{6.70}$$

where

$$\delta = [S/(\rho_0 c_0)]X_0 - \sin kl \cos kl[1 - (R_0^2 + X_0^2)S^2/(\rho_0 c_0)^2]$$

and where $\mathbf{Z}_0 = R_0 + iX_0$. On the other hand the acoustic impedance at the cross–section $S_0(l)$ of the tube of length L is the input acoustic impedance for the tube closed by the rigid cap, see equation (6.43). The matching condition at $x = l$ is

$$\mathbf{Z}(l) = -i\left(\frac{\rho_0 c_0}{S_0}\right)\cot kl \tag{6.71}$$

which leads to the equations,

$$R_0 = 0 \tag{6.72}$$

and

$$\left(\frac{S}{\rho_0 c_0}\right)^2 \left[\left(\frac{S}{S_0}\right)\cot kL \sin^2 kl + \sin kl \cos kl\right] X_0^2$$
$$+ \left(\frac{S}{\rho_0 c_0}\right) X_0 + \left(\frac{S}{S_0}\right)\cot kL \cos^2 kl - \sin kl \cos kl = 0. \tag{6.73}$$

The assumption of the compactness of the system, which can be expressed by the conditions $kl \ll 1$ and $kL \ll 1$, leads to the approximation of equation (6.73) in the form,

$$\left(\frac{S}{\rho_0 c_0}\right)^2 \left[\left(\frac{S}{S_0}\right)\frac{(kl)^2}{kL} + kl\right] X_0^2 + \left(\frac{S}{\rho_0 c_0}\right) X_0$$
$$+ \left(\frac{S}{S_0}\right)\left(\frac{1}{kL} - \frac{kL}{3}\right) - kl - \left(\frac{S}{S_0}\right)\frac{(kl)^2}{kL} = 0. \qquad (6.74)$$

Additionally, as $S/S_0 \ll 1$, the equation (6.74) is reduced to:

$$\left(\frac{S}{\rho_0 c_0}\right)^2 klX_0^2 + \left(\frac{S}{\rho_0 c_0}\right) X_0 + \left(\frac{S}{S_0}\right)\frac{1}{kL} - kl = 0. \qquad (6.75)$$

Consequently, the solution which provides the condition for resonance $(X_0 = 0)$ is

$$X_0 = \frac{\rho_0 c_0}{2S}\left\{\frac{-1 + \sqrt{1 - 4kl[S/(kV) - kl]}}{kl}\right\}$$
$$= \frac{\rho_0 c_0}{2Skl}\left[-1 + \sqrt{1 - 4kl\left(\frac{c_0 S}{\omega V} - \frac{\omega l}{c_0}\right)}\right], \qquad (6.76)$$

where $V = S_0 L$.

Since the Helmholtz resonator has a neck which is short in comparison with the chamber length, the solution (6.76) may be approximated by:

$$X_0 \simeq \frac{\rho_0 c_0}{2Skl}\left[-1 + 1 - \frac{1}{2}4kl\left(\frac{c_0 S}{\omega V} - \frac{\omega l}{c_0}\right)\right] = \rho_0\left[\frac{\omega l}{S} - \frac{c_0^2}{\omega V}\right]. \qquad (6.77)$$

Consequently, the resonant frequency ω_r which is equal to the undamped circular natural frequency for the Helmholtz resonator is determined from the equation,

$$\rho_0\left[\frac{\omega_r l}{S} - \frac{c_0^2}{\omega_r V}\right] = 0$$

which leads to:

$$\omega_r^2 = \frac{c_0^2 S}{lV}. \qquad (6.78)$$

6.8.1 End correction for Helmholtz resonator

The theoretical considerations presented in Section 6.8 has led to the concept of the Helmholtz resonator. It should be noted that equation (6.72), which defines the resistive part of the input acoustic impedance for the Helmholtz resonator, does not incorporate sound radiation losses; nor does the model take into account dissipation phenomena which are, however, for the Helmholtz resonator less significant than radiation losses [1,2,5,6].

Similarly, the equation (6.76), which determines the reactive part of the input impedance, was derived without considering sound radiation from either the outer or inner openings of the resonator neck. If, however, we take into account the sound radiation from the outer termination of the Helmholtz resonator, assuming, for example, that this termination may be approximated by an infinite baffle, the resulting input acoustic impedance Z_0' for the resonator is

$$Z_0' = Z_0 + Z_r/S^2,$$

where Z_r is the radiation impedance of the piston, $Z_0 = R_0 + iX_0$ and $Z_0' = R_0' + iX_0'$. Note that the movement of air at the resonator termination can be compared with the oscillation of a piston. Consequently, see also Section 6.6,

$$R_0' = R_0 + (\rho_0 c_0/S)R_1, \tag{6.79}$$
$$X_0' = X_0 + (\rho_0 c_0/S)X_1, \tag{6.80}$$

where R_1 and X_1 are the piston resistance and reactance functions and S is the cross–sectional area of the resonator neck.

For a Helmholtz resonator with a cylindrical neck of radius a [2]

$$R_1 = (ka)^2/2 \tag{6.81}$$

and

$$X_1 = (8ka)/(3\pi). \tag{6.82}$$

Taking into account equations (6.72), (6.77), (6.79) and (6.80), we have

$$R_0' = \left(\frac{\rho_0 c_0}{S}\right) R_1 \tag{6.83}$$

and

$$X_0' = \rho_0 \left[\frac{\omega l + c_0 X_1}{S} - \frac{c_0^2}{\omega V} \right]. \tag{6.84}$$

The equation (6.84) can be presented in the form

$$X_0' = \rho_0 \left[\frac{\omega[l + (c_0/\omega)X_1]}{S} - \frac{c_0^2}{\omega V} \right] \tag{6.85}$$

or,

$$X_0' = \rho_0 \left[\frac{\omega l_{\text{eff}}}{S} - \frac{c_0^2}{\omega V} \right], \tag{6.86}$$

where

$$l_{\text{eff}} = l + \Delta l \tag{6.87}$$

and, in the considered case,

$$\Delta l = (c_0/\omega)X_1 = X_1/k. \tag{6.88}$$

Consequently, the resonant frequency of the Helmholtz resonator is expressed by:

$$\omega_r^2 = \frac{c_0^2 S}{l_{\text{eff}} V}. \tag{6.89}$$

The above considerations lead to the concept of the apparent *effective length* l_{eff} of the Helmholtz resonator neck. In general the terminations of both ends of the resonator neck (outer and inner openings) determine the end correction $\Delta l = \Delta l_{\text{out}} + \Delta l_{\text{inn}}$, where Δl_{out} and Δl_{inn} denote the end corrections associated with the outer and inner openings, respectively. For example, the end correction Δl_{out} for a cylindrical neck terminated in an infinite baffle is $8a/(3\pi) = 0.85a$, where a is the radius of the neck. This result is obtained from equations (6.82) and (6.88). In some situations it is possible to assume that both ends of the resonator neck are terminated by flanged openings which can be approximated by infinite baffles. Hence, in this case $\Delta l = 2 \times 0.85a = 1.7a$.

Extensive work on resonators have been carried out by Ingard who has established end corrections for different cross–sections of necks [7].

The Helmholtz resonator may be considered as a one degree of freedom system which comprises a spring and a body of a certain mass (mechanical oscillator) [8,9]. In this approach the air in the neck of the resonator can be treated as the vibrating body of the mechanical system and the air in the chamber, compressed and expanded during the oscillations, behaves like the spring [2,10]. The analogy is illustrated in Figure 6.13.

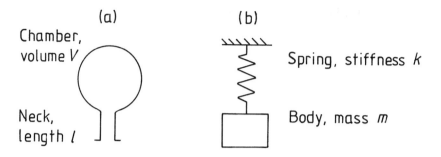

Figure 6.13 Analogy between single degree of freedom mechanical system and Helmholtz resonator.

6.9 Helmholtz resonator as a side branch

A Helmholtz resonator may be placed in a duct as a side branch, as shown in Figure 6.14. In order to calculate the transmission loss provided by the side branch it is necessary to know Z_{sb}, the acoustic impedance at the entrance to the resonator (the input acoustic impedance), see Section 6.7. In Section 6.8 it was shown that the reactive part of the input acoustic impedance for the Helmholtz resonator is expressed by the equation (6.86). The resistive part of the input acoustic impedance is given by the equation (6.83) — see also equation (6.81) — provided that only the sound radiation from the outer opening of the resonator is taken into account.

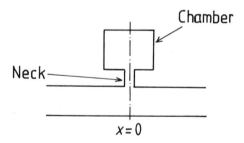

Figure 6.14 Helmholtz resonator as a side branch in a duct.

In general sound radiation from both openings of the resonator neck should be considered, as well as dissipation phenomena in the resonator neck; friction in the neck can be the main dissipative factor. Very often some material is introduced to the resonator to increase dissipation, e.g., a cloth or screen may be placed across the mouth of the resonator to provide damping [11].

Finally, if we denote the resistive part of the input acoustic impedance for the Helmholtz resonator as R, the input acoustic impedance $\mathbf{Z}_{sb}$ may be expressed as

$$\mathbf{Z}_{sb} = R + i \left[\frac{\rho_0 \omega l_{\text{eff}}}{S} - \frac{\rho_0 c_0^2}{\omega V} \right]. \tag{6.90}$$

The sound transmission coefficient for the duct with the Helmholtz resonator as a side branch is determined from the equations (6.66), (6.64) and (6.90) and is given by:

$$\alpha_t = \frac{R^2 + \left[\rho_0 \omega l_{\text{eff}}/S - \rho_0 c_0^2/(\omega V) \right]^2}{\left[R + \rho_0 c_0/(2S) \right]^2 + \left[\rho_0 \omega l_{\text{eff}}/S - \rho_0 c_0^2/(\omega V) \right]^2}. \tag{6.91}$$

Consequently, the transmission loss L_{TL} for the duct with the Helmholtz resonator as a side branch, see equation (6.2), is

$$L_{\text{TL}} = 10 \log \left\{ \frac{\left[R + \rho_0 c_0/(2S) \right]^2 + \left[\rho_0 \omega l_{\text{eff}}/S - \rho_0 c_0^2/(\omega V) \right]^2}{R^2 + \left[\rho_0 \omega l_{\text{eff}}/S - \rho_0 c_0^2/(\omega V) \right]^2} \right\}. \tag{6.92}$$

The maximal transmission loss in the duct with a Helmholtz resonator as a side branch is achieved at the resonant frequencies of the Helmholtz resonator.

6.10 Applications of quarter wave and Helmholtz resonators

6.10.1 Side branch resonator to reduce centrifugal fan noise

Side branch resonators in the form of quarter wave tubes or Helmholtz resonators have been used to reduce the noise from centrifugal fans and blowers [12,13]. The noise from fans and blowers is aerodynamic in origin and comprises a series of tones superimposed on broad band random noise, see Section 8.3.1. The series of tones is formed by a fundamental and its harmonics. The fundamental tone occurs at the blade passing frequency. A centrifugal fan with a resonator placed at the cutoff of the fan casing is shown in Figure 6.15 [12]. The resonator was basically a straight tube with a plug for adjustment of the length. In order to retain, as far as possible, the smooth internal surface of the fan casing the mouth of the resonator was made from 0.5 mm thick sheet steel which was perforated with 43 holes each of 3.1 mm diameter to give an open area of 29%. The fan had 6 blades and when the fan operated at 7500 rev/min the blade passing frequency was 750 Hz. Thus tones would be expected from the fan

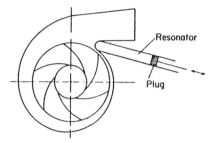

Figure 6.15 Centrifugal fan with a resonator at the cutoff of the fan casing [12].

at 750 Hz, 1500 Hz, etc. A quarter wave tube of the appropriate length is capable of providing attenuation at a fundamental frequency of 750 Hz and at harmonics of 2250 Hz and 3750 Hz, etc.

The spectra of sound from the fan with and without a resonator at the cutoff are shown in Figure 6.16. In this case the measurements were made in the fan outlet duct. The resonator brings about a reduction of over 20 dB at the blade passing frequency of 750 Hz. The length of the resonator corresponded to that of a quarter wave tube. The reduction in level of the third harmonic is additional evidence that the resonator is behaving like a quarter wave tube.

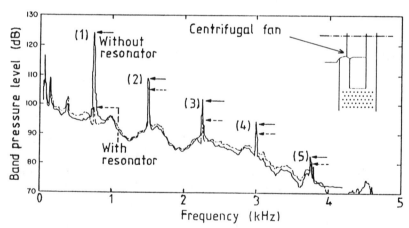

Figure 6.16 Spectra of sound from a centrifugal fan with and without a resonator [12].

In many ways the construction of the resonator under discussion is more similar to a Helmholtz resonator; the holes in the cutoff could be regarded as the multiple necks of a Helmholtz resonator. Note that the value of the resonant frequency of the Helmholtz resonator, when calculated according to equation (6.89) with $\Delta l = 1.7a$, is much greater than 750 Hz. The correction $\Delta l = 1.7a$ is not appropriate when the neck is short (in this case 0.5 mm) [12]. Alster has developed a different formula for the resonant frequencies of Helmholtz resonators [10] and his theory predicts fairly accurately the resonant frequencies in the present case [12].

6.10.2 Helmholtz resonator to reduce compressor intake noise

A second example of the application of a side branch silencer is related to reduction of noise from the intake of reciprocating compressors [14]. A noise problem was caused by the intake of air to two Bellis and Morcom two cylinder double acting compressors which drew 5000 cfm (2500 l/s) of free air through a common intake duct of 24 in (610 mm) diameter. Compressed air was delivered at a pressure of 80 – 90 psig (5.4 – 5.9 bar). The shaft speed of the compressors was 250 rev/min. Most of the noise from the intake was reduced by fitting on the intake a silencer with a design based on two expansion chambers. However, infrasound from the pulsations of the intake air still caused windows in the vicinity of the compressors to vibrate. The fundamental frequency of the intake pulsations was 250/60 or 4.16 Hz. As an expansion chamber type silencer would be too large to reduce the noise at this frequency, a side branch Helmholtz resonator was designed because it takes up less space. The intake and important details of the resonator are shown in Figure 6.17. According to equation (6.89), with $\Delta l = 1.7a$, the resonant frequency f_r of the Helmholtz resonator is given (working in S.I. units) by:

$$f_r = \frac{344}{2\pi} \sqrt{\frac{(\pi/4) \times 0.076^2}{1.22(0.61 + 1.7 \times 0.038)}} = 4.07 \text{ Hz},$$

which is close to the fundamental frequency of the intake pulsations. When fitted the Helmholtz resonator was reported to be successful.

6.10.3 Helmholtz resonators as sound absorbers

Helmholtz resonators have been used as sound absorbers in reverberant spaces, such as churches, for many hundreds of years. Consult Junger [15] for a historical review. Load–bearing concrete blocks are at present commercially available. They provide good sound insulation and also good sound absorption by having a hollow space within the block. This hollow space, which forms the chamber of the Helmholtz resonator, is connected to the reverberant room by means of a narrow slit which acts as the neck.

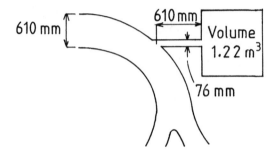

Figure 6.17 Dimensions of side branch Helmholtz resonator for reducing compressor intake noise.

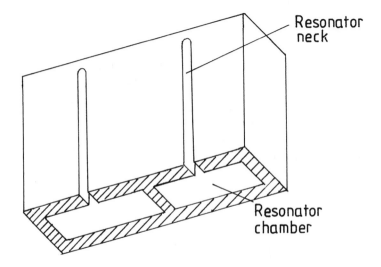

Figure 6.18 Masonry block with internal Helmholtz resonator.

The shape of such masonry blocks is shown in Figure 6.18. An entire wall may be constructed from the blocks.

The values of the sound absorption coefficients for this type of absorber are given in Table 5.1. The dissipation of sound energy originates from friction in the necks of the resonators, as well as being caused by sound absorbing material that is placed in the chamber.

6.10.4 Helmholtz resonators as acoustic filters

Helmholtz resonators play an important role in the system for assisted resonance at the Royal Festival Hall, London [16,17]. The reverberation time of the Royal Festival Hall after its inauguration in 1951 was found to be too small at low frequencies [18]. For most kinds of classical music it was felt that a warmer tone was required, i.e., the reverberation time needed to be increased. In order to increase the reverberation the reflected sound was detected by microphones and relayed through loudspeakers; a system known as assisted resonance.

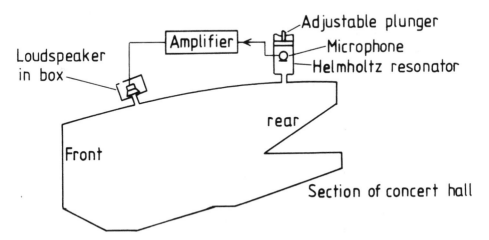

Figure 6.19 Helmholtz resonators used as part of the assisted resonance system at the Royal Festival Hall.

A schematic diagram of the system, as used in the Royal Festival Hall, is shown in Figure 6.19. The system comprises many channels, each with its own microphones, amplifier and loudspeaker. A different frequency is covered by each channel. The microphones (originally of the moving coil type and subsequently condenser) were placed inside the chambers of the Helmholtz resonators. Some form of filter was required to prevent interaction between the channels and, consequently, Helmholtz resonators were chosen in preference to electrical filters because:

(i) they are easy to manufacture,
(ii) they amplify the sound pressure by 30 dB,
(iii) they do not go wrong and
(iv) they are noiseless.

For frequencies above 300 Hz the Helmholtz resonators were too small and had to be replaced by quarter wave tubes. The resonators with the microphones were situated in the ceiling near the front of the hall; only the inlets to the necks of the resonators could be seen. The length of the chamber of each resonator was tuned to the frequency of the channel.

The loudspeakers were placed 15 to 20 m away from the microphones towards the rear of the hall. (In the case of low frequencies the positions of microphones and loudspeakers had to be reversed.) In order to increase the sound output the microphones and loudspeakers were located in boxes and connected to the auditorium by a 150 mm diameter tube which was tuned like a quarter wave tube. This arrangement was only used for frequencies up to 100 Hz.

The gain of the amplifier was adjusted to give the required reverberation time; an increase in the gain lengthens the reverberation time. The reverberation time with and without assisted resonance is shown in Figure 6.20.

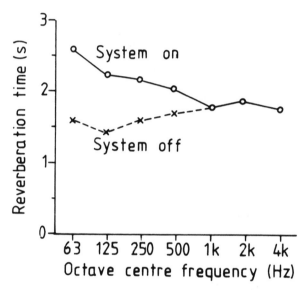

Figure 6.20 Reverberation time in the Royal Festival Hall with and without assisted resonance.

6.11 Expansion chamber silencers with projecting tubes

Real silencers used, for example, with internal combustion engines can have complicated shapes, and for such silencers a systematic design technique is required. In one design method each acoustic element is represented by a two–port network. The input port and output port quantities are related by a square transfer matrix of order two. This approach has been used by many workers, including Igarashi and colleagues [19], Munjal [20] and Lai [21]. The basic technique is widely used in electrical engineering [22]. Another technique is to consider the reflected and transmitted waves at each section of the duct where the acoustic impedance changes [23]. Both techniques can be applied to determine the transmission loss of silencers.

A frequently used type of silencer is an expansion chamber with projecting tubes, as shown in Figure 6.21.

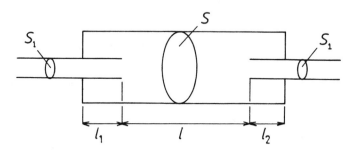

Figure 6.21 Expansion chamber silencer with projecting tubes.

A system equivalent to that shown in Figure 6.21 is presented in Figure 6.22. Both systems are equivalent to each other, since it is assumed that plane waves propagate in the systems and both have the same transmission loss. Note that the side branch ducts of cross–sections $(S–S_1)$, shown in Figure 6.22, should be located in the transient region of the expansion chamber of length l, i.e., at a distance (from the chamber) much smaller than the wavelength.

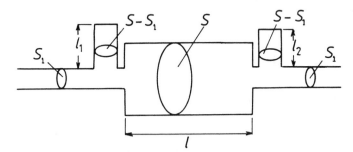

Figure 6.22 Equivalent system for an expansion chamber with projecting tubes.

The part of the expansion chamber adjacent to the projecting tubes, of lengths l_1 and l_2, see Figure 6.21, form separate side branch silencers or quarter wave tubes (as discussed in Section 6.5.1). The expression for the transmission loss is

$$
\begin{aligned}
L_{TL} = 10 \log \Bigg\{ \frac{1}{4} &\left[2 \cos kl - \left(\frac{S - S_1}{S} \right) \sin kl (\tan kl_1 + \tan kl_2) \right]^2 \\
&+ \frac{1}{4} \left[\left(\frac{S}{S_1} + \frac{S_1}{S} \right) \sin kl + \left(\frac{S - S_1}{S_1} \right) \cos kl (\tan kl_1 + \tan kl_2) \right. \\
&\left. - \frac{(S - S_1)^2}{S S_1} \sin kl \tan kl_1 \tan kl_2 \right]^2 \Bigg\}
\end{aligned}
$$

(6.93)

6.12 Acoustic performance of reactive silencers

Reactive silencers are extensively used for reducing the noise from the exhausts of internal combustion engines. Published data is available for a large range of silencers tested on a single cylinder four stroke engine of the type used for motorcycles [24]. The insertion loss of the silencer in this case [24] was obtained firstly by measuring the noise in one third octave bands from the outlet of a straight exhaust pipe 48 in. long (1.22 m) and secondly by taking similar measurements of the noise with the silencer in position. The two measurements are compared to obtain the insertion loss, see Section 6.2 for a description of the method. The engine was run at a speed of 3400 rev/min and at three quarters of full load. The insertion loss, as measured in one third octave bands, is shown in Figure 6.23 for three different silencers whose dimensions are illustrated in the figure. Note that the silencer (a) has three expansion chambers, (b) has two expansion chambers whilst (c) has a single expansion chamber; silencers (a) and (b) also possess projecting tubes.

The corresponding pressure losses for the silencers are shown in Figure 6.24.

6.13 Dissipative silencers

The simplest type of dissipative silencer is a duct which is lined with sound–absorbent material, such as mineral wool or fibre glass. A duct with a rectangular cross–section which is lined to form a silencer is shown in Figure 6.25. The aim of such a silencer is to reduce progressively the amplitude of the sound wave as the wave moves along the duct. The lining of the duct causes sound energy to be dissipated into heat.

6.13.1 Sound attenuation

If it is assumed that there is no wave reflected back to the source by the silencer, then the insertion loss and the transmission loss are the same.

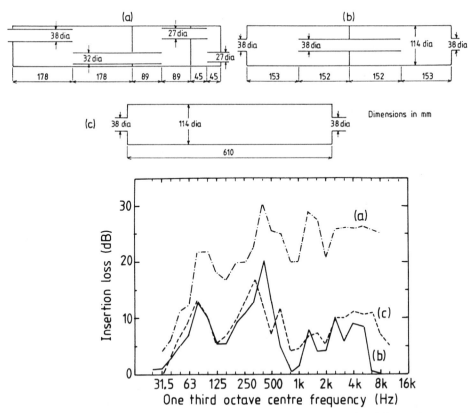

Figure 6.23 Measured insertion loss of silencers used with an internal combustion engine [24].

Mostly, in the case of dissipative silencers, the insertion loss is used to describe the acoustic performance. However, another magnitude often used to describe the sound reduction obtained by a dissipative silencer (or any other component in a duct system) is the *sound attenuation*. The units for attenuation are dB/m.

To appreciate the meaning of the attenuation let us assume that plane, harmonic sound waves propagate in the duct which is lined with sound–absorbent material. Because of the presence of the dissipative material in the silencer, the amplitude decays exponentially with distance x. Hence, in terms of particle displacement, the decay of the sound wave can be written as:

$$\tilde{\mathbf{d}} = \mathbf{A}e^{-mx}e^{i(\omega t - kx)},$$

where m is the sound attenuation (decay) coefficient and $\mathbf{A}e^{-mx}$ is the

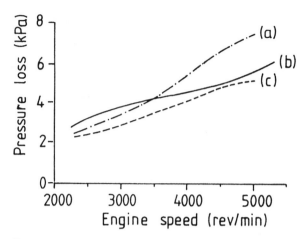

Figure 6.24 Pressure losses for the silencers schematically shown in Figure 6.23.

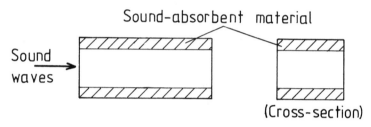

Figure 6.25 A duct lined with sound–absorbent material to form a simple dissipative silencer.

displacement amplitude.

Consequently, the time–averaged sound intensity magnitude changes with distance x as:

$$\overline{I} = \overline{I_0}e^{-2mx}, \tag{6.94}$$

where $\overline{I_0}$ denotes the time–averaged sound intensity at $x = 0$.

Assuming that the time–averaged sound intensities at the entrance and exit of the silencer are $\overline{I_1}$ and $\overline{I_2}$, respectively, we obtain for a silencer of length l_s:

$$\overline{I_2} = \overline{I_1}e^{-2ml_s}$$

and

$$10 \log \left(\frac{\overline{I_1}}{\overline{I_2}} \right) = 2 \times 4.343 \times m l_s = 8.69\, m l_s.$$

Consequently, the sound attenuation is defined as:

$$L_{\text{atten}} = 8.69 m \ \text{dB/m}. \tag{6.95}$$

The sound attenuation is the absolute value of the sound intensity level change per unit distance.

6.13.2 Sabine's empirical formula for sound attenuation

An empirical expression for the sound attenuation achieved by a dissipative silencer has been devised by H. J. Sabine [25]. His expression for the sound attenuation L_{atten}, expressed in units of dB/m, is

$$L_{\text{atten}} = 1.05 \frac{l_p}{S} \alpha^{1.4} \ \text{dB/m}. \tag{6.96}$$

In equation (6.96) S is the area (in m^2) of the cross–section of the interior of the duct, e.g., for the duct schematically shown in Figure 6.25 $S = 2hw$. The length l_p (in m) is that part of the perimeter of the cross–sectional area which is adjacent to sound–absorbent material, e.g., $l_p = 2w$ for the duct shown in Figure 6.25. The sound absorption coefficient α can be found from values listed in Table 5.1.

Equation (6.96) applies to both circular and rectangular ducts. If the duct is rectangular and lined only on two sides, separated by a distance $2h$, as shown in Figure 6.25, equation (6.96) becomes

$$L_{\text{atten}} = 1.05 \frac{\alpha^{1.4}}{h} \ \text{dB/m}. \tag{6.97}$$

The Sabine formula works best for low frequencies or more particularly when $2h/\lambda < 0.1$ (where λ is the wavelength).

An alternative empirical expression is the Piening formula [20,26]:

$$L_{\text{atten}} = 1.5 \frac{l_p}{S} \alpha \ \text{dB/m} \tag{6.98}$$

or,

$$L_{\text{atten}} = 1.5 \frac{\alpha}{h} \ \text{dB/m} \tag{6.99}$$

for the rectangular duct lined on two sides. Another, related expression
which is equivalent to (6.99) is said to be [26]:

$$L_{\text{atten}} = \frac{1.1}{h}(\alpha + \frac{\alpha^2}{2})\,\text{dB/m}. \tag{6.100}$$

Example 6.1

A rectangular duct is lined on both sides by mineral wool of
thickness 50 mm. The separation of the surfaces of the mineral
wool is 100 mm, i.e., $2h = 100$ mm. Use the empirical expressions
(6.97), (6.99) and (6.100) to determine the attenuation in dB/m
at 500 Hz.

Solution From Table 5.1 the value of the sound absorption coefficient of the mineral wool at 500 Hz is found to be 0.4.

$$\text{Expression}(6.97) : L_{\text{atten}} = \frac{1.05 \times 0.4^{1.4}}{0.05} = 5.8 \text{ dB/m}$$

$$\text{Expression}(6.99) : L_{\text{atten}} = \frac{1.5 \times 0.4}{0.05} = 12 \text{ dB/m}$$

$$\text{Expression}(6.100) : L_{\text{atten}} = \frac{1.1}{0.05}(0.4 + \frac{0.4^2}{2}) = 10.6 \text{ dB/m}.$$

6.13.3 Splitter silencers

The Sabine and the Piening formula give some guidance on the geometrical
design of dissipative silencers; for a rectangular cross–section the airway
width of $2h$ should be as small as possible. In addition to the condition
that $2h$ be small a sufficient cross–sectional area should be provided for any
air flow. These requirements can be achieved most readily by introducing
parallel splitters or baffle plates made from sound–absorbent material, as
shown in Figure 6.26.

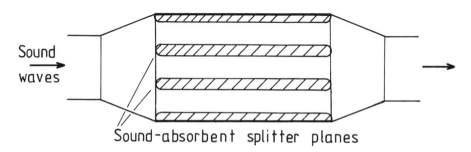

Figure 6.26 Splitter silencer.

The Sabine and Piening formulae are only useful at low frequencies. At high frequencies, when the sound absorption coefficient is unity or approaching unity, these formulae overestimate the sound attenuation value. In practice the attenuation that is achieved is less than predicted. The deficiency can be explained in terms of the tendency of sound at high frequencies to be highly directional so that sound is 'beamed' down the middle of the duct. The sound absorption coefficient α, used in expressions (6.96) to (6.100), is determined in a diffuse field, see Section 5.2. The condition for a diffuse field is clearly not achieved in the case of the highly absorbent duct at high frequencies. Consequently, silencers have been devised which attempt to eliminate the beaming effect by preventing a direct sightline through the silencer [27]. Such silencers maintain a passageway of constant width, but provide a circuitous route for the sound.

6.14 Measurement of insertion loss of dissipative silencers

An appreciation of the meaning of the silencer insertion loss can be gained by considering its measurement. The insertion loss is called the *static insertion loss*, if there is no air or gas flow through the silencer. The insertion loss may be measured by (a) the in–duct method or (b) the reverberation chamber method. The methods described in this section are in accordance with the British Standard [28] and are similar to the American Standard [29].

6.14.1 The in–duct method

The schematic arrangement for this method is shown in Figure 6.27. The silencer under test is placed in the middle of a straight duct. Near the inlet is placed a noise source which can be a loudspeaker or bank of loudspeakers operated in phase and fed with broad band random sound. The inlet duct, situated between the noise source and the silencer, should be either 4 m or $5D$, whichever is the greater, where D is the major dimension of the duct cross–section in the case of a rectangular duct, or the diameter, if the duct is circular. A similar duct (the outlet duct) is placed downstream of the silencer. The duct is terminated anechoically to ensure that no waves are reflected back to the silencer. The anechoic termination is achieved by increasing the cross–section of the duct exponentially and by placing sound–absorbent wedges at the end.

Measurements of the sound pressure are made with a microphone placed at four locations at a section half way along the outlet duct. The four measurement locations for both circular and rectangular ducts are indicated in Figure 6.27.

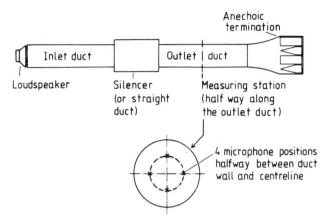

Figure 6.27 Measurement of the insertion loss of a silencer by the in–duct method.

Next, a straight duct is substituted for the silencer and the sound pressures measured at the same locations. Measurements are made with both the silencer–in and the silencer–out with the source operating in the octave bands which have centre frequencies from 63 Hz to 8 kHz. With the silencer in position the sound pressure level based upon the average of the four mean square pressures is $\langle L_1 \rangle$ dB. In the case of the silencer–out the averaged sound pressure level is $\langle L_2 \rangle$ dB. Thus the static insertion loss L_{IL} is given by:

$$L_{\text{IL}} = \langle L_2 \rangle - \langle L_1 \rangle \text{ dB}, \tag{6.101}$$

when the inlet and outlet ducts have the same cross–sectional areas. This insertion loss is equal to the sound power insertion loss, because the averaged mean square pressure at the measurement section is proportional to the time–averaged sound intensity which in turn is proportional to the sound power through a given duct cross–section.

6.14.2 The reverberation chamber method

Another method of measuring the insertion loss makes use of a reverberation chamber. The layout is shown in Figure 6.28. The sound pressure level, averaged in space, for the outlet duct section is determined by averaging several sound pressure levels measured in the space of the reverberation chamber. The averaged sound pressure levels $\langle L_1 \rangle$ and $\langle L_2 \rangle$ are obtained from measurements with the silencer–in and with the silencer–out, respectively. Consequently, the static insertion loss is given by equation (6.101). The positions for the measurements of the sound pressure levels are made according to the suggestions in Section 5.11. The reverberation chamber should have a volume of at least 180 m³.

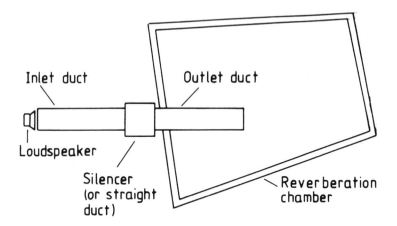

Figure 6.28 Measurement of the insertion loss of a silencer by the reverberation chamber method.

6.15 Effect of air flow

Most of the applications of dissipative silencers involve a flow of a gas which is most likely to be air. When a silencer is used with a fan which is providing air for ventilation purposes the air is generally clean. However, in many industrial applications the air is mixed with dust. Dust and air flow are factors which have to be taken into account in the design, construction and performance of silencers.

Steady flow past any obstruction, such as a silencer, causes aerodynamically generated sound which is propagated in both the upstream and downstream directions. Thus the silencer generates its own sound at the same time as it attenuates external sound. This *self–noise* or *flow-generated noise* might be expected to result in a reduced insertion loss, because of the addition of the self–noise to the noise which is measured downstream. In general this is the case when the flow is in the same direction as the propagation of the sound wave.

Procedures are available to measure the self–noise of silencers [30] and often manufacturers' catalogues quote the sound power levels of the self-noise. For example, in the case of a fan providing ventilation for a room through a silencer and ducting, the self–noise from the silencer and the noise from the fan are considered separately. In estimating the attenuation of the fan noise the *dynamic insertion loss* of the silencer should be used, if available. The dynamic insertion loss is defined, and measured, in a similar manner to the static insertion loss, except that there is an air flow present.

The tendency of the insertion loss of the silencer to be reduced by its self–noise is discussed by Ingard [31]. The values of the dynamic and static insertion losses are also different because of the convection of sound by the flow and the refraction of the sound in the flow. These effects depend upon whether the sound wave is travelling in the same direction as the flow (forward direction), or in the opposite direction to the flow (reverse direction); an intake silencer is an example of the latter situation. The *convection of the sound* by the air flow means that in the forward direction the sound wave is travelling faster in relation to the silencer than in still air. Hence the sound wave 'spends' less time in contact with the absorptive material of the lining. In the reverse direction the speed of the sound wave in relation to the silencer is smaller and there is more time for the sound to be absorbed by the lining. Thus the consequence of the convection effect is to increase the dynamic insertion loss value above the static value in the reverse direction and decrease the dynamic insertion loss value relative to the static in the forward direction. This effect is more pronounced in an air flow with higher velocity.

The *refraction of sound waves* or the change of their propagation directions is explained in terms of the transverse velocity profile of the flow. The velocity of the flow is a maximum at the centre of the silencer, zero at the casing and has some unknown profile within the sound–absorbent lining, as shown in Figure 6.29. In general the spatial gradient of the effective sound velocity, i.e., spatial gradient of the sound velocity in the system connected with the silencer, is responsible for the refraction of the sound waves. A similar phenomenon occurs in the air layer above the ground when the wind velocity changes with altitude. Thus in the silencer the sound wave travelling in the same direction as the flow is refracted towards the absorptive lining and hence the attenuation of the silencer is improved by the presence of the flow, see Figure 6.29a. In the case of the sound wave travelling in the reverse direction, as shown in Figure 6.29b, the sound is refracted towards the centre of the duct away from the lining and the value of the silencer attenuation is reduced. The refraction effect is more significant at high frequencies.

The explanations above of the convection and refraction effects are only qualitative descriptions of how the air flow affects the acoustic per-

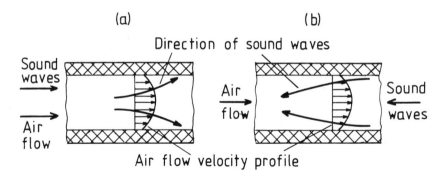

Figure 6.29 Refraction of a sound wave by flow; (a) forward direction and (b) reverse direction.

formance of silencers. The effects have been explained in more detail by Ingard [31]. Pridmore–Brown [32] has provided a mathematical theory to explain the refraction of sound in a duct.

The introduction of a silencer into a duct system increases the pressure loss. The loss of pressure occurs mostly at the entrance and exit of the silencer and can be reduced by rounded fairings at the nose and tail. An approximate, empirical expression for the static pressure loss ΔP for a silencer with an air flow is

$$\Delta P = 0.05 \frac{v^2}{2h} \text{ Pa}, \qquad (6.102)$$

where v is the mean flow velocity in m/s in the airway within the silencer and $2h$ is the airway width in m. As the losses are mostly occurring at the entrance and exit, the losses do not depend so much on the silencer length. Equation (6.102) has been deduced from manufacturers' catalogues. It provides the order of magnitude of the pressure loss. A particular manufacturer's catalogue should be consulted for more accurate data.

As a consequence of the air flow erosion of the sound–absorbing material in the silencer can take place. For flow velocities in the airway above about 5 m/s some protection is required. At low velocities it may be adequate to retain the bulk of the lining material in position by wire mesh. It is even more desirable to give protection by a perforated steel plate about 0.8 mm thick. The amount of open area of the perforated plate is typically about 50%. (See Figure 4.7).

In such applications where it is particularly important that the air should be clean and without fibres of the sound absorbing material, the linings can be covered with a thin plastic sheet such as Melinex. The plastic sheet is applied either in place of the perforated plate or used in addition to the plate. It is important that the sheet is limp and not taut. The sound attenuation at high frequencies is reduced by the presence of the plastic sheet.

In industrial applications where dust is present the plastic sheet can be used also for protection. It is generally best to have a specially designed and constructed silencer so that the splitters and liners can be inspected and removed. If it is possible to remove the splitters, they can be cleaned periodically and in such a case there is no necessity for a plastic sheet. In all events dust should be prevented from clogging up the silencer and rendering it worthless.

6.16 Construction of dissipative silencers

The construction of a typical dissipative silencer with a splitter is shown in Figure 6.30. The acoustic infill is covered by a fireproof cloth and perforated metal. The rounded entry and the evasé exit are designed to minimise pressure losses. The casing is generally of welded steel construction with mounting flanges and stiffeners.

6.17 The acoustic characteristics of typical dissipative silencers

As an example of the characteristics of a dissipative silencer of the type used to silence ventilating fans, performance data of silencers manufactured by the Industrial Acoustics Company, New York, are presented. The results for two types of silencer are illustrated in Figure 6.31. Both silencers have an overall cross–section or face area equivalent to 12 in by 12 in (305 mm × 305 mm). The dynamic insertion losses of both silencers are presented in Figures 6.31a and 6.31b in octave bands for different 'face velocities'. Face velocity refers to the velocity of the air flow at the face of the silencer where the cross–section is 305 mm × 305 mm. The velocity of the air flow within the silencer (the air channel velocity) is greater than the velocity at the silencer face. Results are shown in Figure 6.31a for two silencers, 3S and 7S, of lengths 3 ft (914 mm) and 7 ft (2130 mm) with a 'standard' pressure loss. The air channel width $2h$ is 90 mm. In Figure 6.31b are shown the results for silencers 3L and 7L of the same lengths, but with a lower pressure loss. The lower pressure loss is achieved by having a larger air channel width ($2h = 180$ mm); the acoustic performance is consequently less.

Several conclusions can be drawn in connection with the acoustic performance of the silencers. Firstly, it is only at low frequencies (250 Hz

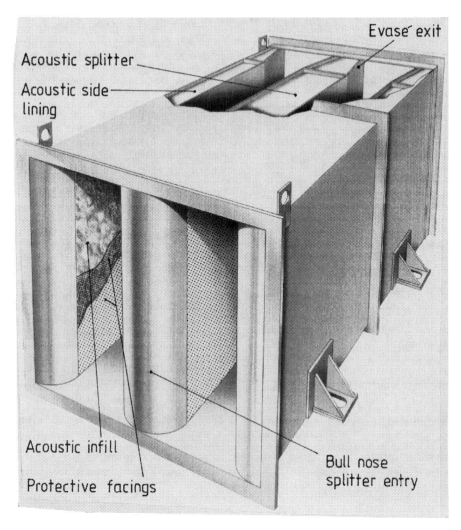

Acoustic splitter

Acoustic side
lining

Evasé exit

Acoustic infill

Protective facings

Bull nose
splitter entry

Figure 6.30 Construction of a dissipative silencer. (By courtesy of AAF–Ltd, a SnyderGeneral Company)

and less) that the dynamic insertion loss is approximately proportional to the length of the silencer; at high frequencies the longer silencers fail to give the expected benefit. Secondly, at frequencies below 1000 Hz the dynamic insertion loss is greater with the air flow and sound waves travelling in the opposite directions; above that frequency the opposite trend is discernable. Thirdly, the maximum acoustic performance occurs at 1000 Hz, apart from

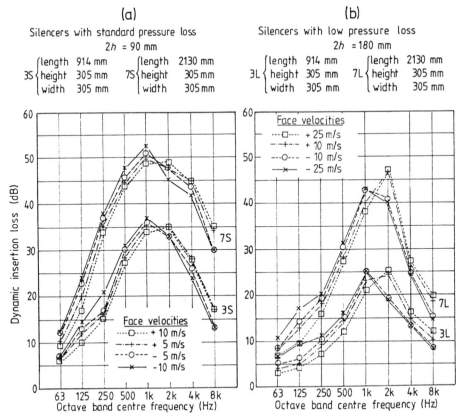

Figure 6.31. Dynamic insertion loss of dissipative silencers for different face velocities(+ in the same direction as the sound waves, − in the opposite direction to the sound waves): (a) standard pressure loss, lengths 3 ft and 7 ft, (b) low pressure loss, lengths 3 ft and 7 ft. (Reproduced with permission of Industrial Acoustics Company Ltd.)

the case of the low pressure loss silencer with the sound waves and the air flow moving in the same direction, see Figure 6.31b.

The sound power levels of the self–noise of the same silencers are shown in Figures 6.32a and 6.32b. The sound power levels in the octave bands do not vary greatly with the octave band centre frequencies.

Finally, the pressure losses for the silencers are shown in Figure 6.33, plotted against face velocity. The pressure loss increases slightly with length of the silencer, but it is the airway width $2h$ which is the most important geometrical factor.

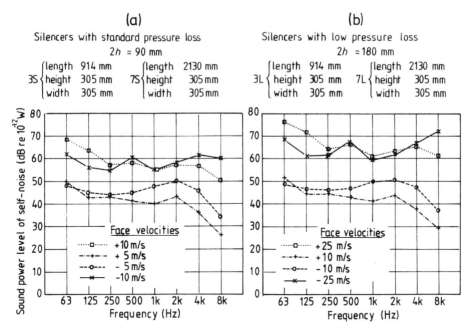

Figure 6.32. Self–noise of dissipative silencers: (a) standard pressure loss, (b) low pressure loss. (Reproduced with permission of Industrial Acoustics Company Ltd.)

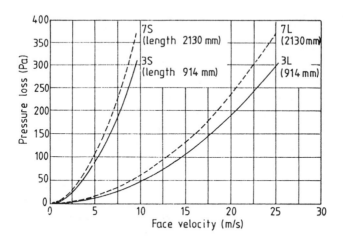

Figure 6.33 Static pressure loss of dissipative silencers. (Reproduced with permission of Industrial Acoustics Company Ltd.)

6.18 Closure

The acoustic design of reactive silencers is well understood; transmission loss and insertion loss can be predicted for most shapes [23]. The transfer matrix technique is particularly suitable for the prediction of the acoustic performance [19]. However, the most important application of reactive silencers is with internal combustion engines and in this case the plane wave acoustic theory does not agree well with measured results [33]. The main reasons for the failure are the presence of a mean flow in the silencer and the fact that the waves from an internal combustion engine are of finite amplitude. Better agreement between theory and measurement can be obtained, if these aspects are taken into account [34].

Dissipative silencers are extensively used in ducts where a fan is the source of noise. In particular they are used in ducts supplying conditioned air or ventilating air to rooms; in these applications it is invariably the case that the silencer is required to give the maximum attenuation in the octave band centred around 125 Hz (or possibly 250 Hz). At such low frequencies dissipative silencers are only moderately effective. Different principles can be used to achieve attenuation at these frequencies. Low frequency attenuation may be attained by the use of membrane absorbers. Silencers using membrane absorbers are lined with thin metal foil. Sound waves excite the vibration of the foil and the conversion of sound to the energy of the foil vibrations represents a reduction in the sound energy available for transmission downstream of the silencer. Usually the metal foil covers a slab of sound absorbing material in the same way as perforated metal covers the infill in a conventional dissipative silencer. It is possible to achieve an insertion loss of 40 dB by combining a dissipative silencer with a silencer based on membrane absorbers. Membrane absorbers have also been combined with Helmholtz resonators [35].

Particularly difficult silencing problems at low frequencies have been overcome by using active silencers [36]. In active silencers loudspeakers create such a sound field in the duct that cancellation is achieved. There are some problems with active cancellation: the loudspeakers have to be robust, especially when used out of doors; if a fault develops the sound emitted can be increased. Nonetheless, the techniques of active silencing are being continually improved and are likely to be extensively used in the future.

The acoustic testing of silencers is an area of some controversy. In the past national standards have differed to some extent [28,29]. There is an increasing tendency to use in air conditioning systems air flows at high velocities, say, above 10 m/s. Thus it is important to know the values of the dynamic insertion loss and the sound power level of the self–noise. Not all standards require these measurements. An international standard is available as a discussion document [37]. With good data on the silencers obtained from the correct tests and with a good technique for predicting the attenuation required [38,39] the most suitable silencer can be selected.

References

[1] S. Temkin (1981). *Elements of Acoustics.* John Wiley, New York.

[2] L. E. Kinsler and A. R. Frey (1962). *Fundamentals of Acoustics,* second edition. John Wiley, New York.

[3] L. E. Kinsler, A. R. Frey, A. B. Coppens and J. V. Sanders (1982). *Fundamentals of Acoustics,* third edition. John Wiley, New York.

[4] A. D. Pierce (1989). *Acoustics, an introduction to its physical principles and applications.* Acoustical Society of America, New York.

[5] P. M. Morse (1948). *Vibration and Sound.* McGraw–Hill, New York.

[6] R. B. Lindsay (1960). *Mechanical Radiation.* McGraw–Hill, New York.

[7] U. Ingard (1953). On the theory and design of acoustic resonators. *Journal of the Acoustical Society of America,* **25** (6), pp 1037–1061.

[8] Y. Rocard (1960). *General Dynamics of Vibration.* Crosby Lockwood, London.

[9] J. S. Anderson and M. Bratos–Anderson (1987). *Solving Problems in Vibrations.* Longman Scientific, Harlow.

[10] M. Alster (1972). Improved calculation of resonant frequencies of Helmholtz resonators. *Journal of Sound and Vibration,* **24** (1), pp 63–85.

[11] L. L. Beranek (1954). *Acoustics.* McGraw–Hill, New York.

[12] W. Neise and G. H. Koopman (1980). Reduction of centrifugal fan noise by use of resonators. *Journal of Sound and Vibration,* **73** (2), pp 297–308.

[13] G. H. Koopman and W. Neise (1982). The use of resonators to silence centrifugal blowers. *Journal of Sound and Vibration,* **82** (1), pp 17–27.

[14] R. Taylor (1972). Silencers for reciprocating compressors. *Proceedings of the British Acoustical Society,* paper 72AE5.

[15] M. C. Junger (1975). Helmholtz resonators in load–bearing walls. *Noise Control Engineering,* 4 (1), pp 17–25.

[16] P. H. Parkin and K. Morgan (1965). "Assisted resonance" in the Royal Festival Hall, London. *Journal of Sound and Vibration,* **2** (1),

pp 74–85.

[17] P. H. Parkin and K. Morgan (1970). "Assisted resonance" in the Royal Festival Hall, London: 1965–1969. *Journal of the Acoustical Society of America*, **48** (5)(part 1), pp 1025–1035.

[18] P. H. Parkin, W. A. Allen, H. J. Purkis and W. E. Scholes (1953). The acoustics of the Royal Festival Hall, London. *Acustica*, **3**, 1–21.

[19] J. Igarashi, M. Toyama, T. Miwa and M. Arai (1955, 59, 60). Fundamentals of Acoustical Silencers, parts 1, 2 and 3. Aeronautical Research Institute, University of Tokyo, report nos 339, 344 and 351.

[20] M. L. Munjal (1987). *Acoustics of Ducts and Mufflers*. John Wiley, New York.

[21] T. C. Lai (1971). Theoretical and experimental attenuation characteristics of acoustic filters. The City University, Department of Mechanical Engineering, Memorandum ML 30.

[22] H. H. Skilling (1965). *Electrical Engineering Circuits*, second edition. John Wiley, New York.

[23] D. D. Davis, G. M. Stokes, D. Moore and G. L. Stevens (1954). Theoretical and experimental investigation of mufflers with comments on engine–exhaust muffler design. NACA report 1192.

[24] C. D. Haynes and R. L. Kell (1964, 65). Engine exhaust silencing. Motor Industry Research Association, reports 1964/3 and 1965/12.

[25] H. J. Sabine (1940). The absorption of noise in ventilating ducts. *Journal of Acoustical Society of America*, **12**, pp 53–57.

[26] H. Schmidt (1976) *Schalltechnisches Taschenbuch*. VDI–Verlag.

[27] T. F. W. Embleton (1971). Mufflers, in *Noise and Vibration Control* (ed. L. L. Beranek). McGraw–Hill, New York.

[28] British Standards Institution (1971) Methods of test for silencers for air distribution systems. BS 4718:1971.

[29] American Society of Testing of Materials, ASTM–E477–80.

[30] U. Ingard, A. Oppenheim and M. Hirschorn (1988). Noise generation in ducts. *American Society of Heating, Refrigeration and Air Conditioning Engineers Transactions*, **74** (1).

[31] U. Ingard (1959). Influence of fluid motion past a plane boundary on sound reflection, absorption and transmission. *Journal of the Acoustical Society of America*, **31** (7) pp 1035–1036.

[32] D. C. Pridmore–Brown (1958). Sound propagation in flow through an attenuating duct. *Journal of Fluid Mechanics*, **4**, pp 393–406.

[33] A. D. Jones (1984). Modelling the exhaust noise radiated from reciprocating internal combustion engines — a literature review. *Noise Control Engineering Journal*, **23** (1), pp 12–31.

[34] R. J. Alfredson and P. O. A. L. Davies (1971). Performance of exhaust silencers components. *Journal of Sound and Vibration*, **15** (2), pp 175–196.

[35] U. Ackermann and H. V. Fuchs (1989). Noise reduction in an exhaust stack of a papermill. *Noise Control Engineering Journal*, **33** (2), pp 57–60.

[36] G. E. Warnacka (1982). Active attenuation of noise — the state of the art. *Noise Control Engineering*, **18** (3), pp 100–110.

[37] International Organization for Standardization (1986). Acoustics — Measurement procedures for ducted silencers. ISO/DIS 7235.

[38] M. A. Iqbal, T. K. Willson and R. J. Thomas (1977). *The Control of Noise in Ventilation Systems, a designers' guide*. E. and F. N. Spon, London.

[39] American Society of Heating, Refrigerating and Air Conditioning Engineers (1987). *ASHRAE Handbook, Heating, Ventilating and Air Conditioning Systems and Applications*; Chapter 52, Sound and vibration control. ASHRAE, Atlanta.

7 The Ear and Hearing Loss

Are you going to come quietly or do we have to use ear plugs?
Spike Milligan.

7.1 Introduction

The ear is a very delicate measuring and detecting organ. At the threshold of hearing of a normal person the displacement of the ear drum is less than 5×10^{-12}m at 3000 Hz, i.e., less than the diameter of an atom of hydrogen [1]. A schematic diagram of the ear is shown in Figure 7.1. The ear is divided into three parts; the external, middle and inner ear.

7.1.1 The external ear

The external or outer ear comprises the visible ear (pinna) and the auditory canal (meatus). The *auditory canal* is about 25 mm long. If the canal is regarded as a duct open at the outside and closed at the ear drum, the fundamental resonant frequency of the canal is about 3500 Hz. At frequencies near to this resonance the ear is most sensitive and most susceptible to damage. The purpose of the *pinna* is to match the impedance of the ear drum to the air. The matching is good at 800 Hz, but poorer at low frequencies. In general the external ear plays a role in determining the sensitivity of the ear as a function of frequency.

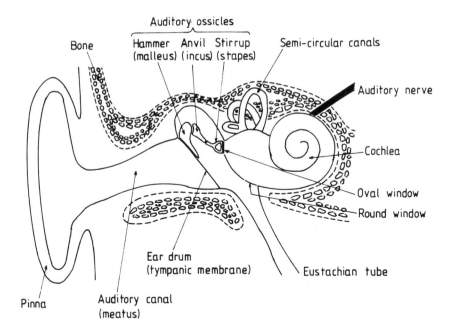

Figure 7.1 Schematic diagram of the ear.

7.1.2 The middle ear

The middle ear is made up of the *ear drum* (tympanic membrane), ossicles and oval window. The *ossicles* are a system of connected bones which act like levers. They provide no appreciable mechanical advantage, but ensure that maximum power is transmitted by matching the impedance at the ear drum on the meatus side with the impedance of the *oval window* as seen from the inner ear. Because of their characteristic shapes the ossicles are called the hammer, anvil and stirrup (or malleus, incus and stapes). The Eustachian tube connects the middle ear to the throat and allows the pressure on either side of the ear drum to be equalised.

Damage to the middle ear mostly affects hearing at low frequencies and is called *conductive deafness*. With very intensive sounds the mode of vibration of the stapes alters and instead of pivoting about the lower right edge of the stirrup (Figure 7.1) it rocks about the main axis of the footplate. In this way the oval window is protected from excessive displacements [2].

7.1.3 The inner ear

The cochlea and the semi–circular canals form the main part of the inner ear. The *semi-circular canals* are associated with our sense of balance and do not directly affect the acoustics of the ear. The *cochlea* is shaped like the shell of a snail.

It is about 8 mm in diameter and when unwound is about 30 mm long. In Figure 7.1 it is shown very much enlarged relative to the other organs, for the purpose of clarity. The cochlea detects, analyses and transmits information to the brain.

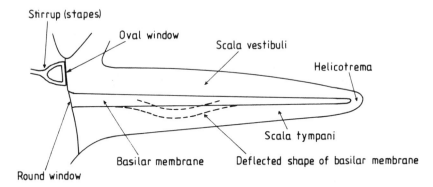

Figure 7.2 Schematic diagram of the unwound cochlea.

A schematic diagram of the cochlea unwound is shown in Figure 7.2. The cochlea is filled with fluid. Pulsations in the fluid, which originated from the oscillation of the oval window, are transmitted along the scala vestibuli and through the scala tympani to the *round window* which is on the boundary with the middle ear. The *basilar membrane* is part of a partition which separates the scala vestibuli from the scala tympani, apart from the connection at the helicotrema. The *organ of Corti* is mounted on the surface of the basilar membrane.

The sensory or *hair cells* are contained in this organ. The pulsations in the fluid within the cochlea cause the basilar membrane to flex rather like a beam in bending. The movement of the membrane stimulates the hair cells which are connected to the brain by the auditory nerve. The hair cells of the inner ear are damaged by continuous exposure to intense noise. The damage cannot be repaired.

The bending waves that travel along the basilar membrane were investigated by von Békésy [3]. He found that for a stimulus of a given frequency the transverse displacement of the membrane depended upon the distance from the oval window. At the high frequencies the maximum amplitudes occur close to the oval window and at the lower frequencies the maxima are closer to the helicotrema, as shown in Figure 7.3. At a given distance the variation of amplitude with frequency has a shape similar to that of a crude bandpass filter. The bandwidth is close to that of a one third octave filter and for this reason, amongst others, one third octave filters have been favoured for acoustic measurements. From experiments on guinea pigs Evans [4] has shown that the nerve fibres which innervate the hair cells are capable of narrower band filtering. Recent work suggests that the basilar membrane and the nerve fibres in the organ of Corti work together in the process of frequency selectivity [5].

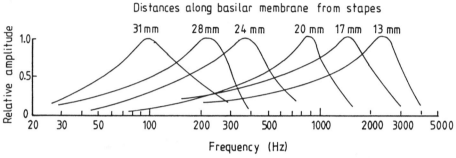

Figure 7.3 Amplitudes of displacement of the basilar membrane as a function of frequency (after von Békésy [3]).

7.2 Types of deafness

Hearing loss may be classified as either conductive or neural deafness. The hearing loss of people is measured with an audiometer (see Section 7.6). The loss of hearing is represented by a *hearing level* at a given frequency. The hearing level, as measured by an audiometer, is relative to a *reference equivalent threshold sound pressure level* (RETSPL). This reference threshold level is similar to, but not exactly the same as, the minimum audible field (MAF) shown in Figure 3.1. A person with good hearing has a zero hearing level, whilst people with a hearing loss have a *hearing threshold level* which is shifted to the higher intensity levels.

7.2.1 Conductive deafness

Conductive deafness arises because of defects in those parts of the external and middle ear which conduct the sound waves to the inner ear. It can occur because of a thickening of the ear drum or a stiffening of the joints of the ossicles. Conductive deafness occurs over the entire frequency range, but particularly at the low and mid frequencies. If the damage is caused by illness or short duration intense sound, it can often be repaired by surgery.

Blockage of the ear drum by wax can cause a 30 dB hearing loss over the entire frequency range.

7.2.2 Neural deafness

Nerve or neural deafness is the type of deafness caused by damage to the hair cells or nerve fibres of the inner ear. Such deafness is irreversible and has no medical remedy. Exposure to high intensity sound for long periods causes deafness of this kind.

Someone who is exposed to intense sound for a short period may suffer a temporary loss of hearing, i.e., a *noise–induced temporary threshold shift* (NITTS), and during this short period the hearing level temporarily increases. His or her hearing will gradually return to normal. However, if the person is subjected continually to further intense noise before regaining the normal threshold the person is likely to suffer eventually a *noise–induced permanent threshold shift* (NIPTS).

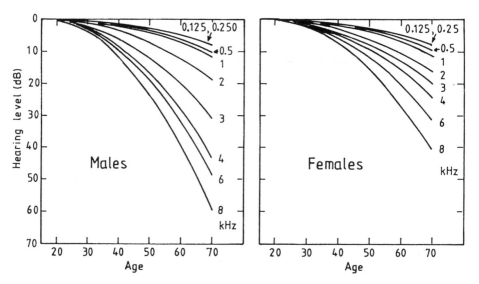

Figure 7.4 Hearing loss with age [6].

Nerve deafness also occurs naturally with ageing, a process known as *presbyacusis* or *presbycusis*. The expected hearing loss that can be expected with age for 50% of an otologically normal group of males and females is shown in Figure 7.4 [6]. The term otologically normal implies in this case that the hearing loss is due only to the ageing process and not to any other factors, such as illness and exposure to high intensity noise.

7.3 Hearing damage criteria

The onset of noise–induced hearing loss, suffered by a worker in a factory, depends upon the intensity of sound and the duration of exposure. Hearing damage criteria normally take the form of a sound level limit which should not be exceeded during a working day of 8 hours. More exactly any criterion has to take account of the variation of the sound level during the working day. As we have seen from Sections 2.8 and 3.7, the equivalent continuous sound level (L_{eq}) is a scale which takes into account both intensity and time, and, as a consequence, is the preferred scale for defining hearing damage criteria.

In a criterion for hearing damage the relationship between sound intensity and time must be decided. There are different views on this relationship. In Europe what is termed the *equal energy principle* is adopted. For the far field the definite integral over time of the square of the A–weighted sound pressure defined as,

$$\mathcal{E} = \int\limits_{0}^{T_0} [p_A(t)]^2 dt, \tag{7.1}$$

is proportional to the product of the A–weighted time–averaged sound energy density and the integration time T_0 and generally has units of Pa^2h. The integration time T_0 is specified normally as 8 hours for a working day. In other words the integral (7.1) is a measure of the sound energy recorded during the working day. For a hearing damage criterion based on the equal energy principle the value of the integral $\mathcal{E}$ should not exceed a constant value, say $\mathcal{E}_0$. It follows from the equal energy principle that the maximal A–weighted sound energy to which workers may be exposed (without hearing damage) during the working day is constant and defined by regulations. The value of the constant $\mathcal{E}_0$ is decided by the level of the A–weighted equivalent continuous sound level based on an 8 hour working day ($L_{eq(8h)}$). For a hearing damage criterion based on an $L_{eq(8h)}$ of 90 dB(A) the constant $\mathcal{E}_0$ is

$$\begin{aligned} \mathcal{E}_0 &= 10^9 \times (2 \times 10^{-5})^2 \times 8 \\ &= 3.2\,Pa^2h. \end{aligned} \tag{7.2}$$

What are the implications of the equal energy principle? The hearing damage criterion discussed so far implies that, if a worker is subjected to a noise with an A–weighted equivalent continuous sound level less than

90 dB(A) over an 8 hour working day, throughout his working life, he is unlikely to suffer noise–induced hearing damage. Of course, there is a risk of damage at 90 dB(A), but any hearing loss criterion has to be a compromise between what is desirable and what is practicable in an economic and technological sense. Suppose the level of noise the worker hears is constant at 93 dB(A) for four hours and the rest of the time is silence. The mean–squared pressure has doubled and the integration time halved. The integral in equation (7.1) becomes

$$\mathcal{E} = 2 \times 10^9 \times \left(2 \times 10^{-5}\right)^2 \times 4$$
$$= 3.2\,\mathrm{Pa^2 h}$$
$$= \mathcal{E}_0 .$$

Similarly, we could consider a constant level of 96 dB(A) for only 2 hours, with silence for the remainder of the 8 hour working day. The value of the integral in equation (7.1) will remain 3.2 Pa^{2}h. In these examples the acoustic energies are equal, even though the sound intensities and exposure times are different.

The equal energy principle is not universally adopted. In the United States of America another definition of $\mathcal{E}$ is used,

$$\mathcal{E} = \int_0^{T_0} [p_A(t)]^{1.2}\, dt. \tag{7.3}$$

The criterion is still based on an A–weighted equivalent continuous sound level of 90 dB(A) for an eight hour working day. From the criterion $\mathcal{E}_0$ can be determined. For example, if a noise exists for only 4 hours, a level of 95 dB(A) is allowed for the four hours.

The difference between the European approach and the American approach is summarised in Table 7.1. The equal energy principle is often referred to as the ISO method (after the International Organization for Standardization) [7] and the American approach as the OSHA method in recognition of its adoption as legislation in the USA, first in the form of the Walsh Healey Act (1969) and subsequently as the Occupational Health and Safety Administration regulations [8].

In the directive of the European Communities [9] noise control measures are expected to be taken when the A–weighted equivalent continuous sound level exceeds 90 dB(A) for the 8 hour day. Additionally, there is a requirement to be aware of the noise at a level of 85 dB(A). More details

TABLE 7.1 Comparison of permissible constant sound levels between ISO (equal energy principle) [7] and OSHA methods [8]

Duration per day hours	Sound level dB(A) ISO	OSHA
8	90	90
6	91	92
4	93	95
3	94	97
2	96	100
1	99	105
$\frac{1}{2}$	102	110
$\frac{1}{4}$	105	115

of the directive are given in the summary in Section 7.16. Since January 1st 1990 the directive has come into force in the Member States of the EC. In the United Kingdom compliance with the directive has taken the form of the Noise at Work regulations [10].

The ISO approach has the advantage of simplicity, relative to the OSHA method. The European method is based upon research into hearing loss, such as that of Burns and Robinson [11], which showed that it is the immission of acoustic energy to the listener's ears which is the factor causing hearing damage. The work of Burns and Robinson [11] is an important source. More recently the results of many researchers and the way in which their work has been incorporated into different standards has been reviewed by Robinson [12]. The American approach was based on a report by the National Academy of Sciences, National Research Council Committee on Hearing, Bioacoustics and Biomechanics (known as CHABA) [13]. CHABA also established hearing damage curves in terms of frequency, as well as intensity and exposure time. The 5 dB increase of sound level each time the exposure time is reduced by a factor of 2 (rather than the 3 dB on the equal energy principle) can be justified by the fact that the ear is able to recover during a rest period.

7.4 Definitions of equivalent continuous sound levels

The European Communities directive [9] introduced a definition of the A–weighted equivalent continuous sound level L_{Aeq, T_e} over a time T_e, which is the daily duration of a worker's exposure to noise, as

$$L_{\text{Aeq},T_e} = 10 \log \left\{ \frac{1}{T_e} \int_0^{T_e} \frac{[p_A(t)]^2}{(2 \times 10^{-5})^2} dt \right\} \text{ dB(A)}. \qquad (7.4)$$

If the equivalent continuous sound level is normalised to T_0, where T_0 is 8 hours, the length of the working day,

$$L_{\text{Aeq},T_0} = 10 \log \left\{ \frac{1}{T_0} \int_0^{T_e} \frac{[p_A(t)]^2}{(2 \times 10^{-5})^2} dt \right\} \text{ dB(A)} \qquad (7.5)$$

and is sometimes indicated by the symbol $L_{\text{eq (8h)}}$.

The European Communities directive [9] introduces the term *daily personal noise exposure* $L_{\text{EP,d}}$. Formally the daily personal noise exposure is defined as:

$$L_{\text{EP,d}} = L_{\text{Aeq},T_e} + 10 \log(T_e/T_0) \text{ dB(A)}. \qquad (7.6)$$

From equation (7.4) the expression for $L_{\text{EP,d}}$ becomes

$$L_{\text{EP,d}} = 10 \log \left\{ \frac{1}{T_0} \int_0^{T_e} \frac{[p_A(t)]^2}{(2 \times 10^{-5})^2} dt \right\} \text{ dB(A)}, \qquad (7.7)$$

which is the same as L_{Aeq,T_0}, the 8 hour equivalent continuous sound level of equation (7.5). Equation (7.7) is quoted in the British regulations [10].

A weekly average $L_{\text{EP,w}}$ of the daily values is found from [9]:

$$L_{\text{EP,w}} = 10 \log \left[\frac{1}{5} \sum_{k=1}^{m} 10^{0.1(L_{\text{EP,d}})_k} \right] \text{ dB(A)}, \qquad (7.8)$$

where $(L_{\text{EP,d}})_k$ are the values of $L_{\text{EP,d}}$ for each of the m working days taken into account during the week.

Example 7.1

A factory operator works for six days during the week. The daily personal noise exposures for the six days are 85, 88, 76, 87, 89 and 91 dB(A). What is the weekly average of the daily personal noise exposures?

Solution:

$$L_{\text{EP,w}} = 10 \log \left[\frac{1}{5} \left(10^{8.5} + 10^{8.8} + 10^{7.6} + 10^{8.7} + 10^{8.9} + 10^{9.1} \right) \right]$$
$$= 88.5 \, \text{dB(A)}.$$

During the eight hour working day the sound level will normally vary. At the completion of the eight hour period a certain amount of acoustic energy will have been received at the measurement point. The equivalent continuous sound level can be regarded as a notional sound level which is constant and which represents the same amount of sound energy associated with the varying sound levels.

7.5 Risk of hearing loss from exposure to noise

At this stage it is worth considering the justification for a regulation based upon a daily personal noise exposure of 90 dB(A). The Health and Safety Executive of the United Kingdom have estimated the likelihood of deafness in workers when exposed to different values of $L_{\text{eq (8h)}}$ during their working lives. Estimates of the percentage of people suffering a hearing handicap when exposed to different levels of noise have been made for *otologically normal* people at 65 years of age and for a typical industrial population at the same age [14,15]. A group is otologically normal if deafness in this case is caused only by noise in the workplace and by ageing and not by illness, use of firearms, etc. A typical sample of the population will have hearing impairment, because of causes other than factory noise and ageing. Hearing loss from other causes is of necessity difficult to estimate; however, estimates have been made for a typical population sample and they are shown in Figure 7.5 [14]. A worker exposed during his working life (from 18 to 65 years) to a daily equivalent continuous sound level of 90 dB(A) has 12% chance of developing a 50 dB hearing loss. Hearing loss, as introduced here, is the arithmetic average of the hearing levels at frequencies of 1, 2 and 3 kHz. A similar exposure leads to a 40% possibility of a 30 dB hearing loss. Someone with a 30 dB loss is likely to suffer some handicap in listening to conversational speech [16]. Thus even if the $L_{\text{eq (8h)}}$ is limited to 90 dB(A), there is still a likelihood of workers exposed to the noise becoming deaf. Note that for an otologically normal group of people the percentage in that group likely to suffer a given hearing loss is less than in a typical population sample.

The likelihood of deafness caused by exposure to noise is difficult to estimate. In Figure 7.5 are illustrated the results of British research.

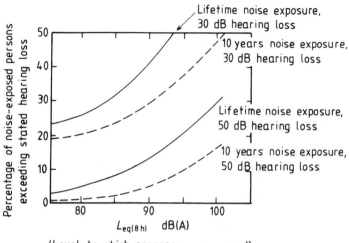

Figure 7.5 Hearing loss in a typical industrial population at 65 years of age after a lifetime's noise exposure and after 10 years noise exposure.

Other studies have given different predictions. For example, the first draft of the Council Directive of the European Communities estimated that the percentage of male workers, who will probably suffer from a definite handicap due solely to noise would be 12% and 28% when exposed to 90 dB(A) and 95 dB(A) $L_{eq\,(8h)}$, respectively, for 40 years [17]. Definite handicap in this case means an average of 30 dB hearing loss at 1, 2 and 3 kHz [17]. These estimates were based on studies by the World Health Organization.

The International Organization for Standardization has published different data [7].

7.5.1 Number of people exposed to noise

It has been estimated that in the European Economic Community a total of 20 to 30 million workers are subjected to a level of noise above 80 dB(A). Half of these are exposed to 85 dB(A) and 6 to 8 millions are exposed to levels of 90 dB(A) or more [17].

In Great Britain it is estimated that in all industries, including manufacturing, construction, agriculture, forestry, quarrying, shipping and transport, there are 1.7 million persons who may be exposed to levels above 85 dB(A) and 630,000 to levels above 90 dB(A) [14,15].

7.6 Audiometry

Hearing loss of people is measured by an audiometer. One of the most common types of audiometer is based on a design by von Békésy [3]. The requirements for pure tone audiometry are outlined in the standards [18,19]. A schematic diagram of an audiometer is shown in Figure 7.6.

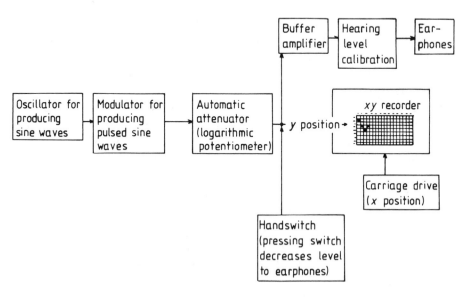

Figure 7.6 Schematic diagram of an audiometer.

This diagram is based on the Brüel and Kjaer audiometer type 1800, a screening audiometer which can be operated by the patient. Shown in the diagram the automatic attenuator consists of a logarithmic potentiometer which has its wiper attached to a pen so that the attenuation of the potentiometer corresponds to the y position of the pen. The patient wears a pair of headphones (earphones) to which are fed pure tones, either continuous or pulsed. Pulsed tones are better, because they cannot be confused with ringing tones that are heard by patients suffering from tinnitus. Initially a tone at 500 Hz is played to the left ear. The level of sound gradually increases. The patient is able to control the level by means of a handset. If he or she presses the button on the handset and keeps the button depressed the level of noise decreases. The level of sound increases when the button is released.

The aim of the operation is for the patient to detect the lowest level of the sound of the tone that he or she can hear. This will be the patient's hearing threshold level. If the hearing is perfect the *hearing threshold level* will coincide with the reference equivalent threshold sound pressure level (RETSPL), and his *hearing level* is registered as 0 dB. The procedure for determining the hearing threshold level is carried out at frequencies of 0.5, 1, 2, 3, 4, 6 and 8 kHz, first on the left and then on the right ear.

The reference equivalent threshold sound pressure level depends upon the earphones used and the way in which the audiometer is calibrated. At this stage some clarification should be given in relation to the different threshold levels that exist. The minimum audible field (MAF), shown in Figure 3.1, is a threshold level of young people with good hearing, obtained when the people are in a free field; the subjects hear pure tones from a loudspeaker in an anechoic chamber and the sound pressure level (SPL) is measured where the listener's head has been. If the same people heard the same sound through earphones, their threshold levels, as measured as a sound pressure level at the ear drum, will generally be greater than the MAF by 5 to 10 dB. The threshold obtained with headphones is referred to as the *minimum audible pressure* (MAP). The values of the MAP differ slightly for different methods of measuring the SPL at the ear. Similarly, the values of RETSPL will differ, because during calibration of the audiometer the values of SPL measured depend upon the way in which the earphone is coupled to an artificial ear, or similar device [19].

The audiometer outlined in Figure 7.6 automatically plots out the hearing level on a chart. Whilst the tone is played into the earphones a pen moves across the paper and traces out the patient's audiogram. Two typical audiograms are shown in Figure 7.7, (a) for a patient with a severe hearing impairment from exposure to noise and (b) for a young person with good hearing. For a patient with perfect hearing the audiogram is ideally a horizontal line through 0 dB. However, the method of operation of the audiometer results in a series of zigzags. The rule when using this type of audiometer is to press the button on the handset whenever the tone is heard through the earphones. Thus, as the sound increases in intensity, the patient eventually hears the tone; this point is represented by A in Figure 7.8. A short time afterwards the patient presses the button; this point represented by B. The difference in time between B and A depends upon the person's reaction time. With the button of the handset depressed the tone becomes fainter, and at C he can hear the tone no longer. A short time afterwards at D the button is released and the sound level increases. For a person with normal, or near normal, hearing the amplitude of the zigzags is a measure of the reaction time, the background noise and the patient's experience or skill in performing the test. The hearing level is

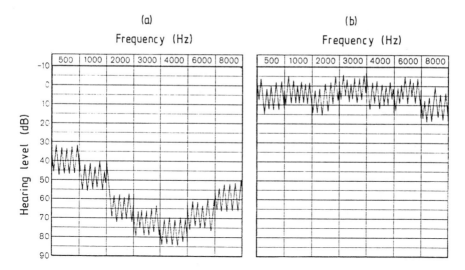

Figure 7.7 Typical audiogram of (a) patient with severe hearing impairment from exposure to noise and (b) a young person with good hearing.

obtained by taking a mean line through the zigzags. Thus in the case of Figure 7.7 for the person with a severe hearing impairment the hearing levels at 1, 2 and 3 kHz are 47, 65 and 75 dB, respectively.

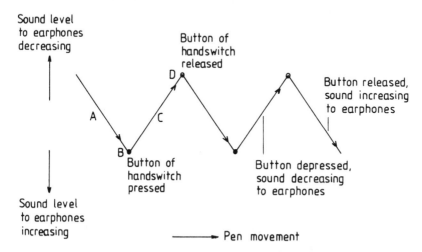

Figure 7.8 Change in the threshold level of an audiogram as the patient controls the button on the handset.

The hearing levels averaged over 1, 2 and 3 kHz is 62 dB for the case of the severe impairment. It is this average hearing loss which is used in Figure 7.5.

7.6.1 Significance of hearing loss

It is of value to put into the context of everyday life the serious hearing impairment characterised by audiogram (a) in Figure 7.7. In Figure 7.9 the curve marked ABC represents the average hearing threshold of young people with normal hearing. Someone with such hearing would have an audiogram similar to (b) in Figure 7.7. The hearing threshold level of the person with a hearing impairment is elevated, as indicated in Figure 7.9. The difference in sound pressure level between the elevated hearing threshold level and the curve ABC is the hearing level, as measured approximately by the audiometer. (Note that the curve ABC of Figure 7.9 is similar to the MAF and not to the RETSPL. The difference is not important for the purpose of this example.)

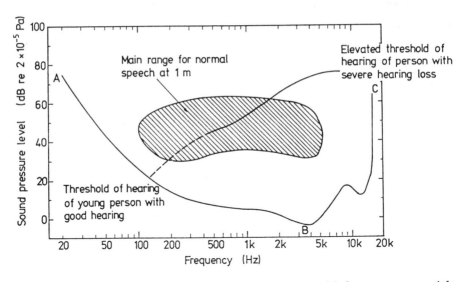

Figure 7.9 Illustration of shift of hearing threshold for a person with severe hearing loss.

Also shown in Figure 7.9 is a shaded area which represents the range of sound pressure levels and frequencies used in normal speech. Clearly some components that make up speech will not be heard by the person with the hearing impairment. Although only a part of the shaded area is cut off by the elevated threshold, the part that is eliminated is important because low intensity, high frequency speech is crucial for *speech intelligibility*. Although speech is made up of vowels and consonants, it is the consonants which are important for intelligibility. As the sound from consonants is in the mid to high frequency range and has a lower level than vowels, the person with the hearing impairment will fail to hear correctly many of the consonants and hence much of ordinary speech will be incomprehensible.

The difficulty in understanding consonants can be appreciated by studying the sound pressure against time of the word 'shift' as shown in Figure 7.10. The final consonant is of low level and will easily be lost to a person with impaired hearing. Also shown in Figure 7.10 are the spectra of the constituent phonemes of the word 'shift'. The power content of the phonemes 'f' and 't' is much greater at high frequencies than for the vowel 'i'.

7.6.2 Recruitment

Hearing aids are, of course, used to help people with impaired hearing. Their purpose is to amplify speech to levels above the hearing threshold level. It might be thought that suitable amplification will enable a deaf person to understand normal speech. Unfortunately, such an amplification will not necessarily solve the problem because of an effect, known as recruitment, whereby the quieter components of the speech tend to disappear completely.

7.6.3 Audiograms of noise–induced hearing loss

An audiogram similar to (a) of Figure 7.7 is shown as (c) in Figure 7.11. Let us assume that the person with this audiogram was exposed to noise with $L_{eq(8h)} = 90$ dB(A) for a working life of 40 years. Hypothetical audiograms for the same person during earlier stages of exposure are shown in Figure 7.11 as (a) and (b). Noise–induced hearing loss invariably occurs at 4 kHz, irrespective of the stage of the impairment. The frequency at which this noise–induced loss occurs is independent of the frequency of the noise to which the sufferer is exposed. The existence of the 4 kHz dip at earlier stages of exposure means that is possible to identify those who are susceptible to noise and likely to develop further hearing loss with continued exposure. Those so identified should be protected from further exposure to noise.

It is clear that audiometry is useful in diagnosing noise–induced hearing loss. For this reason the first draft of the European Communities directive [17] included the proposed regulation that employers had a duty to provide audiometric screening for all workers exposed to an eight hour L_{eq} of 85 dB(A) and above. However, the final directive adopted a regulation

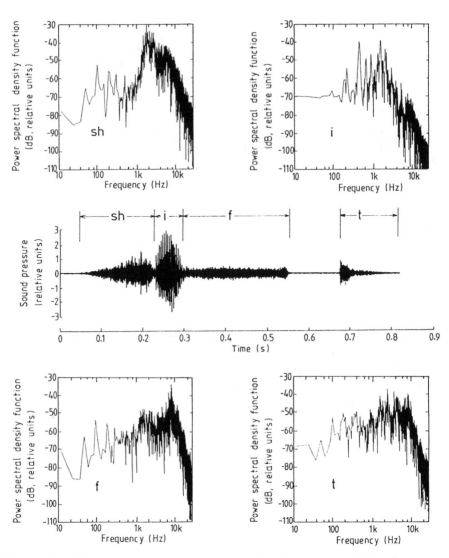

Figure 7.10 Sound pressure against time for the word 'shift' and spectra of constituent phonemes.

which stated that audiometry would be provided for an exposure above 90 dB(A), if requested by the employee [9]. This change was induced, at least in Great Britain, partly by the fear of the cost of providing audiometry to the estimated 1.7 million people exposed to levels of 85 dB(A) or more [20]. The British regulations, introduced in order to comply with

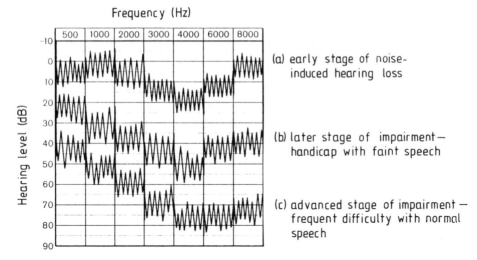

Figure 7.11 Audiograms illustrating the development of noise-induced deafness.

the EC directive, do not mention audiometry [10].

7.7 Measurement and estimation of L_{eq}

In order to estimate the equivalent continuous sound level a sound level meter with fast or slow time weighting or, preferably, an integrating sound level meter is required. If an integrating sound level meter is available it is possible to start the meter in the morning and after 8 hours the $L_{eq(8h)}$ (which will be the $L_{EP,d}$) can be read directly from the meter or digital output. This may not be a convenient method of measurement, and normally it will be sufficient to take sample measurements over a short period of time, such as 60 s. The $L_{EP,d}$ can be estimated from the sample measurements of equivalent continuous sound levels over 60 s (referred to as $L_{Aeq,60s}$ or $L_{eq(60s)}$).

Generally it is needed to know the $L_{eq(8h)}$ to which a particular worker is subjected. A worker who is a machine operator is likely to be in one place for most of the time; his noise will probably vary with time. In a process factory the noise may be mostly steady, but workers will move from place to place and in this case may be subjected to different levels of noise.

7.7.1 Measurement of $L_{EP,d}$ at a fixed position

For a worker who is normally close to his machine during the working

day sample measurements of 60 s duration of the A–weighted equivalent
continuous sound level, obtained with an integrating meter, at hourly
intervals may be sufficient. The integrating meter should be held at arm's
length so that the microphone is close to the operator's head. The Health
and Safety Executive suggest that the microphone should be kept at least
20 mm away from the operator, otherwise errors caused by reflections will
arise [15]. For a sampling time of several minutes the integrating sound
level meter should be set up on a tripod.

When an integrating sound level meter is not available, a simple sound
level meter of grade at least of type 2 can be used. The 'slow' response is
used and visual averaging of the needle on the meter is made.

Example 7.2

The working day in a factory is from 8.00 to 12.00 and 13.00 to
16.30. An integrating sound level meter is used to take samples of
the noise over a 60 s period at a position close to a worker at the
following times: 8.15, 9.15, 10.15, 11.15 and 12.00; 13.15, 14.15,
15.15 and 16.15. The corresponding values of the A–weighted
equivalent continuous sound level $L_{\text{Aeq,60s}}$ are 92, 94, 92, 90 and
82 dB(A); 89, 90, 90 and 87 dB(A). Estimate the daily personal
noise exposure $L_{\text{EP,d}}$ of the worker.

Solution: The calculations required for the estimate are tabulated in Table 7.2.

In the estimate each measurement of $L_{\text{Aeq,60s}}$ has to be regarded as typical of a certain period. The measurement at 8.15
may be regarded as an average value typical of a period t_s of one
hour from 8.00 to 9.00. Thus $L_{\text{Aeq,60s}} = L_{\text{Aeq},\,t_s}$. The measurement at 12 noon is discounted as most machines were switched off
in anticipation of the lunch break. (Or, perhaps, more measurements should have been made to establish how the noise varied
prior to the break.) The measurement at 16.15 is typical of the
final half hour of the working day of $7\frac{1}{2}$ hours.

The mean–squared pressure $\overline{p^2}$ in column 4 is obtained from
the expression,

$$\frac{\overline{p^2}}{(2 \times 10^{-5})^2} = 10^{0.1 L_{\text{Aeq,60s}}} \, \text{Pa}^2. \qquad (7.9)$$

The product $\overline{p^2} t_s$ is provided in column 5. The final column lists

TABLE 7.2 Estimation of $L_{EP,d}$ from sample measurements of $L_{Aeq,60s}$

Time of day	$L_{Aeq,60s}$ dB(A)	t_s (h)	$\overline{p^2}$ (Pa^2)	$\overline{p^2}t_s$ (Pa^{2}h)	f
8.15	92	1	0.632	0.632	0.20
9.15	94	1	1.004	1.004	0.31
10.15	92	1	0.632	0.632	0.20
11.15	90	1	1.004	1.004	0.125
12.00	82	-	-	-	-
13.15	89	1	0.316	0.316	0.10
14.15	90	1	0.400	0.400	0.125
15.15	90	1	0.400	0.400	0.125
16.15	87	1/2	0.200	0.100	0.03
				$\sum f =$	1.22

t_s	Period associated with measurement
$\overline{p^2}$	Mean–squared sound pressure
f	Fractional exposure

f the *fractional exposure* which is given by:

$$f = \frac{\overline{p^2}t_s}{\mathcal{E}_0},\qquad(7.10)$$

where $\mathcal{E}_0$ is 3.2 Pa^{2}h, see equation (7.2). The constant $\mathcal{E}_0$ is the reference sound exposure which is based on an equivalent continuous sound level of 90 dB(A) over an 8 hour period, see definition (7.14). The fractional exposure is the ratio of the acoustic energy received in time t_s to the acoustic energy from a steady noise of 90 dB(A) for 8 hours.

The fractional exposure may also be estimated from the expression,

$$f = \frac{t_s}{8}10^{0.1(L_{Aeq,t_s}-90)}.\qquad(7.11)$$

In the present example the total fractional exposure $\sum f$ is 1.22. The daily personal noise exposure is given, see (7.7) and (7.10), by:

$$L_{EP,d} = 10\log\sum f + 90 \qquad(7.12)$$
$$= 10\log 1.22 + 90$$
$$= 91 \text{ dB(A)}.$$

The daily personal noise exposure is 91 dB(A), to the nearest decibel.

7.7.2 Nomogram method of estimating $L_{EP,d}$

The above type of estimation may be performed more easily with the aid of the nomogram published in [15,21] and redrawn as Figure 7.12. The operation of the nomogram is best explained by an example.

Example 7.3

The sound level is 94 dB(A) for 3 hours, 92 dB(A) for 2 hours and 89 dB(A) for 3 hours. Estimate $L_{EP,d}$.

Solution: On the nomogram a line is drawn joining 94 dB(A) on the left hand scale and 3 hours on the right hand scale. This line intersects the central scale at $f = 1$. Similarly 92 dB(A) at 2 hours gives $f = 0.4$ and 89 dB(A) for 3 hours gives $f = 0.3$. Thus,

$$\sum f = 1 + 0.4 + 0.3 = 1.7.$$

The value of $L_{EP,d}$, corresponding to a fractional exposure of 1.7 is obtained from the central scale, and is 92.5 dB(A). (Note that the nomogram method is approximate.)

7.7.3 Estimation of $L_{EP,d}$ for steady noise and moving worker

In process plants the noise levels are usually constant, but the workers often move around. It is a good technique to measure the A–weighted sound levels at points on a grid in the plant hall, and from these measurements construct a drawing with contours of equal sound level. From information about the movements of a worker during the working day the way in which the sound level varies with time can be determined. The problem of establishing the worker's personal daily noise exposures becomes the same as that outlined in Sections 7.7.1 and 7.7.2.

7.8 Noise dose

The A–weighted equivalent continuous sound level measured over a period T_e has been defined in equation (7.4) which may also be written as

$$L_{Aeq,\,T_e} = 10\log\left(\frac{\mathcal{E}}{T_e(2 \times 10^{-5})^2}\right) \text{ dB(A)}, \qquad (7.13)$$

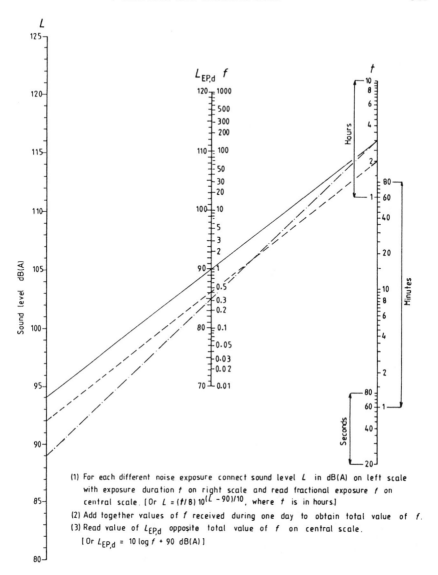

Figure 7.12 Nomogram method for obtaining $L_{EP,d}$.

where $\mathcal{E}$ is the *sound exposure*, and is given by:

$$\mathcal{E} = \int_{0}^{T_e} [p_A(t)]^2 \, dt, \tag{7.14}$$

which is the same expression as (7.1), provided that the exposure time is not longer than 8 hours. For a steady noise $\overline{p^2}$, the mean–squared A–weighted pressure of the sound, is constant and sound exposure linearly depends on exposure time,

$$\mathcal{E} = \overline{p^2} T_e.$$

For a worker subjected to a steady noise level of 90 dB(A) for 8 hours his sound exposure $\mathcal{E}_0$ is 3.2 Pa^{2}h (equation (7.2)).

A sound exposure of 3.2Pa^{2}h represents an upper limit; above this limit the daily personal noise exposure exceeds 90 dB(A), the main EC action level. A sound exposure of 3.2Pa^{2}h is equivalent to a noise dose of 1, a percentage noise dose of 100% or a fractional exposure of 1 (see equation (7.10)). The *noise dose* may be defined as

$$\text{noise dose} = \frac{\int_0^{T_e} [p_A(t)]^2 dt}{3.2} \qquad (7.15)$$

or,

$$\text{noise dose} = \mathcal{E}/\mathcal{E}_0, \qquad (7.16)$$

where $\mathcal{E}_0 = 3.2\,\text{Pa}^2\text{h}$ for the case when the hearing damage criterion is based on 90 dB(A).

$$\text{Percentage noise dose} = \frac{\mathcal{E}}{\mathcal{E}_0} \times 100\%. \qquad (7.17)$$

The concept of noise dose is utilised in an instrument called a *noise dosemeter* or *personal sound exposure meter* [22]. Dosemeters are small personal integrating meters which are intended to be carried by a worker. The microphone moves with the worker so that his personal noise dose is measured. Typically the dosemeter is placed in the worker's breast pocket and the microphone is clipped to his lapel or attached to ear muffs.

The definition of noise dose formulated by equation (7.15) applies only to a hearing loss criterion based on a daily personal noise exposure of 90 dB(A). For a level of 85 dB(A) the 3.2 in the denominator would be replaced by 1.01. It should also be noted that the concept of noise dose, as outlined here, assumes the equal energy principle of hearing damage which is used by the European Communities and the International Organization for Standardization (ISO). A noise dosemeter based on the principle adopted in the USA by the Occupational Safety and Health Act (OSHA) would be different.

It is interesting to consider how the noise dose varies with time in the case illustrated in Example 7.2 in Section 7.7.1 (Table 7.2). The worker wears the noise dosemeter with the microphone on his lapel from 8.00 to 16.30. If it is assumed that the sound level is constant during each hourly period, the noise dose varies during the working day as shown in Figure 7.13. The noise dose increases continuously throughout the day, apart from the lunch hour, held in quiet conditions, when the noise dose remains the same. The final percentage noise dose is 122%. Note that the cumulative fractional exposure was 1.22.

The personal daily noise exposure $L_{EP,d}$ may be found from the expression, see (7.12),

$$L_{EP,d} = 10 \log 1.22 + 90$$

and is 91 dB(A).

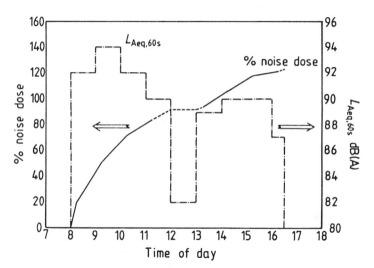

Figure 7.13 Comparison of noise dose and $L_{Aeq,60s}$ for data of Example 7.2.

7.9 Hearing loss from single noise events

In many factories there is considerable noise from forges and presses. Although the impacts are generally repetitive, the noise from each impact can often be identified. Such noise sources can produce single noise events. Provided the peak sound pressure is below a certain level, it is assumed that hearing damage caused by impact noise is the same as that caused by steady noise, and there is no basic difference in the technique for working out the daily personal noise exposure.

With very intense single noise events there is a danger of structural damage to the ear. Normally such damage will not be caused by impacts from forges and presses, but could be caused by cartridge–operated tools or by weapons. For example, permanent damage was incurred by a civilian who fired a Carl Gustav rocket launcher from his shoulder on an experimental test range. In the subsequent court case, in which the operator claimed (and won) damages, the noise level was quoted as '170 dB'. Workers clearly need protection from intense single event noise and the European Communities directive and national regulations [9,10] put an overriding limit of 200 Pa for peak sound pressure; no instantaneous sound pressure in excess of 200 Pa should ever be allowed to reach the unprotected ear. (Note that there may be a temptation to refer to an instantaneous sound pressure of 200 Pa as a sound pressure level of 140 dB. However, this is strictly incorrect, as according to the definition of SPL, equation (1.232), we require the rms value of the sound pressure in the numerator of the argument of the logarithm and not the instantaneous value of the sound pressure.)

A method of measuring the peak, instantaneous sound pressure has been described in Section 2.10. Some sound level meters have the capacity to measure peak pressure.

7.10 Personal ear protectors

It is stated in the directive of the European Communities that, when the daily personal noise exposure of a worker exceeds 85 dB(A), personal ear protectors (hearing protectors) should be made available to workers. When the levels exceed 90 dB(A) hearing protectors must be used. The employer has to supply a range of ear protectors which are suitable for the worker and the working conditions.

Various types are available and they have to be assessed according to:

- comfort
- user's acceptability
- hygiene
- durability
- resistance to chemicals
- cost
- availability.

Ear protectors may be classified as ear plugs (or insert–type protectors) or ear muffs.

7.10.1 Ear plugs

Ear plugs typically provide attenuations of 10 to 15 dB and are appropriate for noise levels in the region of 85 to 100 dB(A).

Ear plugs are sub–divided into prefabricated, disposable and individually moulded types [23].

7.10.2 Prefabricated ear plugs

The prefabricated ear plugs have a preformed shape which fits into the ear canal, as shown in Figure 7.14. An ear plug known as the V–51R is illustrated. It is generally made of soft rubber and has a single flange on the inside. As it is supplied in a range of five sizes, it is possible to select one to fit tightly into most ear canals. These ear plugs are not so comfortable and there is a tendency for wearers to choose, for reasons of comfort, a size which is too small. As the ear canal tends to enlarge when ear plugs are used, it is better from the point of view of sound attenuation to choose the larger size when the size below is too loose. A loss of attenuation of 10 dB can be incurred by poor fitting [23].

Prefabricated ear plugs can be washed easily and reused.

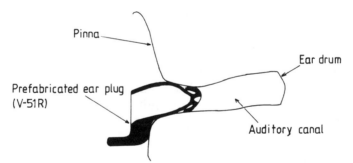

Figure 7.14 Prefabricated ear plug inserted into the ear.

7.10.3 Disposable ear plugs

A commonly used disposable ear plug is made from glass wool, a form of glass fibre with fibre diameter of the order of one micron. The glass wool is formed into a cone by the user and forced into the ear. At the end of the working period the ear plugs are removed and thrown away.

Another type of disposable ear plug takes the form of a short cylinder of foam which is coated with a special chemical. When forced into the ear it expands slowly as it tries to take up its original shape, thus fitting tightly into the ear canal.

Disposable ear plugs can be quite comfortable. The main problem is hygiene; if they are removed and reinserted, there is a danger of dirt in the ear. They are unsuitable for areas with intermittently high noise levels, where workers will wish to take off their hearing protectors from time to time.

7.10.4 Individually moulded ear plugs

Individually moulded or custom–moulded ear plugs are generally made from silicone rubber. The putty–like substance is mixed with a curing agent and pushed into the ear. In a few minutes the rubber sets to the shape of the ear canal. The ear plug can then be removed and is available for reuse. It may be fitted with a short handle for easier insertion and removal.

The custom–moulded ear plug combines the advantages of the prefabricated with the disposable. They can be washed easily and they are comfortable. This type of plug is also more acceptable to the wearer, because it is personalised.

7.10.5 Ear muffs

Ear muffs are formed by two rigid cups which are connected together by a spring–loaded adjustable headband. A soft cushion, attached to the rim of the cup, provides a seal around the ear. In Figure 7.15 there is an illustration of a pair of ear muffs fitted round the ears. The cups should be made from a rigid, dense, non–porous material. The inside of the cup is lined with sound–absorbent material for the purpose of reducing resonances at high frequencies. The ear muff seals are either liquid–filled or foam–filled. The former gives slightly more attenuation, but can leak; the foam–filled seals are more robust.

The advantages of ear muffs are that they are:

    a) issued in one size only
    b) comfortable
    c) hygienic
    d) visible by managers.

Ear muffs have the disadvantages of being:

    a) expensive
    b) bulky
    c) not so robust
    d) difficult to wear with spectacles, long hair or hard hats.

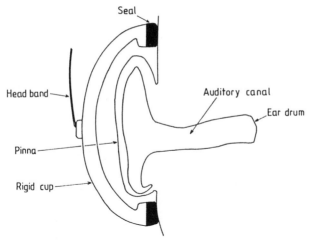

Figure 7.15 Ear muffs fitted round the ears.

A comprehensive list of the advantages and disadvantages of ear muffs and ear plugs is given by Martin [24].

7.11 Attenuation of ear protectors

The method for obtaining the attenuation of ear or hearing protectors, described in standards [25], is based on the subjective measurement of the threshold shift of different subjects whilst wearing the ear protectors in an anechoic room. In the British and international standard fifteen subjects take part twice in the test for each ear protector. The results of the test are expressed as a mean attenuation and a standard deviation in the octave bands with centre frequencies from 125 Hz to 8 kHz. The *assumed protection* is the mean attenuation minus one standard deviation [15]. The measured mean attenuation and the assumed protection for typical V–51R type ear plugs, disposable glass wool plugs, individually moulded ear plugs, disposable expanding foam plugs and foam–sealed ear muffs are shown in Figure 7.16.

7.12 Limitations of the performance of ear protectors

There are four ways in which sound can reach an ear occluded by an ear plug or ear muff. These are [26]:

1) Leaks around the protector — Sound can be transmitted around the ear plug or around the seal of the ear muff. These leaks arise because of poor fitting of the ear protector, or, in the case of ear plugs, because the size is wrong.

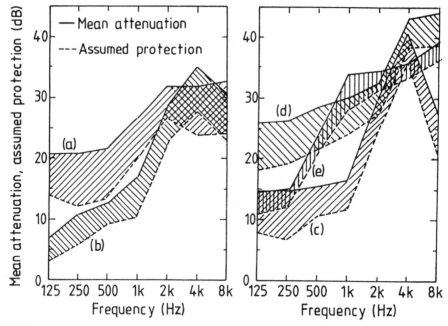

Figure 7.16 Measured mean attenuation and assumed protection for (a) V–51R type ear plugs, (b) disposable glass wool plugs, (c) individually moulded ear plugs, (d) disposable expanding foam plugs and (e) foam-sealed ear muffs.

2) Sound transmission through the protector.
3) Vibration of the ear protector — The ear muff or ear plug has mass and the system has a stiffness which is provided by the entrapped air volume, by the cushions of the ear muff or by the tissue of the ear canal in the case of the ear plug. A spring–mass system is formed which has a certain natural frequency. The attenuation of the ear protector is poor at this natural frequency (and at other natural frequencies which will certainly exist in this in reality multi–degree of freedom system).
4) Bone conduction — It is possible for the ear to receive sound not through the ear canal but by structure–borne sound transmission through the bone of the skull and the tissues of the head. Compared with air conduction, bone conduction is very weak, as can be seen in Figure 7.17.

When the ear is occluded, by an ear plug or ear muff, bone conduction becomes more effective and the threshold is lowered, relative to the open

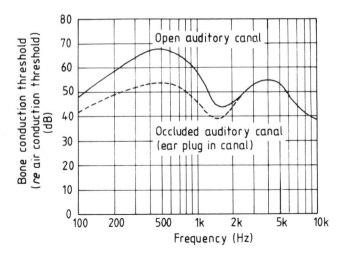

Figure 7.17 Effect of bone conduction.

ear [27]. See also Figure 7.17. The threshold for bone conduction for the occluded ear forms a limit to the performance of any ear protector. Increases in attenuation beyond the bone conduction limit can only be achieved by a helmet, incorporating ear muffs, which covers the skull.

7.12.1 Correct fitting of ear protectors

Correct fitting of ear plugs is important. Martin has reported [23] that poor fitting of V–51R ear plugs reduces the attenuation by 10 dB and increases the standard deviation. In another study [28] workers wearing ear plugs were taken from their workplace and given an audiometric test with their ear plugs in position and untouched. Subsequently they were tested audiometrically without the ear plugs. The difference in the audiograms gave a field attenuation. The results for disposable, expanding foam ear plugs are shown in Figure 7.18. The discrepancy between the laboratory tests and the field tests represents the effect of fitting; in the laboratory tests the fitting of the ear plugs is carefully controlled, whereas in the field tests casual fitting was carried out by the workers. The effect of poor fitting is not normally as bad as in Figure 7.18. In the case of individually moulded ear plugs the discrepancy is much less [28].

7.12.2 Importance of wearing ear protectors throughout noise exposure

It is frequently stated that a ear protector must be worn to be effective [29]. If the ear protector is removed during part of the exposure, the value of the protector is reduced.

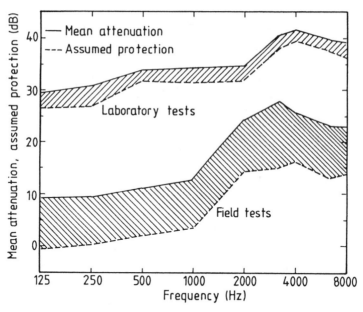

Figure 7.18 Effect of poor fitting of disposable, expanding foam ear plugs.

The effective levels of protection for ear protectors giving 5, 10, 15, 20, 25 and 30 dB protection are shown in Figure 7.19 as a function of time during which the protectors were not worn during the working day (an eight hour period).

Example 7.4

The noise level in a workshop is steady at 105 dB(A). A worker wears ear muffs which reduce the emission at his ears to 90 dB(A). During an 8 hour working day he wears the ear muffs for $7\frac{1}{2}$ hours and removes them for 30 minutes. What is the effective daily personal noise exposure of the worker?

Solution From equation (1.231) $\overline{p_{105}^2}$, the mean–squared A–weighted sound pressure when the sound level is 105 dB(A), is given by:

$$\overline{p_{105}^2} = 3.16 \times 10^{10}(2 \times 10^{-5})^2 \text{ Pa}^2.$$

Similarly,

$$\overline{p_{90}^2} = 1.0 \times 10^9(2 \times 10^{-5})^2 \text{ Pa}^2.$$

The averaged mean–squared sound pressure at the worker's ears during the eight hour day is $\overline{p_{av}^2}$ and is given by:

$$\overline{p_{av}^2} = (\overline{p_{105}^2} \times 0.5 + \overline{p_{90}^2} \times 7.5)/8$$
$$= (3.16 \times 0.5 + 0.1 \times 7.5)10^{10} \times (2 \times 10^{-5})^2/8$$
$$= 0.291 \times 10^{10}(2 \times 10^{-5})^2 \text{ Pa}^2.$$

The effective daily personal noise exposure during the day is:

$$L_{EP,d} = 10\log(0.291 \times 10^{10})$$
$$= 95\,\text{dB(A)}.$$

The worker is obtaining a benefit of only 10 dB instead of 15 dB.

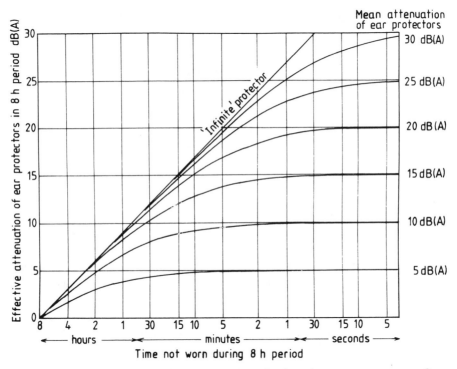

Figure 7.19 Effective levels of protection for hearing protectors as a function of time during which the protectors were not worn during an eight hour period.

7.12.3 Ear plugs and ear muffs combined

Berger [30] has reported that the combination of expanding foam ear plugs with ear muffs gives very good results. See Figure 7.20. The combination of the two ear protectors provides an attenuation which at high frequencies is close to the bone conduction threshold. The bone conduction threshold for the occluded ear is also shown in Figure 7.20 and, of course, it is impossible to achieve attenuations greater than this threshold by means of ear protectors. The results illustrated in Figure 7.20 were obtained from laboratory tests for a particular combination of ear plug and ear muff. Martin [23] and the Health and Safety Commission of the UK [15] report less favourably on dual protection.

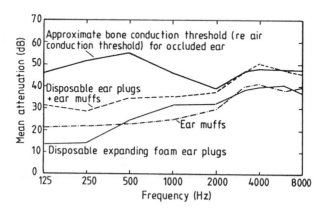

Figure 7.20 Dual hearing protection from expanding foam ear plugs and ear muffs.

7.13 Assessment of protection provided by ear protectors

In order to determine the protection afforded by ear plugs or ear muffs with respect to a given noise, the octave band pressure levels of the noise should first be measured using a sound level meter and octave filter set which covers the octave bands with centre frequencies from 63 Hz to 8 kHz. A sound level meter which satisfied the standards for type 1 should be used, although type 2 would be acceptable; the slow meter response should be chosen. Alternatively, an integrating sound level meter which measures the equivalent continuous sound level may be selected. An integration time of 60 s is normally acceptable for a continuous noise. The method described here is suitable for repetitive impulse noises, including those from forges and presses, but is not suitable for intense, single impulses that occur with gunshots or cartridge–operated tools.

The method is best illustrated by means of the example in Table 7.3. The octave band pressure levels are measured (row 3). In order to use

TABLE 7.3 Estimation of assumed protection level of ear protectors for the case of a typical noise

			63	125	250	500	1k	2k	4k	8k
(1)	Octave band centre frequency	(Hz)	63	125	250	500	1k	2k	4k	8k
(2)	A weighting response	(dB)	−26.2	−16.1	−8.6	−3.2	0	+1.2	+1.0	−1.1
(3)	Octave band pressure level of the noise	(dB)	82	93	100	103	109	109	107	99
(4) (2)+(3)	A-weighted octave band pressure level	(dB)	55.8	76.9	91.4	99.8	109.0	110.2	108.0	97.9
(5)	Mean attenuation of ear muffs	(dB)	12.2	14.7	16.3	27.4	36.8	38.9	40.3	37.5
(6)	Standard deviation	(dB)	4.1	3.3	2.7	3.5	4.2	3.7	2.9	3.8
(7) (5)−(6)	Assumed protection	(dB)	8.1	11.4	13.6	23.9	32.6	35.2	37.4	33.7
(8) (3)−(7)	Assumed protection level in octave band (APL)	(dB)	73.9	81.6	86.4	79.1	76.4	73.8	69.6	65.3
(9) (2)+(8)	A-weighted APL in octave bands	(dB)	47.7	65.5	77.8	75.9	76.4	75.0	70.6	64.2

Linear sound pressure level of the noise (from (3)) =	114 dB
Linear assumed protection level (from (8)) =	89 dB
A-weighted sound level of the noise (from (4)) =	114 dB(A)
A-weighted APL (from (9)) =	83 dB(A)

the method the mean attenuation of the ear protectors, as obtained by an approved test [25], should be available. The standard deviation for the attenuation is also needed. The mean attenuations obtained under careful laboratory conditions are not, as we have seen, typical of the attenuations realised in the factory. To account for this effect a smaller value than the mean attenuation should be assumed. The Health and Safety Commission (UK) recommend that the assumed protection should be the mean attenuation minus one standard deviation. For the hypothetical ear muffs used in the example in Table 7.3 the assumed protection is obtained by subtracting row 6 from row 5 to give row 7. The *assumed protection level* (APL), or the effective noise level at the worker's ears, is obtained by subtracting the assumed protection from the octave band pressure level of the noise (row 3). In Table 7.3 the assumed protection level values in octave bands are given in row 8.

In the example illustrated in Table 7.3 the linear sound pressure level is obtained by 'adding' the band pressure levels in row 3 according to the decibel addition rule, see Section 1.21. The result (to the nearest dB) is 114 dB. Of course, this result should agree with a measurement of sound pressure level with the frequency response on 'linear' on the sound level meter. The corresponding linear level, based upon the assumed protection level of row 8, is 89 dB, and is obtained by 'addition' of the values in row 8. The overall attenuation on a basis of linear frequency weighting is 25 dB (114 – 89 dB).

More realistically, a single figure attenuation can be obtained by basing the subtraction on the A–weighted levels. Firstly, the A–weighting response (row 2) is used to weight the octave band pressure levels of the original noise to give the A–weighted band pressure levels of row 4. Combining the decibel values in row 4 leads to a sound level of 114 dB(A). (The same level as the linear weighting because the sound energy is concentrated in the frequency range from 1 to 2 kHz.) Similarly, the A–weighted assumed protection level is obtained from row 9 and is 83 dB(A). On this A–weighted basis the overall attenuation is 31 dB(A). The ear muffs are effective, because they are providing high attenuation in the frequency range (1 to 2 kHz) where the greatest noise energy is concentrated.

A second example is illustrated in Table 7.4. In this case the noise has a considerable low frequency content; a compressor inlet, for example, could produce such a noise. The same ear muffs are used, but the A–weighted sound level is reduced from 97 dB(A) to an A–weighted APL of 85 dB(A); an attenuation on an A–weighted basis of only 12 dB(A). The reason for the low attenuation is that the energy of the noise is concentrated at frequencies for which the protection provided by the ear muffs is poor.

7.13.1 Approximate graphical method of estimating attenuation

In an approximate graphical method of estimating the attenuation provided by ear protectors, suggested by the Health and Safety Commission [15,21], use is made of charts of sound level against frequency with contours marked with an A–weighted sound level in dB(A), see Figure 7.21. To use these charts the octave band pressure levels are plotted to the same scale. The octave band pressure levels of the noise (from the example of Table 7.3) have been superimposed on the charts in Figure 7.21. The octave

TABLE 7.4 Estimation of assumed protection level of ear protectors for the case of low frequency noise

			63	125	250	500	1k	2k	4k	8k
(1)	Octave band centre frequency	(Hz)	63	125	250	500	1k	2k	4k	8k
(2)	A weighting response	(dB)	−26.2	−16.1	−8.6	−3.2	0	+1.2	+1.0	−1.1
(3)	Octave band pressure level of the noise	(dB)	109	110	101	90	79	70	60	−
(4) (2)+(3)	A-weighted octave band pressure level	(dB)	83.8	93.9	92.4	86.8	79.0	71.2	61.0	−
(5)	Mean attenuation of ear muffs	(dB)	12.2	14.7	16.3	27.4	36.8	38.9	40.3	37.5
(6)	Standard deviation	(dB)	4.1	3.3	2.7	3.5	4.2	3.7	2.9	3.8
(7) (5)−(6)	Assumed protection	(dB)	8.1	11.4	13.6	23.9	32.6	35.2	37.4	33.7
(8) (3)−(7)	Assumed protection level in octave band (APL)	(dB)	100.9	98.6	87.4	66.1	46.4	34.8	22.6	−
(9) (2)+(8)	A-weighted APL in octave bands	(dB)	74.7	82.5	78.8	62.9	46.4	36.0	23.6	−

Linear sound pressure level of the noise (from (3)) =	113 dB
Linear assumed protection level (from (8)) =	103 dB
A-weighted sound level of the noise (from (4)) =	97 dB(A)
A-weighted APL (from (9)) =	85 dB(A)

spectrum touches the contour marked 115 dB(A) at 2000 Hz. Thus according to this approximate method the sound level is 115 dB(A). We note that a more exact value is 114 dB(A). The octave band spectrum of the assumed protection level (row 8 in Table 7.3) is also shown in Figure 7.21. The highest penetration of the contours occurs at 250 Hz where the band pressure level falls between the 80 and 85 dB(A) contours. Interpolation leads to a value of 83 dB(A) for the assumed protection level. Previously the same value of 83 dB(A) was obtained. Thus the approximate method gives an attenuation of 32 dB(A) for the ear muffs with this particular noise. The more exact method gives an attenuation of 31 dB(A).

A similar exercise has been carried out for the example summarised

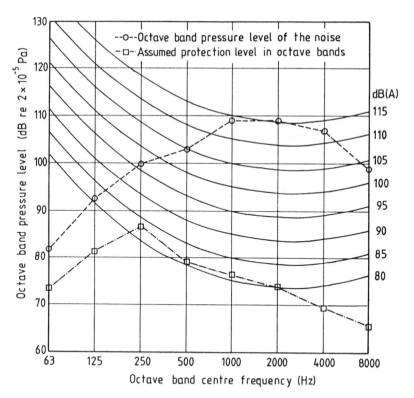

Figure 7.21 Approximate method of estimating assumed protection level for a typical noise.

in Table 7.4, see Figure 7.22. From the chart it can be seen that by this method the sound level of the noise is 99 dB(A) and for the assumed protection level 87 dB(A) is obtained. Thus the attenuation is 12 dB(A). The corresponding values obtained in the more exact analysis (Table 7.4) were 97 dB(A) for the sound level of the noise, 85 dB(A) for the APL and 12 dB(A) for the attenuation. Once again the agreement for the attenuation is good.

On the basis of the two examples considered it is reasonable to conclude that the approximate method works well enough. It is, however, worthwhile to look more carefully at the basis of the method. The contours of different sound levels in Figure 7.21 and 7.22 all have the shape of the inverted A weighting network (see Figure 2.5). The level is adjusted, up or down, depending upon the value assigned to the contour. Each contour also has a shape similar to the equal loudness level con-

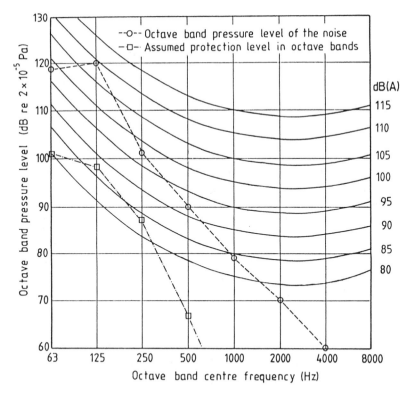

Figure 7.22 Approximate method of estimating assumed protection level for a source of low frequency noise.

tours for low intensity pure tones (40 phons, for example, in Figure 3.1). Consider the contour marked 105 dB(A). Consider also a noise that has octave band pressure levels that follow exactly the 105 dB(A) contour, i.e., the octave band pressure level at 500 Hz is 103 dB, at 1000 Hz it is 100 dB, at 2000 Hz 99 dB, etc. If this sound exists for the eight octave band centre frequencies from 63 Hz to 8 kHz, the A–weighted sound level will be $100 + 10 \log 8 = 109 \text{dB(A)}$, an error of 4 dB(A). (If the frequency of the noise extended below 63 Hz and above 8 kHz and measurements were made in the octave bands with centre frequencies of 31.5 Hz and 16 kHz, the A–weighted sound level would be greater still.) However, a noise with a spectrum that follows exactly the 105 dB(A) contour, or any other such contour, is very unusual.

The method would also give poor results, if one octave band pressure level had a very large value compared to the others, a situation that may occur if a strong pure tone is present. For example, the spectrum of a noise could have a band pressure level of 100 dB at 1000 Hz and low levels at other frequencies; the A–weighted sound level in this case is only 100 dB(A). The approximate method would always give a value of 105 dB(A).

The true A–weighted sound levels could vary from 100 to 109 dB(A), even though the approximate value would always be 105 dB(A).

A more likely situation is for three band pressure levels to have values close to, say, the 105 dB(A) contour and for the other band pressure levels to be lower than the contour. In this case the A–weighted sound level would be $100 + 10 \log 3 = 105$ dB(A), i.e., the value assigned to the contour. In the approximate method of this section it is assumed that noise spectra are similar to the type for which three band pressure levels are in the vicinity of a contour. The approximate graphical method is adequate for most typical industrial noises, but should be avoided if a single, dominant pure tone is present, or if the spectrum shape follows that of the inverted A–weighted network.

7.13.2 The noise reduction rating (NRR)

The methods described previously for the assessment of the protection provided by ear protectors require the measurement of the sound in octave bands. It would be useful to have a simpler method in which the A–weighted sound level alone was measured and the level of noise at the protected ear estimated by subtracting a rating value (which could be stamped on the ear muff) from the A–weighted sound level. The *noise reduction rating* (NRR), devised by the Environmental Protection Agency of the USA, is such a rating. It is a rating for the ear protector and does not depend upon the noise from which protection is sought.

The method of calculation of the noise reduction rating [31] is outlined by means of the example in Table 7.5 which refers to the same ear muffs considered in Tables 7.3 and 7.4. The method assumes pink noise, i.e., a noise which has band pressure levels independent of frequency when measured by a constant percentage filter, such as an octave filter. The level of the pink noise is unimportant; a level of 100 dB is used in row 4 of Table 7.5. In the procedure for estimating the NRR the pink noise is both C–weighted and A–weighted. The effective protection provided by the hearing protectors is taken to be the mean attenuation minus 2 standard deviations. The effective protection is subtracted from the A–weighted pink noise to give the A–weighted protection level in octave bands (row 10). The NRR is the difference between the C–weighted sound level of the pink noise and the sum of the A–weighted sound level at the protected ear and 3 dB. The 3 dB term is a safety factor. In the present example the NRR for the ear muffs is 21 dB.

TABLE 7.5 Estimation of noise reduction rating (NRR) of ear protectors

(1)	Octave band centre frequency	(Hz)	125	250	500	1k	2k	4k	8k
(2)	C weighting response	(dB)	−0.2	0	0	0	−0.2	−0.8	−3.0
(3)	A weighting response	(dB)	−16.1	−8.6	−3.2	0	+1.2	+1.0	−1.1
(4)	Octave band pressure level of hypothetical pink noise	(dB)	100	100	100	100	100	100	100
(5) (2)+(4)	C-weighted octave band pressure level	(dB)	99.8	100.0	100.0	100.0	99.8	99.2	97.0
(6) (3)+(4)	A-weighted octave band pressure level	(dB)	83.9	91.4	96.8	100	101.2	101.0	98.9
(7)	Mean attenuation of ear muffs	(dB)	14.7	16.3	27.4	36.8	38.9	40.3	37.5
(8)	Standard deviation	(dB)	3.3	2.7	3.5	4.2	3.7	2.9	3.8
(9)	2 × Standard deviation	(dB)	6.6	5.4	7.0	8.4	7.4	5.8	7.6
(10) (6)−(7)+(9)	A-weighted protection level in octave bands	(dB)	75.8	80.5	76.4	71.6	69.7	66.5	69.0

C–weighted sound level of the noise (from (5)) =	108 dB(C)
A–weighted sound level for the protected ear (from (10)) =	84 dB(A)
NRR = 108 − 84 − 3 =	21 dB

With the same ear muffs the examples presented in Tables 7.3 and 7.4 resulted in attenuations of 31 dB(A) and 12 dB(A) — evidence that the attenuation depends upon the noise source. The example in Table 7.4 is, however, untypical. The attenuation in the more typical case is in excess of the noise reduction rating. A criticism of the NRR is that it underestimates the attenuation of a given protector [15] and leads to overprotection of workers who would wear ear muffs which are heavier than necessary.

Alternatively, the view could be taken that over–protection is desirable, bearing in mind the limitations of ear protectors as used in a factory.

7.14 Hearing conservation programme

The provision of personal ear protectors to workers can only be a part of a hearing conservation programme. If the daily personal noise exposure is above 90 dB(A), the employers have a duty to reduce levels by noise control techniques, as far as reasonably practicable. It is only after noise control has been attempted that reliance can be put on the use of ear protectors.

Any hearing conservation programme should include the following stages:

(1) *Noise measurement*

    For $L_{EP,d}$ values likely to exceed 85 dB(A) noise surveys should be carried out.

(2) *Evaluate risk to hearing*

    If levels exceed 85 dB(A) workers have to be informed of the risk.

(3) *Carry out noise reduction if levels are above 90 dB(A)*

    Although not obliged to do so, many employers will commence noise reduction at 85 dB(A).

(4) *Monitoring audiometry*

    Many employers will wish to carry out audiometry, even though it is not a statutory obligation in Great Britain, particularly for new employees. New employees should be tested at the start of employment, after 6 months and then annually. Prior to audiometry the patient should have had a rest period from noise of at least 16 hours.

(5) *Distribution and maintenance of ear protectors*

    It is essential that a range of sizes is available and that training is provided in the fitting of the ear protectors. The proper care of the protectors should be emphasised.

7.15 Non–auditory effects of noise

The non–auditory effects of noise have been very difficult to quantify. There is some evidence of physiological effects such as hypertension [32], but the evidence is not conclusive. The effect on human performance is equally controversial. Jansen and Gros [33] report that levels above 90 dB(A) lead to a decrease in the ability to perform activities of any kind. He also pointed out that Kryter [34] took the view that noise had a positive effect on performance, as opposed to Broadbent [35] who claimed that performance was affected negatively by noise.

In the case of sleep there is evidence that road traffic levels of L_{eq} of 35 — 40 dB(A) prevent people from sleeping unless they close the windows. With $L_{eq} = 75$ dB(A) 70% of the residents could not sleep [32].

7.16 Summary of provisions in the EC Council Directive [9]

1) Where the daily personal exposure of a worker to noise is likely to exceed 85 dB(A) or the maximum value of the unweighted instantaneous sound pressure is likely to be greater than 200 Pa, appropriate measures shall be taken to ensure that workers at risk are identified and they or their representatives are fully informed about the noise and the risk to hearing.

2) Where the exposure is likely to exceed 85 dB(A), personal ear protectors must be made available to workers.

3) The risks resulting from exposure to noise must be reduced to the lowest level reasonably practicable, taking account of technical progress and the availability of measures to control the noise, in particular at source.

4) Where the daily personal noise exposure of a worker exceeds 90 dB(A), or the maximum value of the unweighted instantaneous sound pressure is greater than 200 Pa, the reasons for the excess level shall be identified and the employer shall draw up and apply a programme of measures of a technical nature and/or of organisation of work with a view to reducing as far as reasonably practicable the exposure of workers to noise. Workers shall receive adequate information on the excess level and on the measures taken to reduce the noise.

5) Without prejudice to (4) above, where the daily personal exposure of a worker exceeds 90 dB(A) or the maximum value of the unweighted instantaneous sound pressure is greater than 200 Pa, personal ear protectors must be used.

6) Where it is not reasonably practicable to reduce the daily personal exposure of a worker to below 85 dB(A), the worker exposed shall be able to have his hearing checked by a doctor or on the responsibility of a doctor. This means that the worker can ask for an audiometric test if his level is above 85 dB(A).

7) Where a new article (tool, machine, apparatus, etc.) which is intended for use at work is likely to cause for a worker who uses it properly for a conventional eight hour period a daily personal noise exposure equal to or greater than 85 dB(A) or an unweighted instan-

taneous sound pressure the maximum value of which is equal to or greater than 200 Pa, adequate information should be made available about the noise produced in conditions of use to be specified.

8) Where the daily personal noise exposure is likely to exceed 90 dB(A) or where the maximum value of the unweighted instantaneous sound pressure is likely to exceed 200 Pa, information must be given in the workplace and where reasonably practical signs displayed.

7.16.1 The Noise at Work Regulations 1989 [10]

The regulations introduced by the United Kingdom in order to comply with the EC directive have some minor differences in terminology from the directive. A daily personal noise exposure of 85 dB(A) is referred to as the *first action level* and exposure of 90 dB(A) is the *second action level*. The *peak action level* means a peak instantaneous sound pressure of 200 Pa.

The parts of a workplace where the second action level or the peak action level are exceeded are to be designated as ear protection zones and their extent identified. Signs should be used to indicate the boundaries of the *ear protection zone* and within the zone all employees of the company must wear ear protectors.

There is no regulation concerned with audiometry or the monitoring of hearing loss.

The Health and Safety Executive has published a series of Noise Guides which are intended to assist those responsible for organising and supervising the measures required to enable compliance with the regulations [36,37]. The company employees responsible are designated *competent persons*.

7.17 Closure

Interest in noise has burgeoned in recent years because of the imposition of regulations and because employers fear court action (for negligence) by employees who have become deaf at work. Without the fear of legal action little would be done to reduce noise.

In the United Kingdom any worker who believes that his deafness has been caused by noise at work can bring an action for common law negligence against his employers in the civil courts. Negligence is normally defined as an omission to do something that a reasonable person would do. For a long time actions by employees were unsuccessful. In 1969 in the case Down v. Dudley, Coles, Long, Ltd the plaintiff, who suffered permanent loss of hearing through the use of a cartridge–operated ham-

mer, lost the action, because the judge felt that at the time there was
insufficient information available to the defendants to make them aware of
the dangers involved. That knowledge is now available. Furthermore, the
courts recognise that deafness can be predicted from exposure to known
levels of noise, even to the extent of accepting tables published by the
National Physical Laboratory [38]. Employers are now deemed to have an
awareness of the risks and knowledge of the techniques of noise control;
ignorance is no longer an excuse.

In Berry v. Stone Manganese Marine Ltd 1972 a plaintiff was for
the first time awarded damages against his employers for noise–induced
deafness at work. In this case the defendants were found to be negligent
in not providing adequate hearing protection. Although they provided
some ear plugs, ear muffs — which were necessary in this case — were not
supplied. The defence argument, that if ear muffs had been issued they
would not have been worn, was not regarded as acceptable. The case also
established that it is not enough to make available a number of hearing
protectors, but it is necessary to provide an adequate choice and to give
instruction in their fitting and use.

In the case of Berry v. Stone Manganese Marine Ltd the plaintiff
was awarded damages of £ 2500 (reduced to £ 1250 by virtue of the
Limitation Act). In 1991 a man was awarded by the High Court damages
of £ 120,000 for loss of hearing and for tinnitus (ringing in the ears). With
such damages employers will be obliged to act to reduce noise, or be forced
to do so by their insurers.

Deafness develops insidiously over a long period of time. During
this time an employee may have worked for various employers. Who is
responsible? In the case Heslop v. Metalock (Great Britain) Ltd 1981 it
was decided that the final employer was liable, even though the employee
had been subjected to high noise levels by previous employers for longer
periods of time. Such a lack of apportionment of responsibility may seem
unfair, but it does have the advantage of simplicity. An employer would
be well advised to undertake audiometric screening of new employees, at
least for the purpose of assessing his future liabilities.

The question of audiometric screening is controversial. The EC directive stated that employees should be able to request that his hearing
be checked by a doctor. The directive also include a clause that the check
should be carried out in accordance with the national law and practice of
the Member State. The British regulations omit any reference to screening. In Britain it is claimed that every employee has the right to attend
the doctor with whom he is registered and ask for a hearing test; that is

the practice in Britain where there is a well–established National Health Service. It may be that in reality many employed persons, who should have their hearing checked, will find the intended procedure too inconvenient and time–consuming. The arguments against audiometry are that it is expensive and, because there are so many factors which influence hearing loss, that it is inconclusive in indicating whether or not the loss of hearing is caused by noise. However, some Trades Unions are intending to challenge the British government in the European Parliament with respect to this issue.

The European dimension on noise and its control is continually increasing. In addition to the directive on the protection of workers from the risks related to exposure to noise [9] there are also clauses on noise in a directive which relates to machinery [39]. The directive on machinery is concerned with the harmonisation of regulations relating to trade in the Member States and is scheduled to come into force by 31st December, 1992. The directive requires that any machinery causing an equivalent continuous A–weighted sound level in excess of 70 dB(A) at a work station should have the noise level quoted in the machinery's instructions and sales literature. Information should also be given if the C–weighted instantaneous sound pressure exceeds 63 Pa at a work station. When the equivalent continuous A–weighted sound level exceeds 85 dB(A) the sound power level must be quoted. The sound power levels of certain specified machinery must already be quoted (see reference 54, Chapter 5).

Additionally, there is a European Standards Organisation (CEN) which is likely to introduce standards on noise. A standard on ear protectors is expected in 1992.

The steady progress on legislation and its implementation means that the likelihood of workers becoming deaf is reduced. Many years ago in some types of work it was a certainty.

References

[1] T. S. Littler (1965). *The Physics of the Ear*, p 30. Pergamon Press, Oxford.

[2] S. S. Stevens and H. Davis (1983). *Hearing: its psychology and physiology*, p 256. American Institute of Physics, New York.

[3] G. von Békésy (1960). *Experiments in Hearing*. Acoustical Society of America, New York.

[4] E. F. Evans (1986). Recent advances in understanding hearing mechanisms and hearing impairment, in *Noise Pollution* (Scope 24) (eds A. Lara Sáenz and R. W. B. Stephens). John Wiley, Chichester.

[5] S. M. Khanna and D. G. B. Leonard (1982). Basilar membrane tuning in the cat cochlea. *Science*, **215**, pp 305–306.

[6] International Organization for Standardization (1984). Acoustics — Threshold of hearing by air conduction as a function of age and sex for otologically normal persons. ISO 7029–1984. (Also BS 6951:1988.)

[7] International Organization for Standardization (1990). Acoustics — Determination of occupational noise exposure and estimation of noise–induced hearing impairment. ISO 1999–1990.

[8] Department of Labor. Occupational Health and Safety Administration (1974). Occupational Noise Exposure. *Federal Register*, 24 October 1974, pp 37773–37778.

[9] The Council of the European Communities (1986). Council Directive of 12 May 1986 on the protection of workers from the risks related to exposure to noise at work (86/188/EEC). *Official Journal of the European Communities*, No L 137/28. Brussels.

[10] Department of Employment (1989). The Noise at Work Regulations (S.I. 1989/1790). HMSO, London.

[11] W. Burns and D. W. Robinson (1970). *Hearing and Noise in Industry.* HMSO, London.

[12] D. W. Robinson (1987). Noise exposure and hearing: a new look at the experimental data. HSE Contract Research Report No 1/1987. Health and Safety Executive, London.

[13] CHABA (1966). Hazardous exposure to intermittent and steady-state noise. *Journal of Acoustical Society of America*, **39**, pp 451–464.

[14] Health and Safety Commission (1981). Some aspects of noise and hearing loss: notes on the problem of noise at work and report of the HSE Working Group on Machinery Noise. HMSO, London.

[15] Health and Safety Commission (1988). Prevention of damage to hearing from noise at work — Draft proposals for Regulations and Guidance — Consultative Document, HMSO, London.

[16] British Standards Institution (1976). Method of test for estimating the risk of hearing handicap due to noise exposure. BS 5330:1976.

[17] The Commission of the European Communities (1982). Proposal for a Council Directive on the protection of workers from the risks related to exposure to chemical, physical and biological agents at work: noise. *Official Journal of the European Communities*, No C 289/1. Brussels.

[18] International Organization for Standardization (1983). Acoustics — Pure tone air conduction threshold audiometry for hearing conservation purposes. ISO 6189–1983. (Also British Standards Institution

BS 6655:1986.)

[19] International Organization for Standardardization (1985). Acoustics — Standard reference zero for the calibration of pure tone air conduction audiometers. ISO 389–1985. (Also British Standards Institution BS 2497: part 5:1988.)

[20] Confederation of British Industry (1983). Protecting hearing at work. London.

[21] Department of Employment (1972). Code of Practice for reducing the exposure of employed persons to noise. HMSO, London.

[22] British Standards Institution (1983). Personal noise exposure meters. BS 6402:1983.

[23] A. M. Martin (1976). Industrial hearing conservation 1: personal hearing protection. *Noise Control Vibration and Insulation*, Feb 1976, pp 42–50.

[24] A. M. Martin (1982). Occupational hearing loss and hearing conservation, in *Noise and Vibration* (eds R. G. White and J. G. Walker), Ellis Horwood, Chichester.

[25] International Organization for Standardization (1981). Acoustics — Measurement of sound attenuation of hearing protectors – subjective method. ISO 4869–1981. (Also British Standards Institution BS 5108:1983.)

[26] E. H. Berger (1982). Hearing protector performance: how they work and what goes wrong in the real world. EARlog 5.

[27] J. Zwislocki (1957). In search of the bone–conduction threshold in a free sound field. *Journal of Acoustical Society of America*, **29** (7), pp 795–804.

[28] R. G. Edwards, A. B. Broderson, W. W. Green and B. L. Lempert (1983). A second study of the effectiveness of ear plugs as worn in the workplace. *Noise Control Engineering Journal*, **20** (1), pp 6–15.

[29] D. Else (1973). A note on the protection afforded by hearing protectors — implications of the energy principle. *Ann. Occup. Hygiene*, **16**, pp 81–83.

[30] E. H. Berger (1985). Attenuation of ear plugs worn in combination with ear muffs. *Noise and Vibration Control Worldwide*, Nov 1985, pp 264–265.

[31] E. H. Berger (1980). Single number measures of hearing protector noise reduction, EARlog 2.

[32] F. J. Langdon (1986). Noise and health: a brief review of recent research. *Noise and Vibration Control Worldwide*, June 1986, pp 158–162.

[33] G. Jansen and E. Gros (1986). Non–auditory effects of noise: physi-
     ological and psychological effects, in *Noise Pollution* (Scope 24) (eds
     A. Lara Sáenz and R. W. B. Stephens). John Wiley, Chichester.

[34] K. D. Kryter (1970). *The Effects of Noise on Man.* Academic Press,
     New York.

[35] D. E. Broadbent (1971). *Decision and Stress.* Academic Press, Lon-
     don.

[36] Health and Safety Executive (1989). Noise at Work — guidance on
     regulations. Noise guide no 1: Legal duties of employers to prevent
     damage to hearing. Noise guide no 2: Legal duties of designers, man-
     ufacturers, importers and suppliers to prevent damage to hearing.
     HMSO, London.

[37] Health and Safety Executive (1990). Noise at Work — noise assess-
     ment, information and control. Noise guide no 3: Equipment and
     procedures for noise surveys. Noise guide no 4: Engineering control
     of noise, Noise guide no 5: Types and selection of personal ear protec-
     tors. Noise guide no 6: Training for competent persons. Noise guide
     no 7: Procedures for noise testing machinery. Noise guide no 8: Ex-
     emption from certain requirements of the Noise at Work Regulations
     1989. HMSO, London.

[38] M. Dewis (1984). Compensation for deafness. *Occupational Safety
     and Health*, **14** (8), pp 12–14.

[39] The Council of the European Communities (1989). Council Directive
     of 14 June 1989 on the approximation of the laws of Member States
     relating to machinery (89/392/EEC). Official Journal of the European
     Communities, No. L 183/9. Brussels.

8 Noise Sources; Identification and Control

Let there be no noise made,
my gentle friends.
William Shakespeare, *Henry IV, part 2*

8.1 Introduction

Some of the principles of noise control have been described in earlier chapters, for example, sound absorption and insulation in Chapter 4, and silencers in Chapter 6. The purpose of this chapter is to present the possibilities in a more systematic way. Noise control is of interest to machine designers, works engineers and safety officers, amongst others. An incentive for the designer is provided in article 5 of the directive of the Council of the European Communities [1]:

> The risks resulting from exposure to noise must be reduced to the lowest level reasonably practicable, taking account of technical progress and the availability of measures to control the noise, in particular at source.

Similarly, for new factories as a whole there is an obligation to reduce the risk to hearing.

The works engineer or safety officer, operating in an existing factory, has less scope for noise control than the designer, starting afresh. The works engineer has to be content with minor modifications to machines, or noise control techniques, that interfere as little as possible with the smooth running of the factory. He or she has to apply what are often termed band–aid measures.

The techniques used by both the designer and the works engineer have much in common. It is always important to understand the origin of the noise. Any attempt to reduce structure–borne sound with the techniques for controlling air–borne sound, or vice versa, will waste money without bringing any benefit. We consider below some of the different ways in which noise can be produced and how the sources can be identified, before reviewing the possible methods, available to the designer and the works engineer, for reducing noise.

8.2 Noise from vibrating surfaces

The noise caused by vibration can be regarded as the output of a noise–producing system in which the input is a force, torque or pressure (varying with time) which excites the vibrating structure. Not all of the vibrational energy is converted into sound radiated into the far field; the radiation efficiency is a measure of the conversion of vibrational to acoustic energy. The sound output in terms of frequency is the result of the multiplication of three terms, the input, the frequency response of the structure and the radiation efficiency, as shown diagrammatically in Figure 8.1. The sound output can either be in terms of sound power or sound pressure; in the latter case the directivity has to be taken into account.

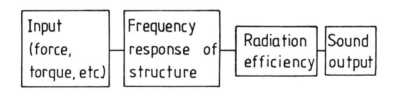

Figure 8.1 Dependence of sound output on input, structural response and radiation efficiency.

8.2.1 System input

The input to the system can be a transient force, for example, from the impact in a punch press, or the input may be the repetitive variation of cylinder pressure, as in a diesel or petrol engine.

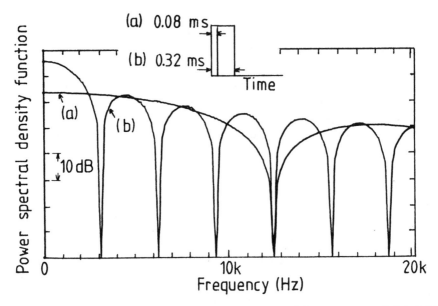

Figure 8.2 Power spectra of rectangular pulses; (a) 0.08 ms and (b) 0.32 ms.

In general transient forces of short duration have a spectral content with a wider bandwidth than a transient of longer duration. This point is illustrated in Figure 8.2, where the power spectra of two rectangular pulses of different widths in the time domain are shown. In the case of a rectangular pulse of duration 0.08 ms the first zero in the power spectrum occurs at 12.5 kHz. A longer pulse of length 0.32 ms has zeros at 3.125, 6.25, 9.375 kHz, etc. Thus the shorter rectangular transient is capable of exciting all the natural frequencies of a structure up to 12.5 kHz. In the same frequency range there would be some natural frequencies that the longer pulse would not excite or only excite weakly.

Rounding the edges of the rectangular pulse has the effect of reducing the amplitudes of the higher frequency content in power spectrum, as shown in Figure 8.3, where a half sine pulse is compared with a rectangular pulse of the same width and height.

In the case of regularly repeated pulses the transform in the frequency domain is controlled by the repetition rate of the pulses and by the shape of an individual pulse. The power spectrum of a series of rectangular pulses is shown in Figure 8.4, where the power spectrum of a single pulse is also shown for comparative purposes. The power spectrum of the series

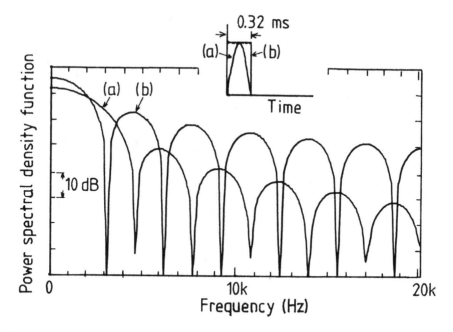

Figure 8.3 Effect on the power spectrum of smoothing a rectangular pulse; (a) rectangular pulse and (b) half sine pulse.

of pulses comprises a set of harmonics (in the frequency domain) of the repetition rate of 781.25 Hz. (The repetition rate of the pulses in the time domain is 1.28 ms, so the harmonics are multiples of 781.25 Hz.) The envelope of the maximum values in power spectrum has the same shape as the power spectrum of a single pulse. The elevation of the power spectrum of the series of pulses by 12 dB, relative to the single pulse, reflects the fact that there are 4 pulses in the series.

The results shown in Figures 8.2 to 8.4 were directly obtained from the output of a Hewlett Packard 5451C Fourier Analyser which is based on the FFT principle, see Section 2.19. In all cases $N = 1024$ and $F_{max} = 100$ kHz; only the first fifth of the spectrum is shown.

8.2.2 Structural response

The structural response, see Figure 8.1, is more a concept than a quantity that can be precisely measured for a complicated machine. What can be measured easily is the frequency response function or transfer function between input and output, for example, between the cylinder pressure

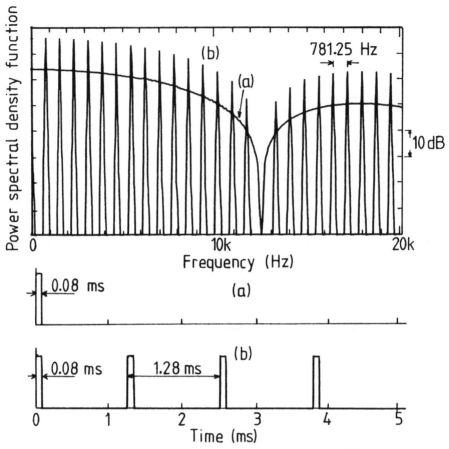

Figure 8.4 Power spectrum of a repetitive pulse.

of the gas above the piston in a single cylinder diesel engine and the velocity of a point on the surface of the exterior of the engine. In vibration testing the transfer function may be obtained by applying a force from a vibrator at one point on a structure and measuring the response in terms of displacement, velocity or acceleration at another point. The transfer function so measured is a property of the structure; it will exhibit, when presented as a function of frequency, a series of peaks which refer to the natural frequencies of the structure. For a given structure and given input position the magnitude of the transfer function at the natural frequencies will depend upon the measurement position of the output. For a particular natural frequency the structure, when vibrating, has certain positions where there is no displacement; these positions are nodes and on

plate–like structures *nodal lines* are formed. At other positions there will
be anti–nodes where the displacements are maximum.

The structural response should be obtained from an average of several
transfer functions. Thus, for the example of the diesel engine, the transfer
function between the cylinder pressure and the velocity at several points
on the engine should be obtained and an average function computed.

The dynamic properties of a structure, as characterised by the transfer
function between output and input at two different points of the structure,
are often the key to noise control. Various advanced techniques, such as
modal analysis [2,3], concentrate on the investigation of the structural
response and the possible modification of the structure in order to reduce
the vibration amplitudes and hence noise. These methods are beyond the
scope of this book. We will concentrate on simpler techniques, such as
vibration isolation and vibration damping, which modify the structure in
a way that reduces the noise levels.

8.2.3 Radiation efficiency

It does not follow that a great deal of sound will be radiated just because
the amplitudes of vibration of a machine component are large. Two plates
can be vibrating with the same average mean–squared velocity, but one
may radiate more sound than the other. To account for the fact that some
bodies radiate sound more effectively a *radiation efficiency* or *radiation
ratio* has been introduced. The radiation efficiency is defined as

$$\sigma_{\mathrm{rad}} = \frac{W_{\mathrm{rad}}}{\rho_0 c_0 S \langle \overline{u^2} \rangle}, \tag{8.1}$$

where W_{rad} is the sound power radiated by a machine component of surface area S. The square of the velocity at a point on the surface of the
component is averaged with respect to time to give $\overline{u^2}$. The time averages are obtained at several points on the surface and a spatial average
$\langle \overline{u^2} \rangle$ is estimated. The product $\rho_0 c_0$ is the specific acoustic resistance
(characteristic impedance) of the air. See Section 1.12.1.

The denominator in equation (8.1) is the sound power radiated by a
rigid, baffled piston at high frequencies. All parts of the piston vibrate in
phase and a beam of plane waves with cross–sectional area S is radiated
[4].

The *critical frequency of wave coincidence* is an important factor in
any discussion of radiation efficiency. Wave coincidence has already been
considered in Section 4.8 in relation to an infinite panel or wall. Wave
coincidence occurs when the projected wavelength of the sound in air, the

trace wavelength, is equal to the wavelength of the bending waves in the panel. The critical frequency occurs when the incident sound waves are running parallel to the panel surface.

In the case of a finite rectangular plate the vibration will tend to occur at a series of natural frequencies. At each natural frequency there will be a grid of nodal lines which divide the plate into smaller rectangles. A particular rectangular area will be vibrating out of phase with the adjacent areas. At frequencies below the critical frequency of wave coincidence, where the wavelength of the bending waves in the plate are less than the wavelength of sound in air, there are cancellation effects with the sound from adjacent rectangular areas and little sound is radiated. At the corners of the plate cancellation is not possible and sound is radiated. For certain modes of vibration strips of the plate along the edges will also radiate sound. In general the cancellation effects in the middle of the plate render the plate an inefficient radiator at low frequencies [5].

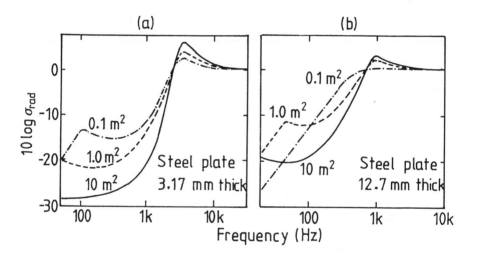

Figure 8.5 Radiation efficiency for square, steel plates in infinite baffles, simply supported along the edges: plate thickness (a) 3.17 mm (b) 12.7 mm [6].

At frequencies above the critical frequency the wavelengths of the bending waves in the plate are greater than the wavelength of sound, the cancellation effects no longer occur and the entire panel radiates sound. In general the radiation efficiency is small at low frequencies and close to unity at high frequencies. At and around the critical frequency of wave coincidence the radiation efficiency can be greater than unity. The exact shape of the radiation efficiency against frequency curves depends upon the type of vibrating body, its dimensions and the method of support. As examples of how the radiation efficiency varies with frequency, theoretical values are shown in Figure 8.5 for square, steel plates, in infinite baffles, simply supported along the edges. The results are presented for plate thicknesses of 3.17 and 12.7 mm and in each case for areas of 0.1, 1.0 and 10 m^2 [6].

8.2.4 Sound output

The sound output which is usually in the form of the sound power level or the sound pressure level in the far field. Sound power is generally the most useful as it gives the integrated effect of the sound field. On the other hand sound pressure level can sometimes be of value because it enables information on the directivity pattern to be retained.

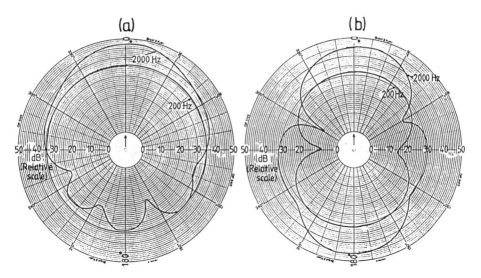

Figure 8.6 Directivity pattern of (a) small loudspeaker box and (b) loudspeaker chassis.

The directivity of a sound source depends upon the product of the wave number k and a maximum dimension D. For small values of the product kD most sound sources of the pulsating type radiate sound almost equally in all directions, see Section 1.17. If surrounded by free space they are a source of spherically symmetric waves, see Section 1.16.1. Such sources are referred to as monopoles or point sources. At larger values of

kD the directivity of the source becomes more complicated. The dependence of the directivity pattern of a source on kD, or on frequency for a given source, is illustrated in Figure 8.6. In Figure 8.6a the source of sound is a small loudspeaker box (height 250 mm, width 190 mm and depth 90 mm) made from a single loudspeaker. The sound pressure level, measured at a fixed point in the far field as the source rotates on a turntable about a vertical axis, is shown plotted against the angle the measurement axis makes with the direction of maximum sound radiation (the axis of symmetry of the loudspeaker chassis). Of the two results presented in Figure 8.6a the sound pressure level at 200 Hz is constant within ±1 dB; at this frequency the loudspeaker box is almost equivalent to a monopole. At 2 kHz the same source is much more directional.

In the case illustrated in Figure 8.6b the sound source is a loudspeaker alone, i.e., just a chassis (of diameter 250 mm) without the box. For both frequencies the sound source is highly directional. At 200 Hz the sound source is similar to a dipole, see Appendix D.

8.3 Aerodynamic noise sources

Aerodynamic noise sources are common in industry, as well as being predominant in aircraft. It is important to identify these sources, as special techniques are required for reduction of their noise.

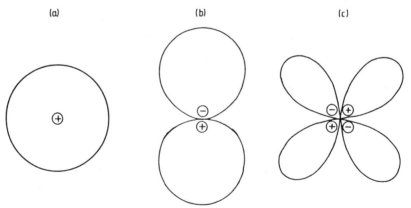

Figure 8.7 Schematic polar plots of the sound intensities in the far field for (a) monopole, (b) dipole and (c) quadrupole (lateral).

Aerodynamic noise sources are frequently classified as *monopole, dipole* or *quadrupole* sources of sound [7]. The higher order sources — the dipole and quadrupole — can be considered as being made up from arrangements of monopoles. (Monopoles, or point sources are the sources of spherical waves discussed in Sections 1.16 and 1.17.) A dipole comprises two monopoles, close together, which are out of phase by 180°, see Appendix D. Cancellation occurs in one direction and addition of the sound field in a direction at right angles. A schematic diagram of the directivity pattern of the sound radiation of a dipole is shown in Figure 8.7b. A quadrupole comprises four monopoles, two of which are in phase and two out of phase. The monopoles can be arranged to form either a longitudinal or lateral quadrupole; a lateral quadrupole is shown in Figure 8.7c (a longitudinal quadrupole has a directivity pattern more like that of a dipole).

Aerodynamic noise monopoles are associated with a fluctuation of mass, dipoles with a fluctuation of momentum (or force) and quadrupoles occur when there is a fluctuation of the momentum flux (or stress). Because of cancellation effects the quadrupole is radiating less sound power than a dipole, and a dipole radiates less sound power than one of its constituent monopoles, if the constituent monopoles are assumed to be of equal strength. In other words the quadrupole is the least efficient of the three sound sources. With aerodynamic sound sources it is velocity which is the most important parameter. The relevant velocity may be the efflux velocity of a jet from a nozzle or the tip speed of rotating fan blades, depending upon the situation. The sound power of the monopole, dipole and quadrupole sources depends upon velocity to the fourth, sixth and eighth power of velocity, respectively.

8.3.1 Dipole sources

Centrifugal and axial flow fans are the most commonly occurring sources of dipole noise. Singing telegraph wires, or Aeolian tones, are another example of dipole sound of aerodynamic origin. A sphere oscillating as a rigid body forms also a dipole source of sound, see Appendix D. The loudspeaker chassis, whose directivity patterns are shown in Figure 8.6b acts like a dipole source at a frequency of 200 Hz. Each side of the diaphragm acts as a separate monopole, but with a phase difference of 180°.

Fan noise The most commonly used fans in industry are of the *centrifugal* or *axial flow* type [8]. Noise from the fans arises because of turbulent flow, either in the oncoming stream or in the blades' wakes. A new type of fan, known as laminar flow fans, avoids producing turbulent flow. This type has been used to replace axial flow fans for cooling purposes [9].

The spectrum of sound from either an axial flow fan or centrifugal fan comprises broad band random noise upon which a series of pure tones are superimposed. (Note that in the context of fan noise these pure tones are often referred to as discrete tones.) The mechanisms of noise generation are similar for both types of fan. In the case of axial flow fans the blades have aerofoil sections. The lift and drag forces on the aerofoils depend upon the direction of the flow onto the leading edge. Turbulence in the approaching stream causes local random variations in the angle of incidence of the flow onto the aerofoils. The lift forces on the blade correspondingly vary in a random way and give rise to random sound which is dipole in origin. This phenomenon of generation of broad band random noise is illustrated in the left part of Figure 8.8 as (a). A second mechanism of random noise generation, illustrated in the right hand part of Figure 8.8, occurs as vortices are shed from the trailing edge of the blade and form a turbulent wake. Consequently, fluctuations in the drag forces on the blade cause corresponding lift fluctuations.

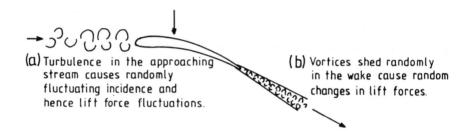

(a) Turbulence in the approaching stream causes randomly fluctuating incidence and hence lift force fluctuations.

(b) Vortices shed randomly in the wake cause random changes in lift forces.

Figure 8.8 Mechanism for the generation of broad band random noise in fans; (a) random fluctuations in lift, (b) vortices in the wake.

Discrete tones can also occur when the turbulent flow approaches the fan blades. Eddies in the flow strike one blade and the next, as the fan rotates, as shown in the upper part of Figure 8.9 as (a). The force on the blade is periodic in the interval of time required for successive blades to pass a fixed point. The reciprocal of this period is the *blade passing frequency* f_p which is given in terms of the number of blades B and the rotational speed N rev/min by the expression,

$$f_p = B \times \frac{N}{60} \text{ Hz.} \qquad (8.2)$$

Normally, harmonics of the blade passing frequency will occur.

Discrete tones can also occur because of obstacle–blade interaction. This is an important source of noise in multi–stage axial flow compressors

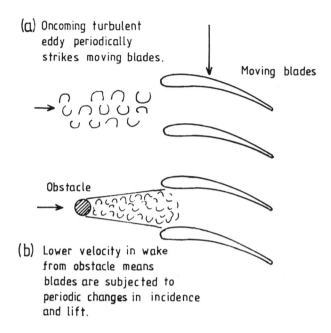

Figure 8.9 Mechanism for the generation of discrete tones in fans; (a) oncoming turbulence, (b) obstacle–blade interaction.

where a rotor is close to a row of stator blades. In the case of axial flow fans of the kind used for ventilation purposes the fan impeller is integral with the shaft of the drive motor, which is normally placed downstream of the impeller. The motor has to be attached to the outer casing by means of struts. These struts (or any obstacle in the flow) play a similar role to stators as far as noise generation is concerned. The mechanism of noise generation is illustrated in the lower part of Figure 8.9. Since the velocity in the wake downstream of the strut is lower than the mean velocity, a cyclical change in the lift forces occurs on the blades as the fan rotates. Thus a periodic force is exerted on the strut and the blades at the blade passing frequency with the result that dipole noise is produced. A similar mechanism occurs with centrifugal fans where the blade wakes interact with the edge of the cutoff. This type of noise can be reduced by increasing the separation between the blades and the obstacles or, in the case of a centrifugal fan, increasing the distance between the impeller tip and the cutoff edge [10].

Previously, Section 6.9.1, it has been shown that tones from centrifugal fans can be reduced by placing the entrance to a quarter wave tube at the cutoff. More commonly the entire noise from fans is reduced by a silencer of the dissipative type.

In order to reduce both discrete tone generation caused by approaching turbulence and broad band random sound all measures that are possible should be taken to prevent the creation of additional turbulence. Most typical flows will be turbulent, but it is possible to prevent unnecessary, additional turbulent eddies by ensuring that as few obstacles and obstructions as possible are upstream of the fan. Correct installation is particularly important in the case of axial flow fans [11]. Any inlet should have a smooth bellmouth instead of an abrupt entrance. Any constriction should be as far upstream as possible, and even then should offer as little obstruction to the flow as possible. For example, a flexible connector should be taut rather than slack. If there is a bend in the duct upstream of the fan, a straight section of duct of at least one duct diameter in length should be installed between the bend and the fan. This final point is illustrated in Figure 8.10. Obstructions not only cause turbulence which excites the fan to generate sound, but also act independently as a source of sound.

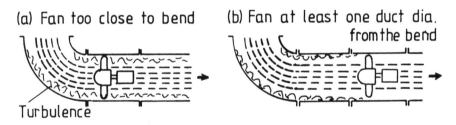

Figure 8.10 Axial flow fan in a duct; (a) incorrect installation (fan too close to the bend), (b) correct installation (fan at least one duct diameter from the bend).

The noise from fans can be kept to a minimum by selecting the correct fan for the duty it has to undertake. The fan manufacturer should be able to supply curves of static pressure against volume flow. The designer of the installation will require a certain volume flow of air for which he can estimate the pressure loss. If the manufacturer also gives information on the total fan efficiency, it is possible to select for the known duty a fan with as great an efficiency as possible. Minimum noise and maximum efficiency approximately coincide. If, however, the fan is subsequently required to operate against a larger pressure drop because, for example, a silencer has been introduced as an afterthought, the noise from the fan will increase. Reducing the static pressure below the design requirement does not increase the noise very much in the case of axial flow fans [12].

8.3.2 Jet noise

Jet noise, which arises from the turbulent mixing of the flow from a nozzle with the surrounding air, has been shown to be sound from quadrupoles [13]. (As we shall describe later, much of the noise from industrial jets is either monopole or dipole.)

Structure of a turbulent jet   A schematic diagram of a subsonic turbulent jet is shown in Figure 8.11. We will refer to a jet of air effluxed from a circular nozzle of diameter D. Any gas is possible and the nozzle could be non–circular.

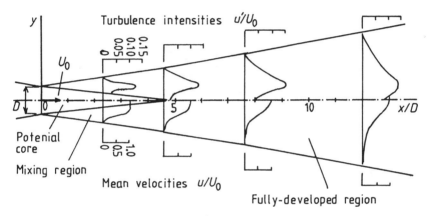

Figure 8.11 Structure of a turbulent jet; (turbulence intensity profiles are shown in the upper part and mean velocity profiles in the lower part).

A *potential cone or core* is formed directly downstream of the nozzle exit and contains an irrotational or laminar flow. The core length is 4 to 5 nozzle diameters and the velocity within the core is that at the nozzle exit U_0. A *mixing region* surrounds the potential core and it is in this region that the nozzle flow induces and mixes with air in the atmosphere. Further downstream the jet spreads out at a wider angle and forms a region of *fully–developed flow* where the velocity profiles are similar. A transition region connects the mixing and fully–developed regions.

In Figure 8.11 the *turbulence intensities* as well as the non–dimensional mean velocities are shown as functions of r for different transverse sections along the jet axis. The turbulence intensity is the ratio of the rms average of the velocity u' to the efflux velocity from the nozzle, U_0. The turbulence intensities have maximum values of about 0.15 and the maxima occur on the surface of a cylinder with a diameter equal approximately to that of the nozzle diameter [14].

The turbulent eddies start to form at the nozzle exit and increase in size as they are convected downstream with a velocity equal to about half that of the jet efflux [15]. The frequency of the sound generated is inversely proportional to the size of the eddies, thus the high frequency sound derives from the mixing region close to the nozzle and the low frequency sound emanates from the fully–developed jet which is well downstream of the nozzle.

Dependence of jet noise on velocity   The sound power W of a circular jet is given by [7]:

$$W = \frac{K\rho_0 U_0^8 D^2}{c_0^5},\tag{8.3}$$

where ρ_0 is the density of air in the atmosphere,
D is the diameter of the circular nozzle,
c_0 is the velocity of sound in the atmosphere and
K is a dimensionless constant which is typically 2.5×10^{-5}.
Equation (8.3) is sometimes referred to as the Lighthill eighth power law.

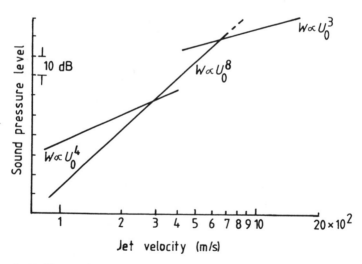

Figure 8.12 Dependence of jet noise on jet velocity (after Bushell [16]).

In deducing equation (8.3) it was assumed that all the noise derived from convected turbulent eddies in a subsonic jet. In some circumstances the eighth power law can prevail, even for supersonic jets, but may break down at low jet velocities. Measurements on jets from a wide range of nozzles, varying from small models to full scale turbojets, have shown that there is not always this velocity dependence. In Figure 8.12 the maximum sound pressure level is shown plotted against jet velocity [16]. The maximum sound level will occur at a certain angle to the jet axis which, although different for each jet, is about 30° or less. Some results show that the eighth power law is obeyed over a wide range of velocities from 100 to 700 m/s. The velocity of the jet depends upon pressure ratio and temperature; the pressure ratio is the ratio of the static pressure at an upstream station where the flow velocity is zero or very low — often in an upstream reservoir — to the atmospheric pressure or the static pressure surrounding the nozzle; temperature is that corresponding to the upstream static pressure. Jets used in a laboratory or workshop are generally cold jets, because they originate from a reservoir which is at room temperature. If the jets discharge from a simple convergent nozzle in which the smallest cross–section is the nozzle exit, the transition from subsonic to supersonic flow occurs when the velocity at the nozzle exit is slightly less than the velocity of sound c_0 in the surrounding atmosphere. The eighth power starts to break down after the transition to supersonic flow occurs. Thus for cold jets the eighth power law holds up to velocities of about 345 m/s.

Once the transition to supersonic flow occurs in a cold jet — by increasing the upstream static pressure — the pattern of sound changes dramatically. The velocity of the jet at the nozzle exit is always equal to the local velocity of sound, but further expansion of the flow takes place downstream of the nozzle exit by means of a series of shock waves within the jet, so that the velocity of the flow downstream — the effective velocity of the jet — is greater than c_0. A new mechanism for sound generation is possible. Vortices from the nozzle lip interact with the shock waves and form a series of monopole sources at the points of interaction [17]. The noise produced in this way is an intense, high–pitched screech at discrete frequencies; it is often referred to as screeching noise. The noise is highly directional and increases rapidly in intensity with increase of the upstream pressure. The eighth power law is not obeyed.

On the other hand, with hot jets — and this is the case with turbojet engines — jet velocities of almost 700 m/s can be attained without the occurrence of the transition to supersonic flow. The eighth power law will apply until this transition occurs or until the turbulent eddies are convected supersonically.

At low velocities the eighth power law is obeyed, provided great care is taken to ensure a smooth flow with low levels of internal noise and turbulence. Such a 'clean' flow can only be achieved in carefully controlled laboratory experiments. In most cases a U_0^4 dependence of sound power on jet velocity will occur and the noise from the jet will be greater.

A dependence of sound power on jet velocity to the power of four indicates that monopole sources are present. The monopole sources are generated upstream of a nozzle exit by obstructions, valves, etc. This type of noise is often referred to as *tailpipe noise* (in the context of jet engines) to distinguish it from true jet noise. It occurs in any duct where obstructions or spoilers are present (hence the name spoiler noise is sometimes used).

At very high velocities of jets the sound power depends on velocity to the power of three. At jet velocities greater than about twice the speed of sound in the atmosphere the turbulent eddies are convected supersonically and a completely different noise mechanism occurs. This type of noise applies to the jets from the Olympus engines of Concorde at take–off (when the jet velocity is 855 m/s (2800 ft/s)) or from rocket engines [18].

Directivity of jet noise   It is characteristic of lateral quadrupoles that the sound is very directional. The stationary quadrupole radiates maximum sound in four directions at 90° to each other, see Figure 8.7c. In the case of jet noise maximum sound would be expected at an angle of 45° to the jet axis, and no sound at 90°. However, the turbulent eddies — the sources of jet noise — are moving, and, as a consequence, the directionality of the sound is modified by a *Doppler shift*. Theoretically, the directivity factor is proportional to $(1 - M_c \cos \theta)^{-5}$, where M_c is the Mach number of the convected turbulent eddies (convection velocity to the velocity of sound c_0 in the atmosphere) and θ is the angle that the measurement axis makes with the jet axis [7].

Some theoretical results are shown in Figure 8.13 for two different eddy convection Mach numbers. These results are relative to a datum at $\theta = 90°$, i.e., at right angles to the jet axis. Measured results often show that the maximum noise level occurs at an angle of 30° to the jet axis.

Frequency content of jet noise   Jet noise from subsonic jets is random in nature and covers a wide frequency range. The frequency f at which the sound energy generated by the jet is maximum depends on jet velocity U_0 and nozzle diameter D. Spectra of jet noise can be normalised in terms of a non–dimensional frequency parameter, the *Strouhal number* S_t, where

$$S_t = fD/U_0. \tag{8.4}$$

The sound power in unit frequency band for cold jets is shown in Fig-

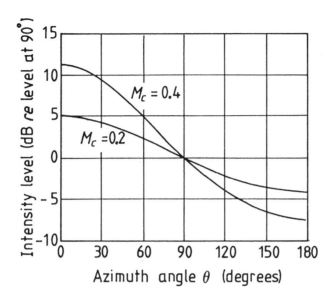

Figure 8.13 Directional distribution of subsonic jet noise; M_c is the Mach number of the convected turbulent eddies.

ure 8.14 as a function of the Strouhal number. As long as the jet efflux velocity is subsonic ($U_0 < c_o$) the spectrum of sound is broad band random and contains no discrete tones. The power spectral density function is greatest when the Strouhal number is about 0.3 to 0.5.

Reduction of jet noise Compressed air is used industrially for a variety of purposes. Wherever compressed air is discharged to atmosphere, jets are created and a noise problem arises. In some applications high velocity, efficient jets have to be used. For example, in machines for sorting potatoes or minerals an air jet may be used to blow away, say, a discoloured potato. In this case the nozzle has to be well designed to produce a jet which exerts the greatest possible force on the potato. The flow process in the nozzle needs to be close to isentropic so that the minimum amount of energy is wasted. For this type of jet the techniques used for silencing the jets from turbojet–propelled aircraft can be attempted.

One method for reducing the noise from high velocity subsonic jets is to change the shape of the nozzle exit from a circle to a series of lobes or smaller circles. Two variants are shown in Figure 8.15. The exact shape of the nozzle exit is not so important as long as the exit cross–section is subdivided into several smaller areas and secondary air can be

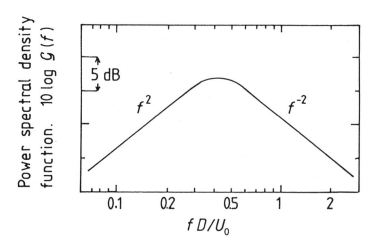

Figure 8.14 Sound power in unit frequency band for jet noise as a function of Strouhal number.

induced between the lobes or smaller circular sections. The changes in cross–section upstream of the exit must be gradual so that thrust losses are minimised. This type of nozzle will give a noise reduction of 3 to 5 dB [19,20].

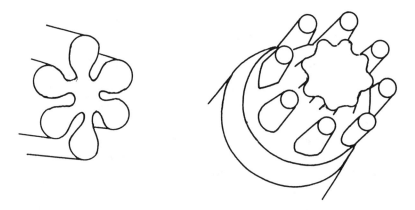

Figure 8.15 Reduction of noise by segmentation of the nozzle exit area.

A second method is to use an ejector, as shown in Figure 8.16a. The most effective way of reducing the noise from air jets is to decrease the jet velocity. The ejector effectively reduces the velocity from the exit of the ejector tube whilst maintaining, or possibly increasing the thrust. A disadvantage of the ejector tube is that it has to be about seven ejector tube diameters in length for maximum noise reduction. It is also possible for the air in the tube to oscillate like the air in an organ pipe and produce discrete tones at the natural frequencies of the ejector tube. Both of these disadvantages can be overcome if the ejector tube is used with the type of lobed nozzle shown in Figure 8.16b. With the lobed nozzle more secondary air is induced through the ejector tube from the atmosphere and mixing of the primary flow from the nozzle with the secondary air occurs in a shorter length. An ejector tube length of three tube diameters is sufficient for maximum noise reduction. Additionally, the organ pipe oscillations are eliminated. A short ejector with a lobed nozzle can give noise reductions of up to 8 dB in sound power level [21]. The ejector tube can also be lined with sound absorbing material (protected from the flow by perforated sheet) to give greater attenuation.

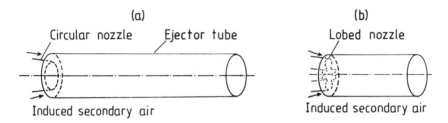

Figure 8.16 Reduction of noise by ejector tubes with (a) circular nozzle and (b) lobed nozzle.

In many applications of compressed air it is not necessary to produce efficient jets. For each application the pressure in the air line should always be reduced to the minimum necessary for the job. This means the introduction of a pressure reducing valve with the correct setting for each airline that discharges to atmosphere.

Many low velocity jets from industrial air lines emit internally generated noise which is monopole in origin. Nozzles incorporating silencers can be fitted to the end of the air line. The noise reduction action of the silencers depends upon sound–absorbent material and an increase in area (and hence lower velocity). A typical silencer nozzle is shown in Figure 8.17.

When larger quantities of air are being discharged a blow–off silencer of the kind shown in Figure 8.18 can be used. Various principles of noise control are applied; the cross–sectional area of the flow is increased, the

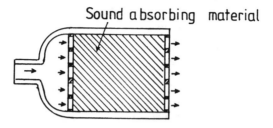

Figure 8.17 Air line silencer.

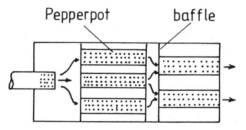

Figure 8.18 Blow–off silencer.

holes in the 'pepperpot' provide acoustic damping, the expansion chambers have a reactive effect and the baffle plates provide sound insulation.

The screeching noise from supersonic jets, described earlier, can be reduced by inserting small teeth into the jet at the nozzle exit.

When using air jets the operator should ensure that the jet is free and not blowing against solid surfaces. Impingement of the jet bodies introduces additional noise sources and can increase the noise by as much as 15 dB.

8.4 Hydrodynamic noise sources

Noise sources in liquid systems can derive from

    (i) pulsations in the flow,
    (ii) turbulence,
    (iii) obstructions,
    (iv) bubbles of air or gas in the liquid, or
    (v) cavitation.

8.4.1 Pulsations in the flow

Hydraulic pumps, such as vane, gear or swash plate types, produce pressure fluctuations of large amplitude which excite vibrations in the pipes containing the flow. The amplitudes of the pulsations can be reduced by placing attenuators in the pipe downstream of the pump. These attenuators are similar in principle to the expansion chamber silencers which have been discussed in Section 6.4 [22,23]. A chamber of large volume is used to smooth out the pulsations.

8.4.2 Turbulence

Turbulence in liquids is not in itself generally found to be a significant noise source, as the flow velocities are low. However, the turbulent pressure fluctuations in the boundary layer adjacent to the wall of the pipe cause random vibration of the pipe and sound is radiated. Turbulence is caused by changes in section, branches and valves, etc., and can be reduced by keeping the flow velocities as low as possible, and by ensuring that bends are rounded or have guide vanes. Specially silenced valves are available which make use of multiple restrictions, tortuous paths or inline silencers downstream of the valves [24,25].

8.4.3 Obstructions

An obstruction in a duct generates both turbulence and sound, whether the fluid is a gas or a liquid. Turbulence decays quickly and is not in itself a significant source of noise in liquids, whereas the sound generated by the obstruction will be transmitted long distances through the pipe. (This source of noise is the equivalent of the spoiler or tailpipe noise discussed in Section 8.3.2 in the context of jet noise).

8.4.4 Bubbles of air in the liquid

The sound of running water is caused by air bubbles in the water which oscillate at their natural frequencies. The frequency f of a bubble is given by [26]:

$$f = \frac{0.0102\sqrt{P_0}}{a} \text{ Hz}, \tag{8.5}$$

where P_0 is the static pressure surrounding the bubble in Pa, and a is the radius of the bubble in m. Thus for a pressure of 10^5 Pa and a bubble radius of 1 mm the natural frequency is 3.2 kHz.

Bubbles form monopole sources of sound when they oscillate. Additionally, sound is created when bubbles form, collapse, divide or coalesce.

8.4.5 Cavitation

In all liquids there are very small voids which are full of gas. When the static pressure in the liquid decreases to that of a threshold pressure, the voids form the nuclei out of which larger bubbles grow. This process of bubble formation is called cavitation. The phenomenon of cavitation occurs in valves, pipe expansions, pumps, turbines, propellers or wherever there is a possibility of low static pressures. The cavitation bubbles grow and eventually collapse. The main source of noise arises from the collapse of the bubbles.

Cavitation can be avoided by ensuring that pressure drops at bends and expansions, etc., are low. In the case of valves special designs are available [27].

8.5 Acceleration noise

Bodies have to possess elasticity to vibrate. Although most bodies are capable of deformation, objects such as small spheres or cylinders are so stiff that their natural frequencies may be above the upper frequency limit of audibility. However, two small spheres, if allowed to strike each other, do emit a sound. This sound does not derive from the vibration of the spheres, but from the acceleration and deceleration during impact. When the spheres collide the area around the point of impact is subjected to an elastic deformation, whilst the rest of the spheres remain effectively rigid. This is the basis of the Hertz theory of contact which allows the impact force, contact time, etc., to be calculated [28]. The contact times are of very short duration — 35 μs in the case of steel spheres of diameter 25.4 mm with a relative velocity of spheres just before impact of 1.52 m/s — and the rate of change of velocity with time is very great. If collision occurs between a moving sphere and a stationary sphere, the impactor is subjected to a deceleration and the impactee to an acceleration. The term acceleration noise is used to describe the type of sound generated by the sudden movement of rigid bodies.

In the case of the impact of two spheres the sound is directional and the peak sound pressure has its greatest value in the direction of the line joining the centres of the spheres. The measured sound pressure against time in the direction of maximum sound radiation is shown in Figure 8.19 for a typical impact between two equal spheres. The way in which sound pressure varies with time can be calculated and reasonable agreement with measurements can be obtained [29].

It is not entirely clear how important acceleration noise is in industrial machines. Richards [30] has shown that the energy emitted as sound can never be greater than 1.5×10^{-4} times the kinetic energy input at impact. The maximum emission occurs only for very short contact times and the fraction of energy emitted as sound decreases sharply as the contact time of the impact increases. Thus acceleration noise is only likely to be important

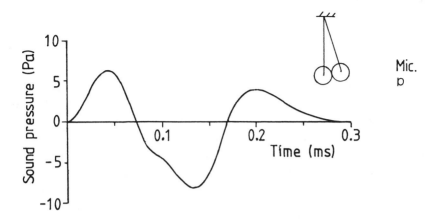

Figure 8.19 Prediction of sound pressure against time for impact of two equal spheres. (Diameters of spheres 25.4 mm, closing velocity 1.5 m/s; sound pressure estimated at a point in line with the sphere centres, 0.375 m from the point of impact.) Calculations performed by S. Hosseini–Hashemi, City University.

in very short metal to metal contact, as, for example, in gears.

8.6 Identification of noise sources

It is essential to identify the source of noise before attempting noise control. It can be a costly mistake to apply noise control in the form of an enclosure to, say, a boiler when the main noise source is the top of a chimney stack. This section is concerned with some of the measures that are used to detect noise sources. Some of the examples have been drawn from the article by Ebbing [31].

8.6.1 Listening to the noise source

The ears are very sensitive organs and can often be used to detect changes of frequency and intensity much more effectively than expensive measuring equipment. Sometimes it is difficult to listen to the noise source, because it is inaccessible. To overcome this problem the microphone can be attached to a hand–held boom and connected via an extension cable to a sound level meter. The output from the sound level meter may be played into headphones. There is sometimes a difficulty in matching the impedance of the sound level meter and the headphones. Electrical circuits designed to overcome this problem are available [31,32].

Easily localised sources, such as bearings, can be examined with a stethoscope.

8.6.2 Source modification

Sir William Bragg in his book based on the Royal Institution lectures gives an example of how the source of noise can be detected by modifications made in the noise producer [33]. In his example a rasping noise was created by a beetle and it was not clear where it came from. A drop of oil was placed on different parts of the beetle in turn until the noise stopped. The rasping noise came from the legs rubbing the body.

It is sometimes possible to eliminate entirely one source in a machine. For example, the noise from the exhaust of an internal combustion engine can be very well silenced and additionally a long exhaust duct can be used so that the outlet is at a large distance from the engine. The exhaust is no longer contributing to the total noise from the engine, and thus the effect of the exhaust on the total noise can be estimated.

An example of source modification was provided by Priede and colleagues [34] who showed that the fan belt pulley of a diesel engine was a significant source of noise. They temporarily removed the pulley and measured the noise both with and without the pulley. The results, obtained with an octave band analyser of the non–contiguous type, are shown in Figure 8.20 and indicate that the pulley is a source of noise. (Any misalignment causes the pulley to act like a piston.) Once it is known that the pulley is a noise source, it becomes a suitable subject for noise control. Vibration isolation of the pulley reduces the noise, see Figure 8.20, although not by as much as total removal. (The technique of vibration isolation is described in Section 8.7.3 and details of the isolated pulley are presented in Figure 8.37.)

8.6.3 Spatial distribution of sound fields

The way in which the sound pressure level varies with measurement position can give useful information on sound sources. As an example, let us consider a burner with a chimney stack of height 21 m. It is not sure whether the noise is radiated directly from the burner or from the top of the stack. Measurements in the far field along the ground at different distances from the burner should be made. If the burner is the main source of noise, we would expect to see a 6 dB decrease in sound pressure level for every doubling of the distance away from the burner. However, if the top of the stack is the main source of noise, we would only expect to experience a decay in sound pressure level of 6 dB per doubling of distance along the ground when measurements are taken some considerable distance away.

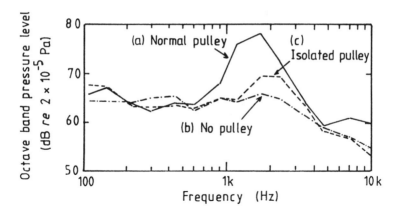

Figure 8.20 Noise from fan belt pulley; (a) unmodified, (b) with pulley removed and (c) reduced by vibration isolation.

Let us take an example with some values of sound power level. Let the top of the chimney stack be a source of spherically symmetric waves of sound power level 100 dB re 10^{-12} W (sound power of 10^{-2} W) and let the burner produce a sound power level of 80 dB (sound power of 10^{-4} W). For simplicity take the acoustic centre of the burner to be 1 m above the ground on the axis of the chimney stack. Measurements are made at a height of 1 m on a hard, reflecting surface. Hemispherical radiation (see equations (5.6) and (5.13)) is assumed for the burner and spherical radiation (see equation (5.5)) for the top of the stack. The mean–squared sound pressure $\overline{p_d^2}$ of the combined field is given by:

$$\overline{p_d^2} = \frac{W_{\text{chimney}}\rho_0 c_0}{4\pi(h^2 + d^2)} + \frac{W_{\text{burner}}\rho_0 c_0}{2\pi d^2},$$

where W_{chimney} is the sound power from the top of the chimney,
         W_{burner} is the sound power from the burner,
         h is the height of the chimney stack above the burner,
         d is the distance from the burner and
         $\rho_0 c_0$ is the specific acoustic resistance
            (taken to be 400 kgm^{-2}s^{-1} in this example).

Thus,

$$\overline{p_d^2} = \frac{400}{2\pi}\left(\frac{10^{-2}}{2(400 + d^2)} + \frac{10^{-4}}{d^2}\right).$$

The sound pressure level from the combined field and the sound pressure levels from the chimney stack and burner alone can be calculated for different values of the distance d. The results of the calculations are shown in Figure 8.21.

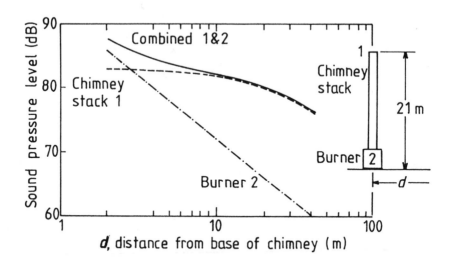

Figure 8.21 Calculated variation of sound pressure level with distance for sound from burner and chimney stack separately and from the two sources combined.

The sound pressure level of the combined field — which should be the same as the measurements — does not show a decay of 6 dB per doubling of distance for small values of d. Only at distances from the burner of more than 40 m is the inverse square law obeyed. If the burner had been the main source of noise, the inverse square law decay would have been obtained for all distances.

Measurements in the field would rarely give such clearcut evidence as the calculations. In reality there would be complications; the sources would be directional, there would be obstacles which reflect sound and atmospheric conditions would have an effect.

Point, line and area sources   It is particularly important to relate the decay of sound pressure level with distance to the type of sound source. The decay of sound pressure level with distance for idealised point, line and area sources is shown in Figure 8.22 [31]. All sources obey the inverse square law decay at a sufficient distance away. With an area source, such as a plate or the side of a diesel engine, the sound level close to the plate is independent of distance. This is the region of the near field of the

sound source (see Sections 1.19.1 and 1.19.2). In reality quite complicated
things may happen in the near field close to the plate; there may be small
amplitude fluctuations in the sound pressure with distance [35].

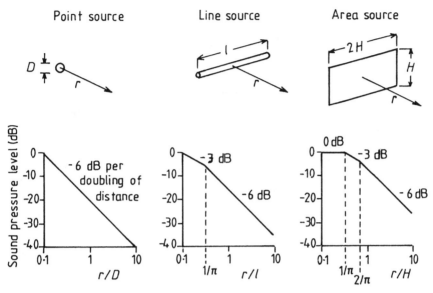

Figure 8.22 Point, line and area sources [31].

8.6.4 Near field sound measurements

In a machine or plant with many separate noise sources it is often useful to
know the sound power levels of the individual sources so that the most significant can be identified. For most industrial purposes A–weighted sound
power levels are sufficient. A–weighted measurements emphasise sound
at frequencies around 1 to 2 kHz, where the wavelengths are relatively
small. In Section 1.19 various conditions are given for the locations of the
near and far fields. In practice a definite boundary between the near and
far fields has to be decided; in many situations, as an approximation, the
boundary may be taken to be half a wavelength, or less, from the source.
In the far field the relationship, $\bar{I} = \overline{p^2}/(\rho_0 c_0)$, between time–averaged
sound intensity and mean–squared pressure may be used. (It can also be
used in that part of the near field called the geometrical near field, see
Section 1.19.2.)

    If A–weighted sound levels are measured at several positions close
to the surface of a machine or component, say, at a distance of 50 to

100 mm away, approximately in the geometrical near field, the time–averaged sound intensity can be related to the mean–squared pressure. The sound power W of the component is equal to the product of the spatial average of the time–averaged sound intensity and the area S of the surface of the component (equation (1.129)). For A–weighted sound power W_A

$$W_A = \frac{\overline{\langle p_A^2 \rangle} S}{\rho_0 c_0},\tag{8.6}$$

where $\langle \overline{p_A^2} \rangle$ is the average of the mean–squared pressures obtained at different positions close to the component. (Note that $\langle\ \rangle$ indicates a spatial average.) Normally, either A–weighted measurements will be taken with a sound level meter switched to the slow response at fixed points on a regular grid in space, or an integrating meter will be slowly traversed, backwards and forwards, over the surface. In terms of the A–weighted sound power level L_{WA} and average sound level $\langle L_A \rangle$ equation (8.6) becomes, approximately,

$$L_{WA} = \langle L_A \rangle + 10 \log S \text{ dB},\tag{8.7}$$

where

$$\langle L_A \rangle = 10 \log \langle \overline{p_A^2} \rangle / (2 \times 10^{-5})^2 \text{ dB}.\tag{8.8}$$

In deriving expression (8.7) it is assumed that $\rho_0 c_0 = 400 \text{ kgm}^{-2}\text{s}^{-1}$. One notes that the sound power level depends not only upon the average sound levels but also upon the area of the component.

The method is illustrated in relation to a centrifugal chiller [31]. As shown in Figure 8.23 the refrigerant is pumped by the compressor 1 to the condenser 11 via a pipe 10 and returns to the compressor via the evaporator 8 and pipe 7. The compressor is driven by a motor 5 through a gearbox 3 with shafts connected by couplings 2 and 4. The ends of the condenser and evaporator are at 9 and 6, respectively.

The A–weighted sound levels, spatially averaged over the component surfaces, are listed in Table 8.1 with the surface areas and the sound power levels estimated from equation (8.7).

The pipe connecting the compressor to the condenser, where the flow is expected to pulsate, ranks first for both sound level and sound power level. It is interesting to note that the second and third most significant sources of sound power are the condenser and evaporator, although the sound levels of these two components are smaller than those of the gearbox and the couplings.

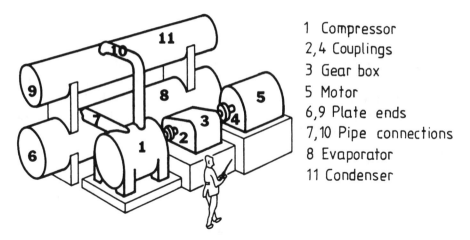

1 Compressor
2,4 Couplings
3 Gear box
5 Motor
6,9 Plate ends
7,10 Pipe connections
8 Evaporator
11 Condenser

Figure 8.23 Diagram of centrifugal chiller.

TABLE 8.1 Noise sources in a centrifugal chiller [31]

Plant component	Spatially averaged sound level dB(A)	Surface area S m²	10 log S	Estimated sound power level dB(A)
1	94	13.6	11.3	105.3
2	98	0.2	−7.0	91.0
3	97	6.8	8.3	105.3
4	98	0.2	−7.0	91.0
5	92	9.1	9.6	101.6
6	91	5.3	7.2	98.2
7	93	13.6	11.3	104.3
8	92	58.3	17.6	109.6
9	95	3.6	5.6	100.6
10	105	5.3	7.2	112.2
11	96	29.2	14.6	110.6

This method of ranking noise sources works quite well for machines with large cylindrical and plane surfaces, but is it not so suitable for machines with small components and voids.

8.6.5 Surface velocity measurements

This method is similar to that of Section 8.6.4, except that, instead of close (to sources) measurements of sound level, velocity transducers (or, more likely, accelerometers with an integrator) are attached to different points on the surface of the machine. From equation (8.1) for a radiation efficiency of unity one obtains

$$W = \rho_0 c_0 \langle \overline{u^2} \rangle S, \qquad (8.9)$$

where $\langle \overline{u^2} \rangle$ is the average of the mean–squared velocity at several positions on the surface of the machine. As equation (8.9) only applies at high frequencies it is appropriate to use the A weighting network on the velocity measurements. Note that it is usually possible to use an accelerometer and integrator with a conventional sound level meter. Attention should be given to the calibration of the velocity measurements.

The method of surface velocity measurements is applicable only if the sources are vibrating surfaces. It has the advantage that it excludes the effect of sound from other sources. It has the disadvantage that the accelerometer has to be attached in turn to several different positions on the surface. In some applications it may be possible to scan the surface by a laser velocity transducer.

8.6.6 Time domain analysis

In machine operations that include impacts it is important to know at which point in the cycle of operations the greatest noise is produced. Measurements in the time domain are often the most useful. A dual beam storage oscilloscope or a dual channel narrow band analyser (such as Brüel and Kjaer types 2034 and 2133) can be used to obtain a trace of sound pressure against time on one channel and to obtain the output from a photocell or force transducer, etc., on the second channel.

An example of the method is illustrated in Figure 8.24 with respect to a small C frame punch press. During the cycle of operations performed by a a punch press there are several possible noise sources, depending upon the type of press used. There will be noise from the clutch, from contact of the punch with the workpiece, blank ejection, etc. [36]. The part of the operation during which the punch is in contact with the material is considered in Figure 8.24. The upper trace (a) is acceleration against time and has been obtained from an accelerometer fixed to the die plate. As most of the noise derives from the transient vibration of the press a trace of sound pressure versus time would have shown similar characteristics. The axial force exerted by the punch was measured by strain gauges placed as close to the end of the punch as possible. The strain gauges can be used to measure force when placed in an ac bridge in a certain way [37]. Force against time is shown as trace (b). The results in Figure 8.24 were obtained with a dual channel Fourier analyser (Hewlett Packard type 5451C).

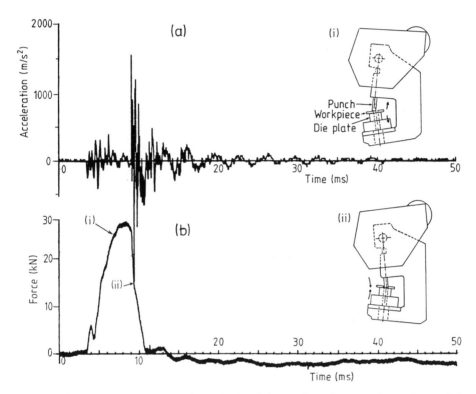

Figure 8.24 Punch press dynamics: (a) acceleration against time; (b) force against time. Schematic diagrams of punch press with (i) increasing force and (ii) decreasing force.

The greatest acceleration values occur when the blank fractures from the workpiece, and not at the initial impact. When impact takes place the material is deformed elastically at first and the frame of the press is relatively gradually displaced. At time referring to the position indicated as (i) on the force–time curve, Figure 8.24b, the deformation of the press is as shown in an exaggerated manner in the inset diagram (i). When fracture occurs at (ii) on the force–time curve, the internal force which has been pushing the C frame apart is suddenly removed and the frame springs impulsively back and after a transient oscillation reverts to its original position. It is the impulsive motion (during the return of the press to its initial position) which is the major source of noise. At the moment referring to (ii) the press takes up a position shown in the inset diagram (ii).

Loom noise has been successfully analysed by considering the variation of sound pressure with time in conjunction with a stroboscope [38].

8.6.7 Narrow band analysis of sound

The subject of narrow band analysis has already been discussed in Chapter 2. Narrow band analysis is particularly useful for locating sources which are dependent upon speed of rotation and give rise to pure tones. For example, a fan may produce a tone at the blade passing frequency (equation (8.2)); thus, if such a pure tone can be located in the spectrum so that its frequency is equal to the product of the number of blades and the rotational speed in rev/s, it is almost certain that the fan is the source of noise.

The presence of a speed–dependent source of noise can sometimes be confirmed by changing the speed of the machine, if it is possible to change the speed. As an example, consider the noise from a pair of meshing gear wheels. Gear noise often includes a tone at the 'tooth passing frequency'. This frequency is equal to the product of the number of teeth and the rotational speed, i.e., very similar in definition to the blade passing frequency of a fan. In the case of gear noise from a single pair of meshing gears of equal size, for which the spectrum is shown in Figure 8.25a, the speed of rotation is 800 rev/min and the number of teeth is 30 for each wheel; the tooth passing frequency is 400 Hz. In the spectrum a discrete tone can be identified at 400 Hz and there is also, amongst others, a very prominent tone at 4288 Hz. (Note that the resolution between lines Δf is about 4.88 Hz).

On changing the speed to 1400 rev/min, the 400 Hz tone disappears, see Figure 8.25b, but a new tone emerges at 703 Hz, which is a frequency close to the new gear tooth passing frequency. The prominent high frequency component remains almost unchanged with speed. The tone at 4288 Hz is caused by the bending vibrations of the wheels which are solid discs. (Note that two frequency components may be seen because the two wheels are not exactly identical.) The same frequency will be recorded, if the wheel is struck and the wheel allowed to ring. Damping materials can be used to reduce this type of noise, see Section 8.7.2.

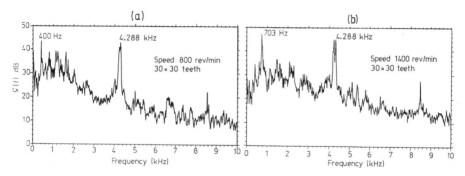

Figure 8.25 Narrow band spectrum of gear noise at a rotational speed of (a) 800 rev/min and (b) 1400 rev/min.

8.6.8 Sound intensity analysis

Sound intensity analysers, described in Section 2.23, can be used to detect noise sources. Intensity meters use a probe with two microphones which is able to measure the gradient of the sound pressure, and hence the particle velocity. If the source under investigation is the side wall of the cylinder block of a diesel engine, the microphone is moved over the surface and average readings are taken. From a knowledge of the area S of the component, or the control surface over which the measurements are made, the sound power can be estimated as the product of the area S and the time–averaged sound intensity.

In an experiment reported by Reinhart and Crocker [39] the sound power levels from different parts of a diesel engine have been measured by the sound intensity method and compared with results from two other techniques. The results are shown in Table 8.2. The surface intensity method is a combination of the approaches described in Sections 8.6.4 and 8.6.5. Lead wrapping is basically the method presented in Section 8.6.2; the engine is wrapped in lead, apart from the component under investigation. The surface intensity method only works if there is a plane surface on which an accelerometer can be mounted; thus some results are omitted from the table. Lead wrapping probably overestimates weak sources, such as the pumps and air cooler, because of transmission of sound from other stronger sources through the lead.

There is quite good agreement for all methods for the high sound power sources, but the sound intensity method is probably more accurate for the less significant sources.

TABLE 8.2 Comparison of overall sound power levels for a diesel engine obtained by the sound intensity, surface intensity and lead–wrapping methods [39]

Engine part	Overall sound power level in dB(A) re 10^{-12}W. Speed 1500 rev/min; torque 542 Nm		
	Sound intensity	Surface intensity	Lead-wrapping
Oil pan	102.7	103.3	102.6
Exhaust manifold, turbocharger, cylinder head and valve covers	101.4	–	101.6
Aftercooler	100.8	101.9	100.6
Engine front	95.0	–	100.0
Oil filter and cooler	91.1	93.4	98.1
Left block wall	97.4	94.6	97.3
Right block wall	94.8	93.3	97.3
Fuel and oil pumps	91.5	–	96.3
Sum of all parts	108.1	–	108.8
Sum of oil pan, aftercooler, oil filter and cooler, and block walls	106.1	106.4	106.7
Bare engine sound power level	108.1	–	109.5

8.7 Noise reduction at the design stage

In Section 8.7 we consider the methods that can be adopted by the designer at the design stage. It is clearly advisable to consider noise control prior to manufacture, although in many situations the measures described here are also applied at the development stage. We will also consider the techniques to be considered when machinery is installed. If such measures are applied as an afterthought they are less successful.

Sometimes a process can be replaced by a new one that leads to less noise. For example, looms in weaving sheds are very noisy, partly because of the impact with the frame of the shuttle carrying the thread. In the case of later designs for man–made fibres it is possible to make use of surface tension to carry the thread in a small jet of water, so considerably reducing noise. The water jet loom is quieter than the conventional loom by about 10 dB(A), because the shuttle and picking arm, responsible for impact noise, have been replaced by the silent water jet. In manipulative machinery impacts are utilised because of the useful effect of the large forces that exist for a short time. Great forces can also be attained by hydraulic rams, with smaller possibility for noise. Hydraulic methods have been devised for sheet piling and for breaking up rocks.

8.7.1 Control of input waveform

Some examples of the power spectra of idealised pulses have been presented in Section 8.2.1. As a principle of noise control it is often noted that the input forces should be spread over as long a time as possible and reduced in amplitude. In general this is an excellent principle. However, as indicated by Figure 8.3, smoothing the waveform reduces the amplitudes of the high frequency components, but does not have much effect on low frequency components.

Priede and his colleagues have made a systematic study of the effect of the input on internal combustion engines and vehicle bodies [34,40,42]. In the case of petrol or diesel engines the input is the cylinder pressure; with vehicle bodies the input is the torque transmitted from the engine to the vehicle body. As an example of Priede's method we will consider the diesel engine.

The rise of cylinder pressure (in a diesel engine) with crank angle is relatively sharp, as can be seen in Figure 8.26a, where the cylinder pressure is shown in a schematic and exaggerated way to emphasise the point to be made. The cycle is repeated every two revolutions of the crank for each cylinder in the four stroke cycle. Spectrum analysis of the cylinder pressure or gas forces leads to spectra which depend upon the rotational speed of the engine, see Figure 8.26b [40]. The effect of an increase in speed is to shift the spectra to the higher frequencies without altering the shape. An increase in speed by a factor of 10 represents a frequency shift of one decade. Note that the spectra shown in Figure 8.26b are idealised. In reality they comprise a series of harmonics with the fundamental frequency which is the repetition rate of the cylinder pressure peak with time; the curves shown in Figure 8.26b are the loci of the harmonics in the frequency domain. The important point to note is that at the high frequencies, the idealised spectra (loci) decrease at a rate of 30 dB per decade.

The idealised acoustical–structural response of the diesel engine is shown in Figure 8.26c. This is a combination of the structural response and radiation efficiency, two component blocks illustrated in Figure 8.1. The acoustical–structural response is obtained by measuring the sound in the far field when a sinusoidal force of constant amplitude is applied to the structure. In reality the acoustical–structural response is a function which varies considerably with frequency; the two straight lines shown in Figure 8.26c are obtained by approximating the peaks of the response [41]. An important point to note in the response is that the maximum value occurs at 1000 Hz. The response shown in Figure 8.26c is typical of many structures which are very stiff. Internal combustion engines and many machine tools are examples of stiff structures with maximum values of acoustical–structure responses around 1000 Hz. Although it might not seem to be an advantage to have an acoustical–structural response determined from measurements, once obtained it can be used to predict the spectrum of sound for different, hypothetical inputs.

The spectrum of the engine noise is the product of the spectrum of the gas forces and the acoustical–structural response; two idealised spectra are shown in Figure 8.26d for speeds of 600 and 6000 rev/min. The spectra show peaks at 1000 Hz, because the acoustical–structural response peaks at that frequency. The spectra are separated by 30 dB, a consequence of the fact that the spectra of cylinder pressure decrease at 30 dB per decade.

The peaks in the noise spectra coincide approximately with the maximum value of the response of the A weighting network, see Figure 2.5. Thus the A–weighted sound level will increase by 30 dB for every increase in speed by a factor of 10, as shown in Figure 8.26e. It follows that the A–weighted time–averaged sound intensity $\overline{I_A}$ varies with engine speed N to the power three.

The approach adopted here can be used to predict how the noise varies with load. The way in which the load influence cylinder pressure is shown in Figure 8.27. The sharp rise in cylinder pressure always occurs, whatever the load. It is this sharp rise which controls the spectrum of cylinder pressure at the higher frequencies. Consequently, the spectrum at the higher frequencies remains largely unchanged. As a result, the effect of load on noise is small; an effect confirmed by experiment [41].

A similar technique has been applied to petrol engines [41]. In the

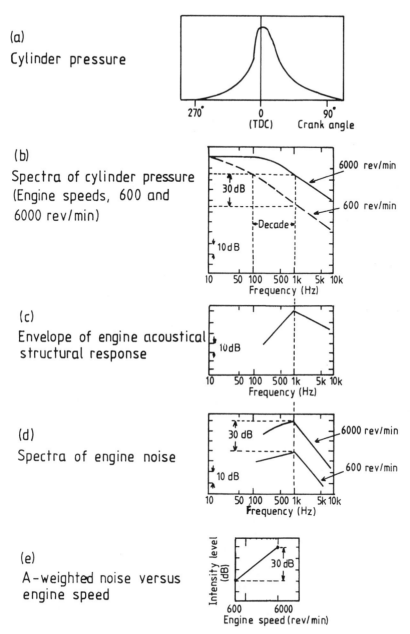

(a)
Cylinder pressure

270°   0   90°
            (TDC)   Crank angle

(b)
Spectra of cylinder pressure
(Engine speeds, 600 and
6000 rev/min)

6000 rev/min
30 dB
600 rev/min
←Decade→
10 dB
10   50 100 500 1k   5k 10k
        Frequency (Hz)

(c)
Envelope of engine acoustical
structural response

10 dB
10   50 100 500 1k   5k 10k
        Frequency (Hz)

(d)
Spectra of engine noise

30 dB
6000 rev/min
10 dB
600 rev/min
10   50 100 500 1k   5k 10k
        Frequency (Hz)

(e)
A-weighted noise versus
engine speed

Intensity level (dB)
30 dB
600   6000
Engine speed (rev/min)

Figure 8.26 Diesel engine noise: (a) cylinder pressure against crank angle, (b) frequency spectra of cylinder pressure, (c) acoustical–structural response of engine, (d) frequency spectra of engine noise and (e) A–weighted intensity level as a function of speed.

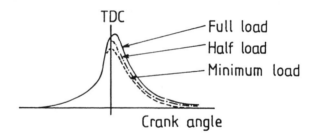

Figure 8.27 The effect of engine load on the cylinder pressure of a diesel engine.

case of petrol engines the variation of cylinder pressure with crank angle is more smooth than in a diesel engine. Consequently, in the frequency domain the harmonics of the cylinder pressure decrease at a rate greater than that of the diesel engine, in fact at 50 dB per decade. The A–weighted time–averaged sound intensity is proportional to engine speed to the power of five. With decreasing load the maximum cylinder pressure decreases. Consequently, for petrol engines the noise decreases with load [41].

In the case of the diesel engine smoothing the input with respect to time would be beneficial. If the increase in cylinder pressure with crank angle could be more smooth, the spectra would decrease at a rate greater than 30 dB per decade and the noise would be reduced. In cases where the structure is much less stiff there may be no benefit in noise reduction to be gained by smoothing the input force against time. A car body provides an appropriate example. The acoustical–structural response of the body of a passenger car has maximum values at frequencies around 100 Hz [41]. In the case of a car body the input which is causing the vehicle to vibrate, and hence radiate sound, is the torque which is transmitted to the body via the engine mounts. The frequency of 100 Hz coincides, in the case of higher operational speeds, with the region of the input spectrum where there is no change in level with frequency. Thus smoothing the torque with respect to crank angle or time would not result in a reduction of noise from the car body.

In Figure 8.28 it is shown how the input torque affects the final noise.

The method developed by Priede and his colleagues has been applied to gear noise [43], fuel injection pumps and others. More recently Richards [44] has elaborated the technique and looked more closely at the individual

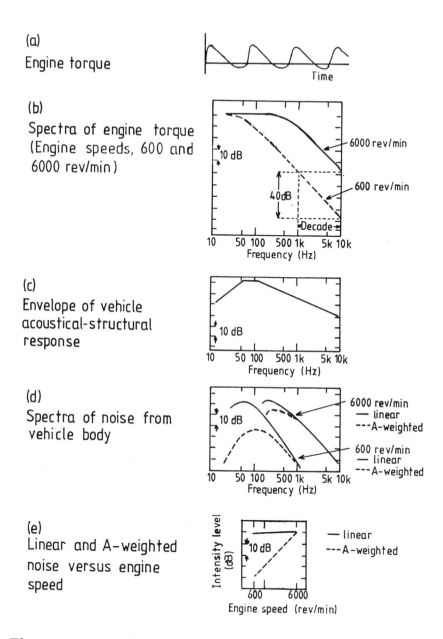

Figure 8.28 Noise from a car body: (a) engine torque against time, (b) spectra of engine torque, (c) acoustical–structural response of car body, (d) frequency spectra of car body noise and (e) A–weighted intensity level as a function of speed [41].

stages. In particular he has separated the structural response from the radiation efficiency. This has enabled him to emphasise the importance of radiation efficiency in the control of noise [6]. As the radiation efficiency depends upon the shape of a machinery component, the designer has the possibility of selecting a component with a low radiation efficiency.

8.7.2 Control of the structural response by damping

The two most common techniques for reducing the structural response — where structural response means the ratio of velocity to force (or mobility) as a function of frequency— are by damping and by vibration isolation. It is also possible to alter the stiffness or the distribution of the mass. Damping will be considered first. The method of manufacture of the structure will influence the amount of damping that is achieved. A riveted or bolted structure achieves some damping by means of *dry friction*. Attached components are able to move relative to one another and the friction forces act to reduce the amplitudes of vibration of the elastic components. In the case of a ship a riveted hull has five times more inherent damping than a welded hull [45]. A welded structure, such as a bedplate to support a machine, has less damping than a cast iron structure.

Material damping   That cast iron provides more damping than steel implies that there is material (internal) damping which depends upon the type of material used for the structure. According to Hooke's law, the force applied to an elastic body is proportional to the deformation (see page 21) or stress is proportional to strain. In reality Hooke's law is not perfectly verified. Even at stresses below the elastic limit, when a force is applied the strain lags behind the stress. There is a hysteresis in the stress–strain relationship when the stress is applied and removed, as during a vibration cycle. (Hence the term hysteretic damping is sometimes used as an alternative expression for material damping.) Hysteresis is a manifestation of the irreversibility of a process.

There are various ways of quantifying material damping. The *specific damping capacity* is defined as

$$\frac{\text{energy dissipated per cycle of vibration}}{\text{total strain energy}}$$

and is normally expressed as a percentage. The *loss factor* is often used to specify material damping and it is the specific damping capacity divided by 2π .

The specific damping capacities for some materials are shown in Table 8.3 [45]. Cast iron can have a high damping capacity, but it depends upon the type. Similarly with aluminium alloys, duralumin provides hardly any damping, whereas hiduminium is quite good. Alloys of manganese and copper combine particularly high damping capacities with good mechanical strength. Unfortunately, manganese–copper alloys have not found many applications. They have been used for spindles [46] and as inserts in the steel bits of pneumatic road drills [47].

TABLE 8.3 Specific damping capacity of different materials

Material	Specific damping capacity %
Steel – mild	1.5
Steel – silver	0.5
Cast iron – high carbon inoculated flake	19
Cast iron – austenitic flake graphite	7
Aluminium alloy – hiduminium	7.5
Aluminium alloy – duralumin	0.3
Manganese–copper – 70% Mn, 30% Cu	42
Nylon 12	31

The specific damping capacity increases with the maximum stress that occurs in the material during the vibration. In Table 8.3 the specific damping capacities refer to a stress of about 30 MPa, apart from nylon for which the stress was 6 MPa.

There is also some damping because of sound radiation. Conversion of mechanical energy into sound must clearly reduce the amplitudes of vibration, but the damping provided in this way is small. In most vibrating structures, unless special measures have been taken the damping derives from the relative motion at the structural joints, material damping or viscous damping at lubricated surfaces.

Mead [48] has pointed out that a great deal of damping comes from the attachment of non–structural items, such as pipes or cables. In the case of a car the steel body has low damping, but the addition of seats, carpets, etc., increases the damping considerably.

Damping by unconstrained and constrained layers It is possible to introduce additional damping to a structure by means of damping layers of visco–elastic material. The damping layers are applied to surfaces which are vibrating in bending. They are particularly effective when applied to large sheets, such as car doors. Car doors tend to ring when given an impact, but the ringing noise can be almost entirely eliminated by the addition of a damping layer.

Damping layers may be either unconstrained or constrained. In the case of unconstrained layers the damping material, which is attached to the metal or alloy base, is subjected to longitudinal extensions and contractions, as shown in Figure 8.29a. For the greatest damping the ratio of the thickness of the damping layer to the base thickness depends upon the ratio of the elastic moduli [49]. For most combinations the optimum condition is achieved when the thickness of the damping material is several times that of the base thickness. In the range of damping material thicknesses that are applied in practice the damping increases with thickness.

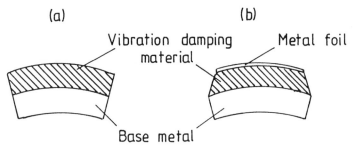

Figure 8.29 Damping with (a) unconstrained layer and (b) constrained layer.

If the base plate vibrates at one frequency and the modal shape corresponding to that frequency is known, it is possible to apply the damping material not to the entire plate but just to the area around the nodes.

The damping effect is dependent on temperature. Thus, if a damping layer is designed to be used at room temperatures, it may not be effective at high operating temperatures.

Constrained damping layers are more effective. The damping material is coated with a thin sheet of metal (often aluminium alloy). Because the damping layer is now sandwiched between two sheets, it is subjected to shear deformation, as shown in Figure 8.29b.

It is also possible to have the damping material sandwiched between two steel sheets. These sandwich sheets are commercially available.

Two applications of constrained damping layers are now presented. The first concerns the wheels of a subway train. When travelling round bends the wheels emitted a loud squealing noise. The noise was reduced by gluing a damping material to the inside of the rim of the wheel and constraining the elastomeric sheet with a thin steel layer. The arrangement is shown in Figure 8.30. The damping material used in this example is called Dyad and is manufactured by the Soundcoat Company. The noise reduction which was achieved is also shown in Figure 8.30.

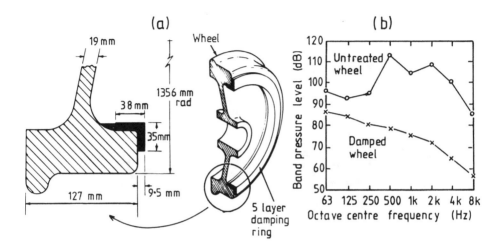

Figure 8.30 Damping of the wheel of a subway train: (a) diagram of wheel with damping material, (b) noise from wheel, with and without damping. (Courtesy of Soundcoat Company, Deer Park, New York)

The second example concerns a space frame which supports an orchestral reflector in a concert hall. The frame was made from several steel tubes connected together by spherical joints. Each tube was 930 mm long, 60.3 mm in diameter and 5 mm thick. A qualitative test was made by allowing a steel ball (of diameter 50.8 mm) to swing through a given distance and strike the tube in the middle. The sound emitted was measured by a microphone and sound level meter. The results are shown in Figure 8.31 (a) for the undamped tube and for (b) the tube covered with a damping layer (apart from a small area where the ball struck the tube).

The damping material, called Dempison, used to obtain the results shown in Figure 8.31b, was of a bitumastic type in the form of a self-adhesive sheet which was about 2 mm thick. It was covered by thin aluminium foil which acted as an anti-abrasive coating and also, to some extent, as the constraining layer.

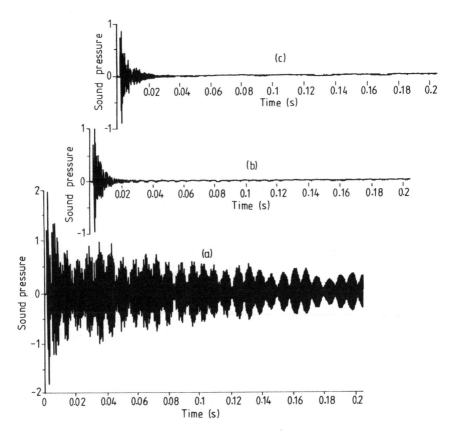

Figure 8.31 Damping of a structural frame component: sound from (a) undamped tube, (b) damped tube and (c) tube filled with sand.

It is interesting to note that when the tube was filled with 2.6 kg of dry sand (grit size 1 to 2 mm) quite good damping was also obtained (Figure 8.31c).

With both the damping layer and the sand any potential ringing of the tube was eliminated.

Damping layers can be used on bins, hoppers and feedbowls in the workshop. If a cutting or hammering process is being carried out on a structure, the structure can be temporarily covered with a damping layer which can be easily removed.

8.7.3 Control of the structural response by vibration isolation

If a vibrating machine is bolted to the floor, it will transmit vibratory forces to the floor which will itself vibrate and radiate structure–borne sound. The transmitted forces can be reduced by supporting the machine on elastic mounts or *vibration isolators*. The subject of vibration isolation of machinery from foundations is extensively covered elsewhere [50–52] and, consequently, only a brief introduction will be given here.

Forced vibrations that are transmitted to the ground can arise because, for example, of cutting forces or the rotation of out–of–balance components. Rotating machinery is normally to some extent out–of–balance and, as it rotates, transmits through the bearings a force which changes with time with a fundamental frequency corresponding to that of rotation. If the rotational speed is N rev/min, f the frequency of rotation is given by $f = N/60$ and is related to the circular frequency ω by equation (1.73). When placed on vibration isolators the machine and the elastic mounts can be regarded as forming a system in which the machine is a rigid body with one degree of freedom (in the vertical direction y), as shown in Figure 8.32. (In reality the system will be more complicated.) As the machine vibrates under the action of the rotating out–of–balance mass, the displacement y of the machine (or body of mass m) from the static equilibrium position is related to the physical parameters of the system by the equation of motion [50], which is

$$m\frac{d^2y}{dt^2} + c\frac{dy}{dt} + ky = m_1 e\omega^2 \sin\omega t, \qquad (8.10)$$

where m is the total mass of the machine,
   c is the damping constant,
   k is the equivalent stiffness of the vibration isolators,
   m_1 is the mass of the rotating parts, and
   e is the distance of the centre of the mass of the
      rotating parts from the axis of rotation.

In equation (8.10) the damping is assumed to be *viscous*, i.e., the damping force is proportional to the velocity of the machine, where the constant of proportionality is the damping constant c. In the case of the machine on the vibration isolators the damping is mostly provided by the isolators. If the machine is mounted on steel helical springs, there is only a small amount of damping. With rubber mounts the damping can be quite considerable. The damping provided by rubber is generally non–viscous, but equivalent–viscous damping is often assumed.

The single degree of freedom system described here has also provided the model for the Helmholtz resonator, see Section 6.8. As with the

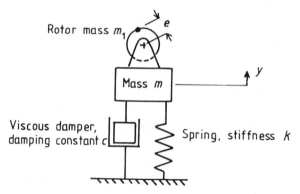

Figure 8.32 Single degree of freedom vibrating system.

Helmholtz resonator, the *undamped natural frequency* f_n of the system is given by:

$$f_n = \frac{1}{2\pi}\sqrt{\frac{k}{m}}.$$

Alternatively,

$$f_n = \frac{1}{2\pi}\sqrt{\frac{1}{\delta}}, \tag{8.11}$$

where δ is the static deflection of the isolators under the weight of the system.

The effectiveness of the vibration isolators is assessed in terms of the *transmissibility*, or *transmission ratio*, which is defined as

$$\frac{\text{amplitude of the force transmitted to the floor}}{\text{amplitude of the out–of–balance force in the machine}}.$$

The transmissibility is a function of the frequency ratio f/f_n and the damping provided by the isolators. The damping is most suitably expressed in terms of the nondimensional *damping ratio* ζ which is the ratio of the damping force to the damping force at critical damping with the same velocity. Critical damping is discussed in vibration texts [50] where it is also shown that for the system shown in Figure 8.32

$$\zeta = \frac{c}{2\sqrt{km}}.$$

The transmissibility TR is given by:

$$\text{TR} = \frac{\sqrt{[1 + (2\zeta f/f_n)^2]}}{\sqrt{[\{1 - (f/f_n)^2\}^2 + (2\zeta f/f_n)^2]}}. \tag{8.12}$$

The way in which TR varies with the ratio f/f_n for different values of ζ is shown in Figure 8.33. For the amplitude of the force transmitted to the floor to be reduced to a value less than that for the excitation force the ratio f/f_n must be greater than $\sqrt{2}$. For ratios of f/f_n less than $\sqrt{2}$ the transmitted force is greater than the excitation force. Quite good vibration isolation is achieved with low damping for f/f_n ratios above two. In order to achieve a ratio of 2 or more the static deflection δ may have to be quite large, if the excitation frequency f is small. In other words a soft spring is needed; a requirement that may be difficult to realise in practice.

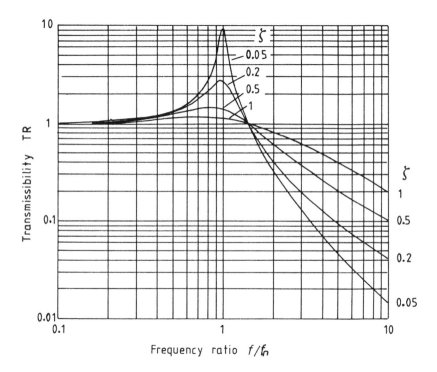

Figure 8.33 Vibration isolation; transmissibility of a single degree of freedom vibrating system for different damping ratios.

In many situations the speed of operation of a machine is changing and thus for a given system with isolators the transmissibility also changes. Vibration isolation which is adequate at a certain speed may be inadequate at a lower rotational speed. In general the out–of–balance force (as a function of time) can consist of not only the fundamental but also higher harmonics. In some cases vibration isolation may give an unacceptably high value of transmissibility at the fundamental frequency but may provide good isolation at the higher harmonics of the out–of–balance force.

Vibration isolation for noise control   The above discussion is concerned with the prevention of the transmission of vibratory forces to foundations; the same principles can be used to provide isolation of a machine or instrument from a vibrating floor. Vibration isolation has, however, quite general application. The following example illustrates the advantages of using resilient mounts (or vibration isolators) rather than rigid attachments. The example concerns an outboard motor attached to the stern of a boat [41]. Normally, the motor is attached rigidly to the boat so that the forces on the motor casing are transmitted directly to the hull which is caused to vibrate and become a source of structure–borne sound. The noise can be reduced by attaching the motor to a cover by means of resilient mountings; the cover is rigidly attached to the boat, as shown in Figure 8.34a. Noise reductions of 10 dB and more are achieved over a wide range of one third octave bands (see Figure 8.34b). It is interesting to note that the low frequency sound is increased, presumably because the mounts are too stiff to provide isolation of the vibratory forces at the fundamental firing frequency of the engine and at low frequencies of the harmonics.

A further example concerns panels attached to the framework of a machine. Panels on, say, a machine tool improve the appearance, but can result in an increase of the radiated sound, because of the transmission of vibration from the frame to the panel. A scheme for the resilient connection of a panel is shown in Figure 8.35. In this case, although the panel is bolted to the main structure, there is no direct metal to metal contact between the frame and the panel. Every vibrational transmission path is interrupted by an elastic component.

In Section 8.6.2, in connection with the location of noise sources, an example was presented of a pulley of an engine which was found to be a source of noise. It was observed that the noise was reduced when the pulley was removed. Once it is known that the pulley is a source of noise, vibration isolation becomes a possible remedy. The design of the pulley with a vibration isolator is shown in Figure 8.36. The noise was reduced by about 8 dB at 1600 Hz [34], as shown in Figure 8.20b.

Installation of vibration isolators and pipe hangers   It is quite common to bolt a machine to a heavy bedplate or concrete plinth and support the entire system on vibration isolators. The increase in mass of the vi-

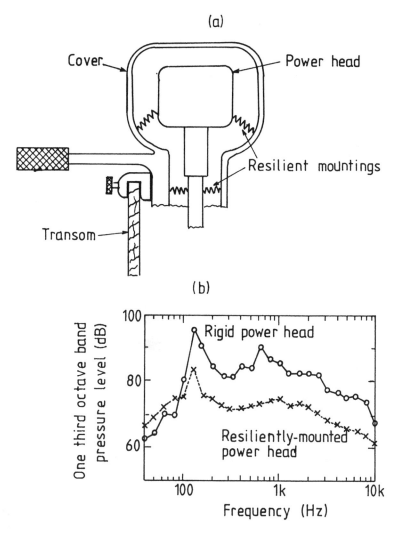

Figure 8.34 Vibration isolation applied to the outboard motor of a boat [41].

brating system due to the addition of the bedplate or plinth results in lower amplitudes of vibration. It is particularly advantageous to restrict the vibratory displacements when pipes and conduits are attached to the machine. Any pipe has to be secured to a wall or ceiling and, if the pipe is vibrating with the machine, the vibrations are transmitted to the wall and

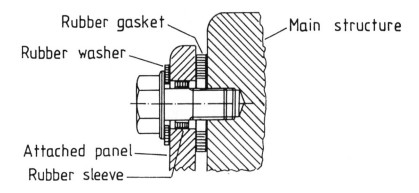

Figure 8.35 Vibration isolation of a panel on a machine.

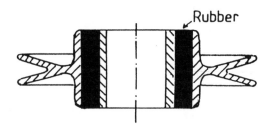

Figure 8.36 Scheme of a vibration isolator for a fan belt pulley [34].

ceiling which become sources of sound. In addition to the heavy bedplate supporting the machine it is advisable to support the pipe with flexible pipe hangers which provide vibration isolation between the pipe and the ceiling.

An arrangement that may be used with rotating or reciprocating machinery is shown schematically in Figure 8.37. The pipes are attached to the ceiling by means of flexible pipe hangers. Although not shown on the diagram, it would be useful to have an additional bend downstream of the flexible pipe hangers so that the final direction of the pipe is still horizontal. In this way the floating pipe has the possibility of three types of motion, one translational and two rotational.

Further information on vibration isolating couplings is given by Smith [53] and general information on vibration isolation by Fry [52].

8.7.4 Modification of radiation efficiency

As the radiation efficiency of machinery components is small at low fre-

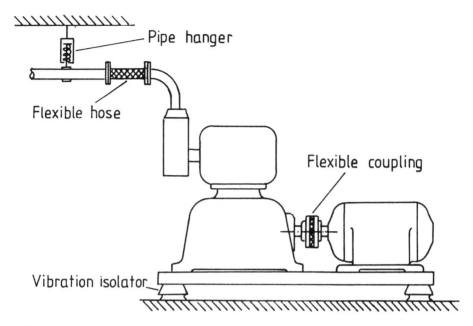

Figure 8.37 Vibration isolation of a machine and attached pipework.

quencies, it seems reasonable to ensure that excitation frequencies are as low as possible. However, raising the natural frequencies of structural vibrations by increasing the stiffness is sometimes suggested as a means of noise control [54]. At low frequencies there can be large amplitude excitation forces which causes the sympathetic vibration of the structure. Stiffening the structure can increase the natural frequencies of the structure above the frequency of the excitation forces, thus reducing the vibration of the machine. If the radiation efficiency, both before and after stiffening, is of the order of unity, the radiated sound will be reduced by stiffening.

Perforations in a plate or panel reduce the radiation efficiency and, provided the panel is not required for sound insulation, replacement of a solid panel by a perforated version will reduce the radiated sound.

Information on radiation efficiencies is given by Richards and his colleagues [6,55].

8.8 Enclosure

Enclosure is a commonly used form of noise control in the workplace. It is also possible to incorporate enclosure into the design.

8.8.1 Enclosure as part of the design

Enclosure has been incorporated into the design of internal combustion engines [56–58], electric motors, rotary screw compressors and diesel–driven air compressors for roadworks [59,60]. The problems with enclosure as part of the design arise from the existence of structure–borne sound and the provision of adequate ventilation. A schematic diagram of a system for a portable air compressor is shown in Figure 8.38 [59].

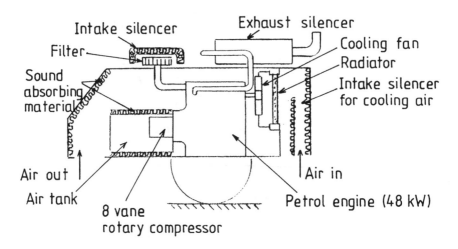

Figure 8.38 Scheme for the enclosure of a portable air compressor.

Air is required to cool the petrol engine and to drive the rotary vane compressor. The combined unit is enclosed within one shell. The passageways for the inflowing and outflowing air are lined with sound absorbing material (forming in effect a silencer, see Chapter 6). There is also a silencer on the engine intake. The engine of the version described here was attached to the shell by hard engine mounts, but normally vibration isolators would be used. The sound from the enclosed compressor unit was 76 dB(A) when measured at a distance of 7 m away in accordance with a standard test [59]. The sound reduction measures of the kind adopted in this example would normally result in an attenuation of about 10 dB(A) with respect to an unenclosed unit.

It is common practice to include a heavy rubber mat to the bulkhead separating the engine compartment from the passenger compartment of a vehicle. The use of a barrier in this way cannot prevent either sound travelling round the barrier or the radiation of structure–borne sound. Complete enclosure of a vehicle engine and gearbox has been attempted. In this way noise reduction of 10 dB(A) can be achieved [56].

8.8.2 Enclosure in the workplace

Some of the principles relating to enclosures have already been discussed in Section 4.10. The first point to be aware of with respect to enclosures is that they will only attenuate air–borne noise. Structure–borne sound should be eliminated as much as possible prior to the erection of the enclosure. Thus a vibrating machine should be mounted on vibration isolators (see Section 8.7.3). Any pipes from the machine should not be rigidly attached to the enclosure, but should be supported by vibration isolating hangers. Metal to metal contact has to be avoided in locations where pipes pass through the enclosure. Air gaps need to be minimised, as they result in a sound transmission coefficient of unity and poor acoustic performance of the enclosure. Careful design is required for places where cables pass through the enclosure; a rubber seal may need to be incorporated.

Similarly, ducts for fresh air and exhaust gases have to be adequately silenced so that they do not form a weak part in the acoustic transmission path. Details of the ventilation requirements for enclosures can be found elsewhere [61].

With enclosures there is a problem of access for both personnel and materials. An effective, although expensive, solution is to have a telescopic–type of enclosure which can slide away to give easy access to the machine within. Such an enclosure is shown in Figure 8.39. In situations where a smaller attenuation is adequate it is possible to use heavy plastic curtains in the form of vertical strips which hang from a rail. It is very much preferable for the enclosure formed by the curtains to have a roof, otherwise the curtains form no more than a barrier.

8.9 Noise reduction by factory modification

Enclosure and the introduction of damping materials are measures that can be adopted in the workplace to reduce noise, often as an afterthought. These techniques are directed towards a particular machine or process. Measures that can be applied to reduce the levels of noise in general in a factory include the creation of graduated noise zones, use of barriers and functional absorbers.

8.9.1 Noise reduction by changing factory layout

In the free field of a sound source the sound intensity level decreases by 6 dB for every doubling of the distance away from the source, provided

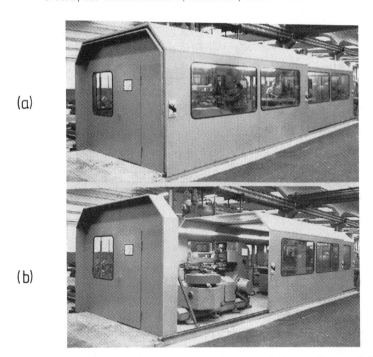

(a)

(b)

Figure 8.39 Telescopic enclosure in use in a workshop; shown (a) closed and (b) open. (Courtesy G and H Montage GmbH.)

that the free field is also the far field. Thus workers should be kept as far away from noise sources as possible. Although this is not generally practicable, certain noise sources, such as fan extracts and compressed air discharges, can often be sited away from personnel.

In order to ensure that the noisiest machines are separated from most people the technique of segregation or zoning is used. With this technique the noisiest machines are placed together in one area and the quietest machines furthest away. In between are placed machines of intermediate noise levels. In an example where this method was successfully applied the noisiest machines, which were semi–automatic, were placed in an area of the factory where they could be separated from the rest by a barrier or screen [62]. Within the quieter area the noise levels were reduced by more than 20 dB(A).

Noise sources which are placed against a wall or in a corner produce a greater sound intensity in the free field than the same source positioned in the middle of the room, see Section 5.3. In a factory, where there are rows of machines with aisles for pedestrians between the rows, it is better to have the aisles near the wall and two rows of machines immediately adjacent to each other rather than rows of machines next to the wall [63]. Noisy machines should be kept away from the walls as much as possible.

8.9.2 Noise reduction by boundary absorption

The sound in the direct or free field can be reduced to some extent by zoning for those workers who are sufficiently far from the noise of the machines. Additionally, the reverberant levels from the reflected sound can be reduced by introducing extra sound absorbing material into the room. Generally, mineral wool or fibre glass is introduced into the ceiling area or, less often, placed on the walls or on screens. The absorbers, in the shape of flat slabs or cylinders, are often called *functional absorbers*.

The reverberant sound pressure level in the diffuse field depends upon the room constant R_c, see equations (5.1) and (5.17). If additional material is introduced which increases the room constant from $R_{c,\,\text{before}}$ to $R_{c,\,\text{after}}$, the noise reduction ΔL in the diffuse field is:

$$\Delta L = 10 \log(R_{c,\,\text{after}}/R_{c,\,\text{before}}) \text{ dB}.$$

Further, if the room constant is assumed to be equal to the absorption $S\overline{\alpha}$, the noise reduction ΔL becomes

$$\Delta L = 10 \log(T_{\text{before}}/T_{\text{after}}) \text{ dB}, \tag{8.13}$$

where T_{before} is the reverberation time for the room before the introduction of the sound absorbing material and T_{after} is the reverberation time with the material present. The reverberation time is given by the Sabine formula, equation (5.22).

Addition of absorption to a factory hall is usually only beneficial if the room is 'live' and has little absorption before treatment. As large quantities of sound absorbing material are expensive, it is advisable to check that its introduction is worthwhile. Firstly, the reverberation time is measured (see Section 5.9 for the method) and T_{before} is obtained. The sound absorption coefficient α of the sound–absorbing material is obtained from the manufacturer's data. The total sound absorbing area S_{ab} is also estimated. In the case of flat slabs suspended from the ceiling both sides of slabs absorb sound and the area used should be twice that of the surface area of one side of the slab. In the case of blankets or acoustic tiles on the walls the area of one side should be only considered. The notional reverberation time T_{ab} for the hypothetical room with only the sound–absorbent material present may be estimated from equation (5.22) to be

$$T_{ab} = \frac{0.161V}{S_{ab}\alpha_{ab}}, \qquad (8.14)$$

where V is the volume of the room. Thus the reverberation time after treatment T_{after} is given by:

$$\frac{1}{T_{after}} = \frac{1}{T_{before}} + \frac{1}{T_{ab}}. \qquad (8.15)$$

The noise reduction ΔL can be estimated according to expression (8.13).

Normally, noise reduction of not more than 3 to 5 dB can be expected from the introduction of sound absorbing material. However, such reductions are often worthwhile, particularly as high levels of reflected sound are subjectively unpleasant. Reductions of 8 to 10 dB(A) are sometimes claimed [64].

In bottling halls the sound level of the diffuse field is very high, particularly if the hall is tiled. Sound levels can be reduced by functional absorbers, provided that hygiene requirements are satisfied. An example of their use is shown in Figure 8.40. In this case cylindrical absorbers of mineral wool are used (manufactured by Grünzweig and Hartmann Montage GmbH). The absorbers were suspended at a height of 2.5 m over the entire floor area. The absorption — the product $S_{ab}\alpha_{ab}$, see equation (8.14), — was increased by 1600 m^2. The sound level in the diffuse field was reduced by 8 dB(A).

It should be noted that, whilst increasing the amount of surface absorption in a room is very effective in reducing the level of the diffuse sound field, it does not bring a benefit to the worker who is in the direct field of his own machine. Care has to be taken to ensure that absorbers suspended from the ceiling do not interfere with the lighting and other services. Also lightweight structures must not be loaded excessively.

8.9.3 Use of barriers

Barriers or screens in a factory can be moveable and cause little obstruction. The noise reduction which can be achieved by barriers is limited and is much less than that from full enclosures. The main effect of a barrier is to bring about diffraction of the sound [65]. In the case of low frequency sound the top of the barrier acts like a new source of sound which radiates in all directions; the noise reduction provided by the barrier is small. With

Figure 8.40 Reduction of reverberant sound levels in a bottling hall by the introduction of functional absorbers. (Courtesy of G and H Montage GmbH.)

high frequency sound the barrier casts the acoustic equivalent of a shadow and an attenuation is achieved on the opposite side of the barrier.

When used in a workshop the barrier should have a lining of sound absorbing material facing the noise source and should be used in conjunction with functional sound absorbers on the ceiling above the source, as shown in Figure 8.41. The mineral wool or fibre glass on the screen should be at least 50 mm thick and the screens at least 2 to 2.5 m high.

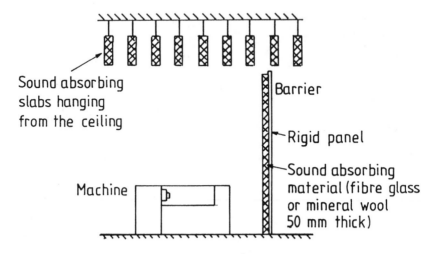

Figure 8.41 Barrier in use in a workshop.

8.10 Closure

Current legislation provides the incentive for noise control. Further incentives could come from the threat of civil action by workers or from the requirements of insurance companies.

It is essential to have a noise control programme. Such a programme will vary with different situations, but would typically comprise the following steps.

(1) Carry out a noise survey.
(2) Identify the areas where the sound levels are greater than (a) 90 dB(A) and (b) 85 dB(A).
(3) Identify the personnel who have a daily noise exposure level in excess of (a) 90 dB(A) and (b) 85 dB(A).
(4) Ensure that all machines are operating correctly and are maintained in good working order; adequate lubrication should be provided and there should be no loose components.

(5) Consider local noise control measures, such as changing the position of machines, creation of separate noise zones or the introduction of barriers.

(6) Consider whether noise from material handling can be reduced; reduce the height from which the material is dropped, apply vibration damping material to bins and hoppers, allow material to drop onto heavy, wear–resistant rubber and stop unnecessary clatter and banging.

(7) Ensure that vibration isolation is provided for machines and pipes.

(8) Consider increasing the amount of absorption in the room by hanging functional absorbers from the ceiling.

(9) Consider enclosing machines with either a simple, factory–built enclosure to give approximately 20 dB(A) attenuation or a commercially made enclosure to give attenuation of 30 dB(A), or more.

(10) If daily noise exposure levels of some workers are still in excess of 85 dB(A), initiate a hearing conservation programme.

This book has attempted to give a systematic approach to the basics of acoustics, as applied to noise and its control. There is no reason to regard noise control as a mystery. Some years ago an advertisement for an acoustics consultant implied that prodigious talents were required.

Personable, well–educated literate individual with college degree in any form of Engineering or Physics to work for a small firm. Long hours, no fringe benefits, no security, little chance for advancement are among the inducements offered. Job requires wide knowledge and experience in physical sciences, materials, construction techniques, mathematics and drafting. Competence in the use of spoken and written English is required. Must be willing to suffer personal derision from peers in more conventional jobs and slanderous insults from colleagues. Job involves frequent physical danger, trips to inaccessible locations throughout the world, manual labour and extreme frustration from lack of data on which to base decisions. Applicant must be willing to risk personal and professional future on decisions based on inadequate information and complete lack of control over acceptance of recommendations by clients. Responsibilities for work are unclear and little or no guidance offered. Authority commensurate with responsibility is not provided either by firm or its clients!

We hope this book goes some way towards making the subject of noise control more acceptable.

References

[1] The Council of the European Communities (1986). *Official Journal of the European Communities*, No L 137/28, 24.5. 1986, Brussels.

[2] D. J. Ewins (1984). *Modal Testing: theory and practice*. Research Studies Press, Letchworth.

[3] K. A. Ramsey and M. Richardson (1976). Making effective transfer function measurements for modal analysis. *Proceedings of the 22nd Annual Meeting of the Institute of Environmental Sciences*, Philadelphia.

[4] L. E. Kinsler and A. R. Frey (1962). *Fundamentals of Acoustics*, second edition, p 171. John Wiley, New York.

[5] I. L. Ver and C. I. Holmer (1971). Interaction of sound waves with solid structures, in *Noise and Vibration Control* (ed. L. L. Beranek). McGraw–Hill, New York.

[6] E. J. Richards, M. E. Westcott and R. K. Jeyapalan (1979). On the prediction of impact noise, II: ringing noise. *Journal of Sound and Vibration*, **65** (3), pp 419–451.

[7] H. S. Ribner (1964). The generation of sound by turbulent jets. *Advances in Applied Mechanics*, **8**, pp 103–182. Academic Press, New York.

[8] W. C. Osborne (1977). *Fans*, second edition. Pergamon Press, Oxford.

[9] Health and Safety Executive (1983). *100 practical applications of noise reduction methods*, p 6. HMSO, London.

[10] W. Neise (1976). Noise Reduction in centrifugal fans: a literature survey. *Journal of Sound and Vibration*, **45** (3), pp 375–403.

[11] B. B. Daly (1978). *Woods Practical Guide to Fan Engineering*. Woods of Colchester, Colchester.

[12] I. Sharland (1972). *Woods Practical Guide to Noise Control*. Woods of Colchester, Colchester.

[13] M. J. Lighthill (1963). Jet noise. *American Institute of Aeronautics and Astronautics Journal*, **1** (7), pp 1507–1517.

[14] J. C. Laurence (1956). Intensity, scale and spectra of turbulence in the mixing region of a free, subsonic jet. NACA report 1292.

[15] P. O. A. L. Davies, M. J. Fisher and M. J. Barratt (1963). The characteristics of the turbulence in the mixing region of a round jet. *Journal of Fluid Mechanics*, **15**, pp 337–367.

[16] K. W. Bushell (1971). A survey of low velocity and coaxial jet noise with application to prediction. *Journal of Sound and Vibration*, **17** (2), pp 271–282.

[17] A. Powell (1953). On the mechanism of choked jet noise. *Proceedings of the Physical Society*, **66**B, pp 1039–1056.

[18] J. E. Ffowcs Williams (1968). Jet noise at very low and very high speed. Proceedings of the AFOSR–UTIAS Symposium, Toronto, University of Toronto Press.

[19] F. B. Greatrex (1959). Noise suppressors for Avon and Conway engines. *American Society of Mechanical Engineers*, preprint 59–AV–49.

[20] F. B. Greatrex and D. M. Brown (1959). Progress in jet engine noise reduction. *Advances in Aeronautical Sciences*, **1**, pp 364–392, Pergamon Press, Oxford.

[21] W. D. Coles, J. A. Mihaloew and E. E. Callaghan (1958). Turbojet engine noise reduction with mixing nozzle ejector combinations. NACA TN D–60.

[22] J. Kane, T. D. Richmond and D. N. Robb (1965–66). Noise in hydrostatic systems and its suppression. Marine applications of fluid power. *Proceedings of the Institution of Mechanical Engineers*, London, **180** (32), pp 36–52.

[23] R. J. Whitson (1980). The measured transmission loss characteristics of some hydraulic attenuators. Paper C382/80, Research Project Seminar on Quieter Hydraulics, Institution of Mechanical Engineers, London.

[24] C. B. Schuder (1971). Control valve noise–prediction and abatement. *Proceedings of the Purdue Noise Control Conference*; Noise and Vibration Control Engineering, Lafayette, Indiana, pp 90–94.

[25] G. Reethof (1977). Control valve and regulator noise generation, propagation and reduction. *Noise Control Engineering*, **9** (2), pp 74–85.

[26] D. Ross (1976). *Mechanics of Underwater Noise*. Pergamon Press, New York.

[27] L. S. Goodfriend (1979). Fluid–flow systems; plumbing, piping, steam, in *Handbook of Noise Control* (ed. C. M. Harris), second edition. McGraw–Hill, New York.

[28] K. L. Johnson (1985). *Contact Mechanics*. Cambridge University Press, Cambridge.

[29] L. L. Koss and R. J. Alfredson (1973). Transient sound radiated by spheres undergoing elastic collision. *Journal of Sound and Vibration*,

27 (1), pp 59–75.

[30] E. J. Richards, M. E. Westcott and R. K. Jeyapalan (1979). On the prediction of impact of noise source, I: acceleration noise. *Journal of Sound and Vibration*, **62** (4), pp 547–575.

[31] C. E. Ebbing (1974). Diagnostic tests for locating noise sources: classical techniques (part I). *Noise Control Engineering*, **3** (1), pp 30–36.

[32] B. Fader (1981). *Industrial Noise Control*. John Wiley, New York.

[33] W. Bragg (1920). *The World of Sound*. 1930 reprint, G. Bell, London.

[34] T. Priede, A. E. W. Austen and E. C. Grover (1964–65). Effect of engine structure on noise of diesel engines. *Proceedings of the Institution of Mechanical Engineers*, London, **179** (2A) (4), pp 113–144.

[35] T. F. W. Embleton (1971). Sound in large rooms, in *Noise and Vibration Control* (ed. L. L. Beranek). McGraw–Hill, New York.

[36] J. R. Bailey, J. A. Daggerhart and N. D. Stewart (1975). A systems approach for control of punch press noise. *American Society of Mechanical Engineers*, paper 75–DET–49.

[37] C. C. Perry and H. R. Lissner (1962). *The Strain Gage Primer*, second edition. McGraw–Hill, New York.

[38] A. J. Hillier (1974). Loom noise and its control. *Noise Control and Vibration Reduction*, Nov. pp 312–317.

[39] T. E. Reinhart and M. J. Crocker (1982). Source identification on a diesel engine using acoustic intensity measurements. *Noise Control Engineering*, **18** (3), pp 84–92.

[40] T. Priede (1961). Relation between form of cylinder pressure diagram and noise in diesel engines. *Proceedings of the Institution of Mechanical Engineers* (Automobile Division), London, **175**, pp 63–77.

[41] T. Priede (1969). Noise and engineering design. *Proceedings of the Philosophical Transactions of the Royal Society*, **263** A, pp 461–480.

[42] T. Priede (1971) Origins of automobile vehicle noise. *Journal of Sound and Vibration*, **15** (1), pp 61–73.

[43] B. L. Clarkson (1972). The social consequences of noise. *Proceedings of the Institution of Mechanical Engineers*, London, **186**, pp 97–107.

[44] E. J. Richards (1981). On the prediction of impact noise, III: energy accountancy in industrial machines, *Journal of Sound and Vibration*, **76** (2), pp 187–232.

[45] D. Birchon (1964). High damping alloys; parts 1 and 2. *Engineering Materials and Design*, **7** (9), pp 606–608, and **7** (10), pp 692–696.

[46] M. A. Satter, B. Downs and G. R. Wray (1969–70). Reduction of noise at the design stage: a specific case study. *Proceedings of the Institution of Mechanical Engineers*, London, **184** (1) (33), pp 593–

614.

[47] E. J. Richards (1966). Noise consideration in machines and factories. *The Chartered Mechanical Engineer*, June, pp 266–273.

[48] D. J. Mead (1982). Vibration control (I), in *Noise and Vibration* (eds R. G. White and J. G. Walker). Ellis Horwood, Chichester.

[49] E. E. Ungar (1971). Damping of panels, in *Noise and Vibration Control* (ed. L. L. Beranek). McGraw–Hill, New York.

[50] J. S. Anderson and M. Bratos–Anderson (1987) *Solving Problems in Vibrations*. Longman Scientific, Harlow.

[51] C. E. Crede and J. E. Ruzicka (1976). Theory of vibration isolation, in *Shock and Vibration Handbook* (eds C. M. Harris and C. E. Crede), second edition. McGraw–Hill, New York.

[52] A. Fry (ed.) (1988). *Noise Control in Building Services*. Pergamon Press, Oxford.

[53] T. J. B. Smith (1977). Vibration–isolating couplings and links. *Noise Control Vibration and Insulation*, Jan, pp 7–11.

[54] R. H. Lyon (1978). Noise reduction by design — an alternative to machinery noise control. *Proceedings in Internoise 78*, San Francisco, California (ed. W. W. Lang), pp 31–44.

[55] R. K. Jeyapalan and E. J. Richards (1979). Radiation efficiencies of beams in flexural vibration. *Journal of Sound and Vibration*, **67** (1), pp 55–67.

[56] D. T. Aspinall (1967). Reducing the noise of motor vehicles. *Noise Control*; Symposium of the Production Engineering Research Association of Great Britain, 26–27 April, 1967.

[57] J. W. Tyler (1979). The TRRL quiet heavy vehicle project. *Proceedings of the Institution of Mechanical Engineers* (Automobile Division), London, **193**, pp 137–147.

[58] H. Drewitz and M. Stiglmaier (1989). Noise reduction on trucks, of 6 to 10 t gvt through engine encapsulation. *Noise Control Engineering Journal*, **33** (1), pp 5–10.

[59] W. N. Patterson (1975). Quieting portable air compressor. *Noise Control Engineering*, **5** (1), pp 41–47.

[60] L. Janssen (1989). Noise reduction analysis of a rotary air compressor for enclosure design. *Noise Control Engineering Journal*, **33** (2), pp 61–66.

[61] R. K. Miller and W. V. Montone (1978). *Handbook of Acoustical Enclosures and Barriers*. The Fairmont Press, Atlanta.

[62] M. S. Langley (1984). The noise and vibration consultant. *Noise and Vibration Control Worldwide*, July, pp 180–182.

[63] Anon. (1982). *Noise Control; Principles and Practice*, p 65. Brüel and Kjaer, Naerum, Denmark.

[64] D. Banks (1975) Noise reduction by the use of suspended noise absorbers and acoustical treatment. Noise Control *Vibration and Insulation*, Oct, pp 305–308.

[65] Z. Maekawa (1968) Noise reductions by screens. *Applied Acoustics*, **1**, pp 157–173.

Appendices

Appendix A

Acoustic energy balance equation

For isentropic flow the law of conservation of energy can be expressed by equation (1.27) or equation (1.28), see also remarks concerning equation (1.25). Combining the mass conservation equation,

$$\frac{\partial \rho}{\partial t} + \text{div}(\rho \vec{u}) = 0 \tag{A.1}$$

and the momentum conservation equation for an ideal fluid [1–5]

$$\rho \frac{d\vec{u}}{dt} = -\text{grad}\, P \tag{A.2}$$

with equation (1.28), leads to a different form of the energy conservation equation for an ideal fluid:

$$\frac{\partial}{\partial t}\left(\frac{1}{2}\rho u^2 + \rho \epsilon\right) + \text{div}\left[\rho \vec{u}\left(\frac{1}{2}u^2 + h\right)\right] = 0, \tag{A.3}$$

where $\vec{u} = \vec{u}(x, y, z, t)$ is the vector of the particle velocity at a certain point in space and time (fluid velocity vector). In the above equation:

$u = |\vec{u}|$ is the magnitude of the fluid velocity vector;

ρ is the fluid density, $\rho = \rho(x, y, z, t)$;

P is the fluid pressure, $P = P(x, y, z, t)$;

ϵ is the specific internal energy, i.e., the internal energy per unit mass;

h is the specific enthalpy, hence $h = \epsilon + P/\rho$.

The term $\frac{1}{2}\rho u^2 + \rho \epsilon$ denotes the energy of the unit fluid volume; $\frac{1}{2}\rho u^2$ is the kinetic energy for this volume and $\rho \epsilon$ is its internal energy.

To understand the meaning of both terms in equation (A.3), let us integrate equation (A.3) over a fluid volume V, which is fixed in space (i.e., does not depend upon time). The result is

$$\frac{\partial}{\partial t}\int_V \left(\frac{1}{2}\rho u^2 + \rho \epsilon\right) dV = -\int_V \text{div}\left[\vec{u}\left(\frac{1}{2}\rho u^2 + \rho \epsilon + P\right)\right] dV. \tag{A.4}$$

433

On the base of Gauss's theorem [6,7] the volume integral on the right hand side of equation (A.4) can be converted into a surface integral. Hence,

$$\frac{\partial}{\partial t} \int_V \left(\frac{1}{2}\rho u^2 + \rho\epsilon\right) dV = -\oint_S \vec{u}\left(\frac{1}{2}\rho u^2 + \rho\epsilon\right)\cdot d\vec{S} - \oint_S P\vec{u}\cdot d\vec{S}. \quad (A.5)$$

The term on the left side of the equation (A.5) is the rate of change of the fluid energy in the volume V. The term on the right hand side denotes energy flux flowing out through the surface S surrounding this volume V. It contains two terms concerning the flux of kinetic and internal fluid energy and the work made in unit time by the pressure forces on the fluid in the volume V, respectively. The expression $\vec{u}\left(\frac{1}{2}\rho u^2 + \rho\epsilon + P\right) = \rho\vec{u}(\frac{1}{2}u^2 + h)$ is the energy flux density vector, the magnitude of which is equal to the amount of energy passing in unit time through the unit surface perpendicular to the particle velocity vector $\vec{u}$.

Let us consider the fluid flow generated by sound propagation in an undisturbed medium. In this case the mean flow is reduced to zero, since for an undisturbed medium: $\vec{u}_0 = 0$ and ρ_0, P_0, ϵ_0 and h_0 have constant values. Sound wave motion causes only a small departure from the state of fluid at rest. Hence, $\rho = \rho_0 + \Delta\rho$, where $|\Delta\rho|/\rho_0 \ll 1$; $P = P_0 + p$, where $p \ll \gamma P_0$ (see equation (1.50)); $\vec{u} = 0 + \vec{u}$, where $|\vec{u}|/c_0 \ll 1$. Since $P = P_0 + p$ the energy conservation equation (A.3) is reduced to:

$$\frac{\partial}{\partial t}\left[\frac{1}{2}\left(\rho u^2 + \rho\epsilon\right)\right] + \mathrm{div}\left[\left(\frac{1}{2}\rho u^2 + \rho\epsilon + \rho_0\frac{P_0}{\rho_0}\right)\vec{u}\right] = -\mathrm{div}(p\vec{u}). \quad (A.6)$$

The energy conservation equation for a sound wave can be obtained from equation (A.6), taking into account the fact that sound is a small disturbance propagation and that sound propagation is an isentropic process. The term $\rho\epsilon$, see equation (A.6), where $\epsilon = \epsilon(\rho, s) = \epsilon(\rho)$, may be expanded as a function of the density disturbance $\Delta\rho = \rho - \rho_0$, namely,

$$\rho\epsilon = \rho_0\epsilon_0 + \left\{\left[\frac{d(\rho\epsilon)}{d\rho}\right]_s\right\}_{\rho=\rho_0}\Delta\rho + \frac{1}{2}\left\{\left[\frac{d^2(\rho\epsilon)}{d\rho^2}\right]_s\right\}_{\rho=\rho_0}(\Delta\rho)^2 + \dots \quad (A.7)$$

For the isentropic process equation (1.26) is reduced to:

$$d\epsilon = -PdV = \frac{P}{\rho^2}d\rho \quad (A.8)$$

or,

$$\left(\frac{d\epsilon}{d\rho}\right)_s = \frac{P}{\rho^2}.$$

(A.9)

Therefore,

$$\left[\frac{d(\rho\epsilon)}{d\rho}\right]_S = \frac{P}{\rho} + \epsilon = h.$$

(A.10)

Since,

$$Tds = dh - \frac{1}{\rho}dP,$$

(A.11)

see equation (1.25), for the isentropic process there is the relation,

$$dh = \frac{1}{\rho}dP.$$

(A.12)

Hence,

$$\left(\frac{dh}{d\rho}\right)_s\left(\frac{d\rho}{dP}\right)_s = \frac{1}{\rho}$$

(A.13)

and from equations (1.41) and (A.12)

$$\left\{\left(\frac{dh}{d\rho}\right)_s\right\}_{\rho=\rho_0} = \frac{c_0^2}{\rho_0}.$$

(A.14)

Differentiation of equation (A.10) leads to:

$$\left\{\left[\frac{d^2(\rho\epsilon)}{d\rho^2}\right]_s\right\}_{\rho=\rho_0} = \left\{\left(\frac{dh}{d\rho}\right)_s\right\}_{\rho=\rho_0} = \frac{c_0^2}{\rho_0}.$$

(A.15)

Using equations (A.10) and (A.15) in equation (A.7), we obtain

$$\rho\epsilon = \rho_0\epsilon_0 + h_0\Delta\rho + \frac{1}{2}\frac{c_0^2}{\rho_0}(\Delta\rho)^2 + \dots$$

(A.16)

Putting equation (A.16) into (A.6) and neglecting all terms of third order magnitude and smaller, since sound is a small disturbance propagation in an elastic medium, we obtain:

$$\frac{\partial}{\partial t}\left[\frac{1}{2}\rho_0 u^2 + \rho_0\epsilon_0 + h_0\Delta\rho + \frac{1}{2}\frac{c_0^2}{\rho_0}(\Delta\rho)^2\right] +$$
$$\text{div}\left\{\vec{u}\left[\left(\epsilon_0 + \frac{P_0}{\rho_0}\right) + h_0\Delta\rho\right]\right\} = -\,\text{div}(p\vec{u})$$

(A.17)

or,

$$\frac{\partial}{\partial t}\left[\frac{1}{2}\rho_0 u^2 + \frac{1}{2}\frac{c_0^2}{\rho_0}(\Delta\rho)^2\right] +$$

$$\frac{\partial}{\partial t}(h_0\Delta\rho + h_0\rho_0) - \frac{\partial}{\partial t}\left(\rho_0\frac{P_0}{\rho_0}\right) +$$

$$\text{div}\left[\vec{u}(\rho_0 + \Delta\rho)h_0\right] = -\text{div}(p\vec{u}), \tag{A.18}$$

where $\partial P_0/\partial t = 0$. Finally, taking into account the continuity equation,

$$\frac{\partial\rho}{\partial t} + \text{div}(\rho\vec{u}) = 0, \tag{A.19}$$

we derive from (A.18) the *acoustic energy balance equation,*

$$\frac{\partial}{\partial t}\left[\frac{1}{2}\rho_0 u^2 + \frac{1}{2}\frac{c_0^2}{\rho_0}(\Delta\rho)^2\right] = -\text{div}(p\vec{u}). \tag{A.20}$$

The term $\frac{1}{2}\rho_0 u^2 + \frac{1}{2}c_0^2(\Delta\rho)^2/\rho_0$ is the acoustic energy density E_a and contains, respectively, the kinetic and potential energy density of the sound wave. In general both terms are not equal to one another, except in the case of a plane, freely progressing wave. For a plane, freely progressing sound wave, see equations (1.47) and (1.82),

$$u^2 = \frac{c_0^2(\Delta\rho)^2}{\rho_0^2} \tag{A.21}$$

therefore,

$$\frac{1}{2}\rho_0 u^2 = \frac{1}{2}\frac{c_0^2}{\rho_0}(\Delta\rho)^2. \tag{A.22}$$

It should be noted that the sound energy density of the sound wave is a second order quantity. The acoustic energy balance equation possesses only second order quantities. On the other hand the linearised acoustic equations, which lead to the sound wave equation, are accurate only for first order quantities.

Let us integrate the equation (A.20) over the fluid volume V which is fixed in space. The integration leads to:

$$\frac{\partial}{\partial t}\int_V\left[\frac{1}{2}\rho_0 u^2 + \frac{1}{2}\frac{c_0^2}{\rho_0}(\Delta\rho)^2\right]dV = -\int_V\text{div}(p\vec{u})dV \tag{A.23}$$

or,

$$\frac{\partial}{\partial t} \int_V \left[\frac{1}{2}\rho_0 u^2 + \frac{1}{2}\frac{c_0^2}{\rho_0}(\Delta\rho)^2 \right] dV = - \oint_S p\vec{u} \cdot d\vec{S}. \qquad (A.24)$$

Hence,

$$\frac{\partial}{\partial t} \int_V \left[\frac{1}{2}\rho_0 u^2 + \frac{1}{2}\frac{c_0^2}{\rho_0}(\Delta\rho)^2 \right] dV = - \oint_S p\vec{u} \cdot \vec{n}dS, \qquad (A.25)$$

where $\vec{n}$ is a unit vector normal to dS, the infinitesimal element of the surface S. By convention $\vec{n}$ is the outward normal. S is the surface surrounding the volume V. The term on the left side is the rate of change of the acoustic energy in the volume V. The expression on the right side of the equation (A.25) describes the net acoustic energy flow (per unit time) out of the volume V. In other words it is the acoustic energy flux through the surface S.

It should be stressed that the acoustic energy balance equation holds in any sound source-free region of space.

Appendix B

B1 Active and reactive intensities of sound

Since the particle velocity phasor can be resolved into the active and reactive components, which are in phase and in quadrature with the sound
pressure phasor, respectively, the instantaneous sound intensity can be
presented as the sum of the active and reactive (instantaneous) sound
intensities, namely,

$$\vec{I} = \vec{I}_{ac} + \vec{I}_{rc}. \tag{B.1}$$

To prove this statement and to give a physical interpretation of the active
and reactive intensities, let us consider an arbitrary monochromatic sound
field. An arbitrary sound wave may be built up by superposition of plane
waves [8]. In terms of sound pressure the monochromatic sound wave may
be described as follows:

$$\widetilde{\mathbf{p}} = p_0(\vec{r})e^{i[\omega t + \Phi_0(\vec{r})]} \tag{B.2}$$

or,

$$\widetilde{\mathbf{p}} = p_0(\vec{r})e^{i\Phi(\vec{r},t)}, \tag{B.3}$$

where $\vec{r}$ is the position vector (radius vector); $\vec{r} = x\vec{e}_x + y\vec{e}_y + z\vec{e}_z = r\sin\theta\cos\psi\vec{e}_x + r\sin\theta\sin\psi\vec{e}_y + r\cos\theta\vec{e}_z$, see Figure 1.15. The functions
$\Phi(\vec{r},t)$ and $\Phi_0(\vec{r})$ are the phase and initial phase, respectively. The mo-
mentum conservation equation (Euler's equation) in vector form:

$$\frac{\partial \vec{u}}{\partial t} = -\frac{1}{\rho_0}\nabla\left\{\mathrm{Re}(\widetilde{\mathbf{p}})\right\} \tag{B.4}$$

in combination with equation (B.2) leads to an expression for the particle
velocity vector $\vec{u}$,

$$\vec{u} = -\frac{p_0(\vec{r})}{\rho_0\omega}\nabla\Phi_0(\vec{r})\cos\left[\omega t + \Phi_0(\vec{r})\right] - \frac{\nabla p_0}{\rho_0\omega}\sin\left[\omega t + \Phi_0(\vec{r})\right]. \tag{B.5}$$

The first term is the active part of the particle velocity vector, the second
term is its reactive part, i.e.,

$$\vec{u} = \vec{u}_{ac} + \vec{u}_{rc}. \tag{B.6}$$

Since the (instantaneous) sound intensity vector, see equation (1.186), is

$$\vec{I} = \mathrm{Re}(\widetilde{\mathbf{p}})\,\mathrm{Re}(\widetilde{\vec{\mathbf{u}}}) = [\mathrm{Re}(\widetilde{\mathbf{p}})]\vec{u} \tag{B.7}$$

then

$$\vec{I} = -\frac{p_0^2(\vec{r}) \, \nabla \, \Phi_0(\vec{r})}{\rho_0 \omega} \cos^2 \left[\omega t + \Phi_0(\vec{r})\right] - \frac{\nabla p_0^2(\vec{r})}{4\rho_0 \omega} \sin 2 \left[\omega t + \Phi_0(\vec{r})\right]. \quad \text{(B.8)}$$

The sound intensity is the sum of two vectors, namely, the active sound intensity vector,

$$\vec{I}_{ac} = -\frac{p_0^2(\vec{r}) \, \nabla \, \Phi_0(\vec{r})}{\rho_0 \omega} \cos^2 \left[\omega t + \Phi_0(\vec{r})\right] \quad \text{(B.9)}$$

and the reactive sound intensity vector,

$$\vec{I}_{rc} = -\frac{\nabla p_0^2(\vec{r})}{4\rho_0 \omega} \sin 2 \left[\omega t + \Phi_0(\vec{r})\right]. \quad \text{(B.10)}$$

The sound wave fronts, see Figure B.1, are at different instants surfaces of constant phase, $\Phi(\vec{r}, t) = \omega t + \Phi_0(\vec{r}) = \text{const}$. Hence, the active intensity vector is perpendicular to the sound wave front, see equation (B.9), and its magnitude is proportional to the magnitude of the spatial gradient of the phase, $\nabla\Phi(\vec{r})$.

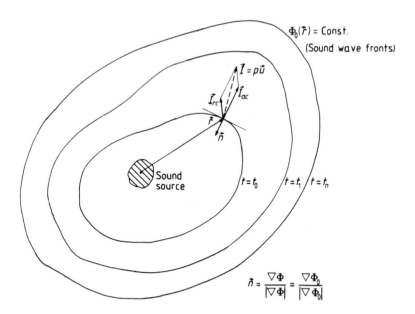

Figure B.1 Family of constant phase surfaces (sound wave fronts), $\Phi(\vec{r}, t) = \text{constant}$.

Since $p_{rms}^2 = p_0^2(\vec{r})/2$, see also equation (1.143), we can conclude from equation (B.10) that the reactive intensity vector and the spatial gradient of the mean–squared pressure, ∇p_{rms}^2, are unconformally collinear, see Figure B.2.

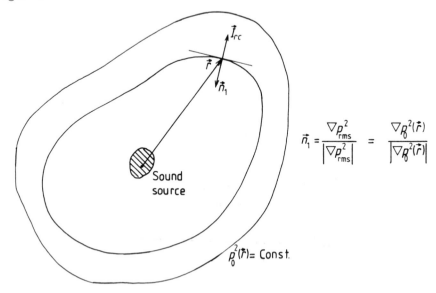

$$\vec{n}_1 = \frac{\nabla p_{rms}^2}{\left|\nabla p_{rms}^2\right|} = \frac{\nabla p_0^2(\vec{r})}{\left|\nabla p_0^2(\vec{r})\right|}$$

Figure B.2 Family of constant mean–squared pressure surfaces for an arbitrary sound field at a certain instant.

It follows from equation (B.8) that the time–averaged or mean acoustic intensity $\overline{\vec{I}}$ is equal to the time–averaged active sound intensity $\overline{\vec{I}_{ac}}$, since $\overline{\vec{I}_{rc}}$ is equal to zero. Consequently,

$$\vec{I} = 2\overline{\vec{I}}\cos^2\left(\omega t + \Phi_0\right) + \vec{I}_{rc}. \tag{B.11}$$

In the case of spherical waves from a point source, see equations (1.168), (1.191) and (B.10), the active intensity vector is

$$\vec{I}_{ac} = \frac{A^2}{\rho_0 c_0 r^2}\cos^2\left(\omega t - \vec{k}\cdot\vec{r}\right)\vec{e}_r \tag{B.12}$$

and the reactive intensity vector,

$$\vec{I}_{rc} = \frac{A^2}{2\rho_0\omega r^3}\sin 2\left(\omega t - \vec{k}\cdot\vec{r}\right)\vec{e}_r, \tag{B.13}$$

where $r = |\vec{r}|$.

The amplitude of the reactive intensity for a spherical wave diminishes proportionally to the inverse of r^3, whereas the amplitude of the active intensity decreases proportionally to the inverse of r^2. Hence, the reactive intensity manifests itself mainly near the sound source.

B2 Complex intensity of sound

The momentum conservation equation in the complex vector representation may be written down as follows:

$$\frac{\partial \vec{\tilde{u}}}{\partial t} = -\frac{1}{\rho_0} \nabla \tilde{p}, \tag{B.14}$$

where $\vec{u} = \mathrm{Re}(\vec{\tilde{u}})$. Insertion of the expression for $\tilde{p}$, equation (B.2), into equation (B.14) leads to the expression for $\vec{\tilde{u}}$, namely,

$$\vec{\tilde{u}} = -\frac{p_0(\vec{r})}{\rho_0\omega} \{\nabla \Phi_0 - i \nabla \ln p_0\} e^{i[\omega t + \Phi_0(\vec{r})]}. \tag{B.15}$$

The complex intensity is defined as [9,10],

$$\vec{I}_c = \frac{1}{2} \left(\tilde{p}\vec{\tilde{u}}^* \right). \tag{B.16}$$

Thus, for a monochromatic sound field

$$\vec{I}_c = -\frac{p_0^2(\vec{r})}{2\rho_0\omega} \{\nabla \Phi_0 + i \nabla \ln p_0\}. \tag{B.17}$$

The formula (B.17) can be presented in another form:

$$\vec{I}_c = \overline{\vec{I}_{ac}} + i\vec{I}_{r0}, \tag{B.18}$$

where $\vec{I}_{r0} = -\nabla p_0^2/(4\rho_0\omega)$, see equation (B.10).

On the other hand the (instantaneous) sound intensity $\vec{I} = \vec{I}_{ac} + \vec{I}_{rc}$ can be expressed in terms of the complex intensity. From equations (B.9) and (B.10) we have

$$\vec{I} = -\frac{p_0^2(\vec{r}) \nabla \Phi_0(\vec{r})}{2\rho_0\omega} [1 + \cos 2(\omega t + \Phi_0)] - \frac{\nabla p_0^2}{4\rho_0\omega} \sin 2(\omega t + \Phi_0) \tag{B.19}$$

or,

$$\vec{I} = \overline{\vec{I}_{ac}}\,[1 + \cos 2(\omega t + \Phi_0)] + \vec{I}_{r0}\sin 2(\omega t + \Phi_0). \tag{B.20}$$

Using the complex exponential representation, we obtain instead of formula (B.20) the expression for $\vec{I}$ as a function of the complex intensity, namely,

$$\vec{I} = \mathrm{Re}\left\{\left[\overline{\vec{I}_{ac}} + i\vec{I}_{r0}\right]\left[1 + e^{-2i(\omega t + \Phi_0)}\right]\right\} \tag{B.21}$$

or,

$$\vec{I} = \mathrm{Re}\left\{\mathbf{\vec{I}}_c\left[1 + e^{-2i(\omega t + \Phi_0)}\right]\right\}. \tag{B.22}$$

Appendix C

C1 Radiation from a harmonically pulsating sphere

The simplest acoustic radiator of spherical waves is a pulsating sphere. As was mentioned in Section 1.17, the concept of a simple sound source originates from the properties of the sound field generated by a pulsating sphere. Many radiators to a certain approximation behave like a pulsating sphere.

Let us consider a sound field generated by a pulsating sphere embedded in an infinite medium. To determine the sound field of the pulsating sphere we must find the solution of the wave equation which fulfils the boundary conditions. Since the sphere — of radius a — is located in an infinite medium, there is no reflected wave, hence the solution should have the form of a wave outgoing from the sound source. Additionally, the boundary condition is assumed in the form of a surface velocity distribution. The particle velocity of the fluid adjacent to the sphere is equal to the velocity of the sphere surface. Since the considered problem is characterised by spherical symmetry, its solution in terms of a (radial) particle velocity may be expressed by equation (1.179), see also equation (1.170). Taking into account formula (1.170), we can rewrite equation (1.179) in the form,

$$\tilde{\mathbf{u}}_r(r,t) = \frac{\mathbf{A}e^{i(\omega t - \vec{k}\cdot\vec{r} + \phi)}}{\rho_0 c_0 r \cos\phi}, \tag{C.1}$$

where ϕ is defined by equation (1.177).

The constant $\mathbf{A}$ is determined from the boundary condition which is specified for the fluid particle velocity on the surface of the sphere. Since the boundary condition written in phasor convention is as follows:

$$\tilde{\mathbf{u}}_r(a,t) = U_0 e^{i\omega t}, \tag{C.2}$$

where it is assumed that U_0, the amplitude of the particle velocity on the surface of the sphere, is small, such that $U_0/\omega \ll a$, and that also $U_0 \ll c_0$, then

$$U_0 e^{i\omega t} = \frac{\mathbf{A}e^{i(\omega t - ka + \phi_a)}}{\rho_0 c_0 a \cos\phi_a}, \tag{C.3}$$

where

$$\phi_a = \phi(a) = \arccos\left\{\frac{ka}{\sqrt{(1 + k^2 a^2)}}\right\} = \arctan\left(-\frac{1}{ka}\right). \tag{C.4}$$

Therefore, from equation (C.3), we obtain:

$$\mathbf{A} = U_0 a \rho_0 c_0 \cos \phi_a e^{i(ka - \phi_a)} \qquad (C.5)$$

and the solution (C.1) can be presented as

$$\tilde{\mathbf{u}}_r(r, t) = U_0 \left(\frac{a}{r}\right) \frac{\cos \phi_a}{\cos \phi} e^{i[\omega t - k(r-a) + \phi - \phi_a]} \qquad (C.6)$$

or,

$$\tilde{\mathbf{u}}_r(r, t) = U_0 \left(\frac{a}{r}\right)^2 \left(\frac{1 + k^2 r^2}{1 + k^2 a^2}\right)^{1/2} e^{i[\omega t - k(r-a) + \phi - \phi_a]}. \qquad (C.7)$$

Formula (C.7) for the phasor of the radial particle velocity may be presented in another form, namely,

$$\tilde{\mathbf{u}}_r(r, t) = \frac{Q}{4\pi r^2} \left(\frac{1 + k^2 r^2}{1 + k^2 a^2}\right)^{1/2} e^{i[\omega t - k(r-a) + \phi - \phi_a]}, \qquad (C.8)$$

where $Q = 4\pi a^2 U_0$ is the strength of the sound source (pulsating sphere) or the amplitude of the volume velocity across the surface of the sphere.

The sound field generated by the pulsating sphere in terms of the sound pressure phasor is

$$\tilde{\mathbf{p}}(r, t) = \left(\frac{a}{r}\right) U_0 \rho_0 c_0 \cos \phi_a e^{i[\omega t - k(r-a) - \phi_a]}. \qquad (C.9)$$

The equation may be presented as

$$\tilde{\mathbf{p}}(r, t) = \frac{i \rho_0 \omega Q}{4\pi r (1 + k^2 a^2)^{1/2}} e^{i[\omega t - k(r-a) - \phi_a - \pi/2]}. \qquad (C.10)$$

Either comparing formulae (1.178) and (1.210) with the formula (1.206), or directly combining equations (1.208), (C.5) and (1.206) leads to the equation

$$\mathbf{Q}_p = \left[\frac{Q}{(1 + k^2 a^2)^{1/2}}\right] e^{i(ka - \phi_a - \pi/2)},$$

which is equivalent to the equations

$$Q_p = |\mathbf{Q}_p| = \frac{Q}{(1 + k^2 a^2)^{1/2}} \quad \text{and} \quad \beta = ka - \phi_a,$$

since $\mathbf{Q}_p = Q_p e^{i(\beta - \pi/2)}$, see also equation (1.170). From the above equations we can conclude that the point sound source with strength of magnitude $Q_p = Q/(1 + k^2 a^2)^{1/2}$ is equivalent to the pulsating spherical sound source of radius a and of strength Q. We can also deduce that for $ka \ll 1$, i.e., for the compact pulsating sphere of strength Q, $Q_p = Q$ and $\beta = \pi/2$. This means that the compact pulsating sphere generates an identical sound field to the sound field from a point source of the same strength, namely, $\mathbf{Q}_p = Q_p = Q$.

In the sound field generated by a pulsating sphere the instantaneous sound intensity vector, as well as the active and the reactive sound intensity vectors, see Appendix B, are directed radially from the sound source, see Figure C.1.

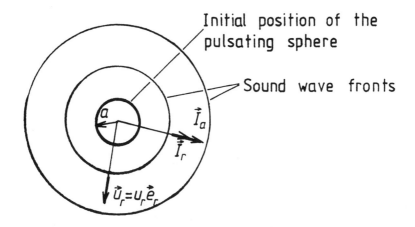

Figure C.1 The sound field emitted from a pulsating sphere.

From the formulae (1.186), (C.6) and (C.9), or directly from (B.9), (B.10) and (C.9) we obtain for the active and the reactive intensity vectors the following equations:

$$\vec{I}_{ac} = U_0^2 \left(\frac{a}{r}\right)^2 \rho_0 c_0 \cos^2 \phi_a \cos^2 [\omega t - k(r - a) - \phi_a] \vec{e}_r \qquad (C.11)$$

$$\vec{I}_{rc} = \frac{U_0^2}{2kr} \left(\frac{a}{r}\right)^2 \rho_0 c_0 \cos^2 \phi_a \sin 2 [\omega t - k(r - a) - \phi_a] \vec{e}_r. \qquad (C.12)$$

If we define I_{a0} as the amplitude of the active intensity magnitude, then

$$\vec{I}_{ac} = I_{a0} \cos^2 [\omega t - k(r - a) - \phi_a] \vec{e}_r \qquad (C.13)$$

$$\vec{I}_{rc} = \frac{I_{a0}}{2kr} \sin 2 [\omega t - k(r - a) - \phi_a] \vec{e}_r. \qquad (C.14)$$

Hence, we can conclude that the reactive intensity plays a more important role in the vicinity of a sound source and its amplitude decreases with distance r from the sound source centre as $1/r^3$.

Finally, the time-averaged intensity $\bar{\vec{I}}$ at a certain distance r from the centre of the pulsating sphere is

$$\bar{\vec{I}} = \frac{1}{2}U_0^2\left(\frac{a}{r}\right)^2\rho_0 c_0\frac{k^2a^2}{1+k^2a^2}\vec{e}_r \tag{C.15}$$

or,

$$\bar{\vec{I}} = \frac{Q^2\rho_0\omega k}{32\pi r^2(1+k^2a^2)}\vec{e}_r. \tag{C.16}$$

Integration of equation (C.16) over a closed area enclosing the pulsating sphere leads to a formula describing the sound power W_{ps} emitted by the pulsating sphere of radius a.

$$W_{ps} = \frac{Q^2\rho_0\omega k}{8\pi(1+k^2a^2)} \tag{C.17}$$

or,

$$W_{ps} = \left(\frac{Q^2\rho_0\omega^2}{8\pi c_0}\right)\frac{1}{1+\omega^2a^2/c_0^2}. \tag{C.18}$$

The last relation indicates that for an assumed source strength the sound power emitted by the pulsating sphere, which fulfils the compactness condition $ka = \omega a/c_0 \ll 1$, is proportional to ω^2.

It should be noted that the compactness condition denotes that the sound source dimension is much smaller than the wavelength of the sound produced by the sound source.

We can also deduce from equation (C.15) that for constant U_0 and for the case when $ka \ll 1$ the time–averaged intensity is

$$\bar{\vec{I}} = \frac{1}{2}U_0^2\left(\frac{a}{r}\right)^2\rho_0 c_0 k^2a^2\vec{e}_r = \frac{1}{2}U_0^2\left(\frac{a}{r}\right)^2\frac{\rho_0}{c_0}a^2\omega^2\vec{e}_r. \tag{C.19}$$

Equations (C.19) imply that the time–averaged intensity is proportional to the fourth power of the source radius, and to the square of the frequency of sound wave emitted by the source. Hence, a small sound source (with respect to the wavelength of sound) is a much poorer radiator of sound energy than an even slightly larger sound source characterised by the same amplitude of surface velocity U_0.

To give a physical interpretation of the different capabilities of sound sources to radiate sound energy, let us consider a pulsating sphere in an elastic medium as a mechanical vibrator (oscillator) with viscous damping.

C2 Pulsating sphere as a vibrator with viscous damping

Let us assume that a sphere of radius a pulsates in a vacuum under an applied harmonic force $\tilde{\mathbf{F}}(t) = F_0 e^{i\omega t}$. The equation of motion of the pulsating sphere in vacuum, written in phasor convention in terms of the displacement of the sphere surface $\tilde{\mathbf{d}}_a$, is

$$\mathcal{M}\frac{d^2}{dt^2}(\tilde{\mathbf{d}}_a) + \mathcal{R}\frac{d}{dt}(\tilde{\mathbf{d}}_a) + \mathcal{K}\tilde{\mathbf{d}}_a = F_0 e^{i\omega t}, \tag{C.20}$$

where $\mathcal{M}$ is the mass of the vibrator

  $\mathcal{R}$ is the effective mechanical damping constant of the vibrator,

and $\mathcal{K}$ denotes the effective stiffness of the vibrator.

For the case when the pulsating sphere is suspended in an elastic medium, e.g. air, the equation of motion (C.20) must be replaced by another one, namely,

$$\mathcal{M}\frac{d^2}{dt^2}(\tilde{\mathbf{d}}_a) + \mathcal{R}\frac{d}{dt}(\tilde{\mathbf{d}}_a) + \mathcal{K}\tilde{\mathbf{d}}_a = F_0 e^{i\omega t} - \tilde{\mathbf{F}}_a, \tag{C.21}$$

where $\tilde{\mathbf{F}}_a$ denotes the force that the sphere exerts on the medium during the pulsation. The force $\tilde{\mathbf{F}}_a$ is called also the *radiation reaction*, since the medium must exert on the sphere, in agreement with Newton's third law, a force opposite in sense, but of the same magnitude, as the force $\tilde{\mathbf{F}}_a$. The force which the sphere of a radius a exerts on the surrounding fluid is

$$\tilde{\mathbf{F}}_a(t) = 4\pi a^2 \tilde{\mathbf{p}}(a, t), \tag{C.22}$$

where the sound pressure at the sphere surface $\tilde{\mathbf{p}}(a, t)$ is given by formula (C.9). Hence,

$$\tilde{\mathbf{F}}_a(t) = 4\pi a^2 U_0 \rho_0 c_0 \cos\phi_a e^{-i\phi_a} e^{i\omega t} \tag{C.23}$$

and, finally, with the aid of equations (C.2) and (C.4) we obtain

$$\tilde{\mathbf{F}}_a = 4\pi a^2 \tilde{\mathbf{u}}_a \rho_0 c_0 \left(\frac{k^2 a^2}{1 + k^2 a^2} + i\frac{ka}{1 + k^2 a^2} \right), \tag{C.24}$$

where $\tilde{\mathbf{u}}_a$ denotes $\tilde{\mathbf{u}}_r(a,t)$. Equation (C.24) may be presented as

$$\tilde{\mathbf{F}}_a = \frac{4\pi a^4 \rho_0 \omega k}{1 + k^2 a^2} \tilde{\mathbf{u}}_a + \frac{4\pi a^3 \rho_0}{1 + k^2 a^2} \frac{d\tilde{\mathbf{u}}_a}{dt} \tag{C.25}$$

or,

$$\tilde{\mathbf{F}}_a = \frac{4\pi a^4 \rho_0 \omega k}{1 + k^2 a^2} \frac{d}{dt}(\tilde{\mathbf{d}}_a) + \frac{4\pi a^3 \rho_0}{1 + k^2 a^2} \frac{d^2}{dt^2}(\tilde{\mathbf{d}}_a). \tag{C.26}$$

Inserting the expression for $\tilde{\mathbf{F}}_a$ from equation (C.26) into equation (C.21), we obtain:

$$\left(\mathcal{M} + \frac{4\pi a^3 \rho_0}{1 + k^2 a^2} \right) \frac{d^2}{dt^2}(\tilde{\mathbf{d}}_a) + \left(\mathcal{R} + \frac{4\pi a^4 \rho_0 \omega k}{1 + k^2 a^2} \right) \frac{d}{dt}(\tilde{\mathbf{d}}_a) + \mathcal{K}\tilde{\mathbf{d}}_a = F_0 e^{i\omega t} \tag{C.27}$$

or,

$$(\mathcal{M} + m_a) \frac{d^2}{dt^2}(\tilde{\mathbf{d}}_a) + (\mathcal{R} + R_a) \frac{d}{dt}(\tilde{\mathbf{d}}_a) + \mathcal{K}\tilde{\mathbf{d}}_a = F_0 e^{i\omega t}, \tag{C.28}$$

where $m_a = 4\pi a^3 \rho_0/(1 + k^2 a^2)$ and $R_a = 4\pi a^4 \rho_0 \omega k/(1 + k^2 a^2)$.
    Finally, remembering that

$$\frac{d}{dt}(\tilde{\mathbf{d}}_a) = i\omega \tilde{\mathbf{d}}_a \quad \text{and} \quad \frac{d^2}{dt^2}(\tilde{\mathbf{d}}_a) = -\omega^2 \tilde{\mathbf{d}}_a$$

we may write

$$-\omega^2 (\mathcal{M} + m_a) \tilde{\mathbf{d}}_a + i\omega (\mathcal{R} + R_a) \tilde{\mathbf{d}}_a + \mathcal{K}\tilde{\mathbf{d}}_a = F_0 e^{i\omega t}. \tag{C.29}$$

The term $m_a = 4\pi a^3 \rho_0/(1 + k^2 a^2)$ may be interpreted as a certain inertial term which increases the effective mass of the vibrator. Since the inertial force $m_a d^2(\tilde{\mathbf{d}}_a)/dt^2$ is $\pi/2$ radians out of phase with respect to the surface velocity, we can conclude that this term is not responsible for any net energy transport.
    The term $\mathcal{M} + m_a$ is called the *effective mass* [11]. The term m_a is the *attached mass* and it represents the inertia of the medium surrounding a pulsating sphere whose surface is accelerating.
    The term $R_a = 4\pi a^4 \rho_0 \omega k/(1 + k^2 a^2)$ increases the effective damping of the vibrator and, since it is in phase with the surface velocity, it

results in net sound energy transport (sound radiation). This term is also described as dissipative. Treating the medium, which surrounds the sphere as a reference system, we can interpret the energy dissipated by a pulsating sphere as a manifestation of the medium 'resistance' to sound radiation. Hence, the expression $R_a = 4\pi a^4 \rho_0 \omega k/(1 + k^2 a^2)$ defines the so–called *radiation resistance* for a pulsating sphere.

The sound power generated by a pulsating sphere may be described as a function of the sound radiation resistance; insertion of $Q = 4\pi a^2 U_0$ in equation (C.17) leads to:

$$W_{ps} = \frac{4\pi a^4 \rho_0 \omega k}{1 + k^2 a^2} \frac{U_0^2}{2} \qquad (C.30)$$

or,

$$W_{ps} = R_a \frac{U_0^2}{2}. \qquad (C.31)$$

The expression $X_a = \omega m_a = 4\pi a^3 \rho_0 \omega/(1 + k^2 a^2)$ is called the *radiation reactance* for the pulsating sphere, since the inertial force $m_a d\tilde{u}_a/dt$ appears as a reaction to the sphere surface acceleration.

The complex quantity,

$$Z_{ra} = R_a + i\omega m_a = R_a + iX_a, \qquad (C.32)$$

is named as the *radiation impedance* for the pulsating sphere of radius a.

It should be noted that the concept of impedance exists not only in acoustics but also in mechanics and electrical engineering. Vibrating mechanical systems and ac circuits are described by the same type of differential equations [12–14].

Equation (C.32) may be presented in another form as

$$Z_{ra} = \frac{4\pi a^4 \rho_0 \omega k}{1 + k^2 a^2} + i\frac{4\pi a^3 \rho_0 \omega}{1 + k^2 a^2} \qquad (C.33)$$

or,

$$Z_{ra} = 4\pi a^2 \rho_0 c_0 \frac{k^2 a^2}{1 + k^2 a^2} + im_0 \frac{3\omega}{1 + k^2 a^2}, \qquad (C.34)$$

where $m_0 = (4/3)\pi a^3 \rho_0$ is the mass of the fluid displaced by the sphere.

At very high frequencies, when $a \gg \lambda$ or $ka \gg 1$, the radiation resistance of a pulsating sphere approaches its maximum value, namely

$4\pi a^2 \rho_0 c_0$, which is the product of the sphere surface and $\rho_0 c_0$, the characteristic resistance of the fluid (called also the specific acoustic resistance of a plane wave). At high frequencies, when $ka \gg 1$, the radiation reactance X_a is very small in comparison with the radiation resistance, since $X_a/R_a \rightarrow 1/(ka) \ll 1$. Hence, for high frequencies the sound radiation, manifested by a high rate of dissipation of energy from the sphere, is the dominating phenomenon. It should be noted that, if $ka \gg 1$, the pulsating sphere emits the same amount of energy per unit area as a plane surface, all points of which vibrate in phase.

On the other hand, at low frequencies, when $ka \ll 1$, the radiation reactance X_a dominates the radiation resistance R_a since $R_a/X_a \rightarrow ka \ll 1$. The radiation reactance approaches in this case the value of $3m_0\omega$. Hence, we can formulate the conclusion that a pulsating sphere is a more efficient sound radiator at high frequencies than at low frequencies, see also Appendix C1.

The expression for the radiation impedance of a pulsating sphere may be presented in another form, namely as a ratio of a radial force exerted on the sphere surface $S = 4\pi a^2$ by a sound field to the radial velocity of the sphere surface.

We can prove that the radiation impedance for a pulsating sphere, see equations (C.33), (C.9) and (C.6), fulfils the relation,

$$\mathbf{Z}_{ra} = 4\pi a^2 \, \frac{\widetilde{\mathbf{p}}(a)}{\widetilde{\mathbf{u}}_r(a)} \qquad (C.35)$$

or,

$$\mathbf{Z}_{ra} = S \frac{\widetilde{\mathbf{p}}(a)}{\widetilde{\mathbf{u}}_r(a)} = \frac{\widetilde{\mathbf{F}}_a}{\widetilde{\mathbf{u}}_r(a)}. \qquad (C.36)$$

The last relation (C.36) has a more global meaning, since it can be applied to any acoustic system, not only to a pulsating sphere. It may be treated as a general definition of radiation impedance.

Appendix D

Dipole sound sources

D1 Definition of dipole

Let us imagine two point sound sources of identical strength, but opposite in phase, situated at points P_1 and P_2 at a distance d apart, see Figure D.1.

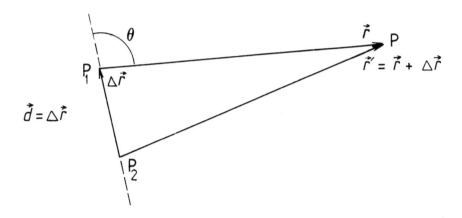

Figure D.1 Two point sound sources close together.

Both sound sources radiate sound simultaneously and create a sound field which is the superposition of the sound fields generated independently by each point source. The sound pressure $\widetilde{p}$ at the point P is the sum of the sound pressures, say $\widetilde{p}_1$ and $\widetilde{p}_2$, caused by sound radiation from the sound sources at points P_1 and P_2, respectively. Hence,

$$\widetilde{p} = \widetilde{p}_1 + \widetilde{p}_2 = \widetilde{p}_p(\vec{r}) - \widetilde{p}_p(\vec{r} + \Delta\vec{r}), \tag{D.1}$$

where $\widetilde{p}_p$ denotes the sound pressure generated by a point source, see equation (1.210). The minus sign in front of the term $\widetilde{p}_p(\vec{r}+\Delta\vec{r})$ indicates that the point sound source at point P_2 is π radians out of phase in relation to the sound source at P_1. Taking into account the formulae (D.1) and (1.210), we obtain:

$$\widetilde{p}(\vec{r}, \Delta\vec{r}) = -\frac{i\omega\rho_0 \mathbf{Q}_p}{4\pi} e^{i\omega t} \left(\frac{e^{-ik|\vec{r}+\Delta\vec{r}|}}{|\vec{r}+\Delta\vec{r}|} - \frac{e^{-ik|\vec{r}|}}{|\vec{r}|} \right), \tag{D.2}$$

where $k = |\vec{k}|$ and $\mathbf{Q}_p$ is the strength of the point source.

An acoustic dipole is a mathematical idealisation of the pair of point sources characterised by the same strength, but opposite in phase. Thus, an acoustic dipole is the limit configuration of two point sources for which the distance $\vec{d} = \Delta\vec{r} \to 0$, while $Q_p \to \infty$ in such a way that the vector $\vec{\mathbf{Q}}_d = \mathbf{Q}_p\vec{d}$ is finite. The vector $\vec{\mathbf{Q}}_d$ is called the dipole strength. Hence, the sound pressure in a dipole field at the point P, see equation (D.2), is given by

$$\tilde{\mathbf{p}}_d = \lim_{\Delta\vec{r}\to 0} \tilde{\mathbf{p}}(\vec{r}, \Delta\vec{r}) = \lim_{\Delta\vec{r}\to 0}\left[-\frac{i\rho_0\omega\mathbf{Q}_p}{4\pi}e^{i\omega t}\left(\frac{e^{-ik|\vec{r}+\Delta\vec{r}|}}{|\vec{r}+\Delta\vec{r}|} - \frac{e^{-ik|\vec{r}|}}{|\vec{r}|} \right) \right],$$
$$\text{(D.3)}$$

where $\Delta r = |\Delta\vec{r}| = |\vec{d}|$, or

$$\tilde{\mathbf{p}}_d = \lim_{\Delta\vec{r}\to 0}\left[-\frac{i\rho_0\omega\mathbf{Q}_p}{4\pi}e^{i\omega t}\Delta r\left(\frac{e^{-ik|\vec{r}+\Delta\vec{r}|}}{|\vec{r}+\Delta\vec{r}|} - \frac{e^{-ik|\vec{r}|}}{|\vec{r}|} \right)/\Delta r \right]. \qquad \text{(D.4)}$$

Equation (D.4) can be presented in the form,

$$\tilde{\mathbf{p}}_d = \lim_{\Delta r\to 0}\left(-\frac{i\rho_0\omega\mathbf{Q}_p}{4\pi}e^{i\omega t}\Delta r \right)\lim_{\Delta r\to 0}\left[\left(\frac{e^{-ik|\vec{r}+(\Delta r)\vec{e}_d|}}{|\vec{r}+(\Delta r)\vec{e}_d|} - \frac{e^{-ik|\vec{r}|}}{|\vec{r}|} \right)/\Delta r \right],$$
$$\text{(D.5)}$$

where $\vec{e}_d = \Delta\vec{r}/\Delta r = \vec{d}/|\vec{d}|$. Introducing the notation

$$f(\vec{r}) = \frac{e^{-ik|\vec{r}|}}{|\vec{r}|},$$

we obtain from equation (D.5) the following:

$$\tilde{\mathbf{p}}_d = \lim_{\Delta r\to 0}\left(-\frac{i\rho_0\omega\mathbf{Q}_p}{4\pi}e^{i\omega t}\Delta r \right)\frac{df(\vec{r})}{d\vec{e}_d}, \qquad \text{(D.6)}$$

where $df(\vec{r})/d\vec{e}_d$ denotes the directional derivative of $f(\vec{r})$ at point P in the direction of $\vec{e}_d$. From equation (D.6) we have

$$\tilde{\mathbf{p}}_d = \lim_{\Delta r\to 0}\left(-\frac{i\rho_0\omega\mathbf{Q}_p}{4\pi}e^{i\omega t}\Delta r \right)\vec{e}_d \cdot \nabla f \qquad \text{(D.7)}$$

or,

$$\tilde{\mathbf{p}}_d = \lim_{\Delta r\to 0}\left(-\frac{i\rho_0\omega\mathbf{Q}_p}{4\pi}\Delta r e^{i\omega t} \right)\vec{e}_d \cdot \vec{e}_r\frac{d}{d\vec{e}_r}\left(\frac{e^{-ik|\vec{r}|}}{|\vec{r}|} \right) \qquad \text{(D.8)}$$

or,

$$\widetilde{\mathbf{p}}_d = -\frac{i\rho_0\omega}{4\pi}e^{i\omega t}\vec{\mathbf{Q}}_d \cdot \vec{e}_r \frac{d}{dr}\left(\frac{e^{-ikr}}{r}\right), \tag{D.9}$$

where $\vec{e}_r = \vec{r}/|\vec{r}|$, $r = |\vec{r}|$, $\vec{e}_d\Delta r = \vec{d}$ and $\mathbf{Q}_p\vec{d} = \vec{\mathbf{Q}}_d$. However, as $\vec{\mathbf{Q}}_d \cdot \vec{e}_r = \mathbf{Q}_d \cos\theta$, see Figure D.1,

$$\widetilde{\mathbf{p}}_d = -\frac{i\rho_0\omega}{4\pi}e^{i\omega t}\mathbf{Q}_d \cos\theta\frac{d}{dr}\left(\frac{e^{-ikr}}{r}\right) \tag{D.10}$$

or,

$$\widetilde{\mathbf{p}}_d = \frac{i\rho_0\omega}{4\pi r^2}\mathbf{Q}_d \cos\theta(1 + ikr)e^{i(\omega t-kr)}. \tag{D.11}$$

Formula (D.11) is the expression for the complex sound pressure field of an acoustic dipole. The presence of the term $\cos\theta$ indicates that the sound field depends on the direction of radiation. The sound pressure achieves maximum values along the axis of the dipole ($P_1\,P_2$ in Figure D.1) and zero values along the direction perpendicular to the dipole axis. However, the sound wave front (surface of constant phase) has a spherical shape.

D2 Particle velocity in a dipole field

Putting the expression for the sound pressure (D.11) into equation (B.14), we can obtain an expression for the vector of the particle velocity in the acoustic dipole field. Since the sound pressure $\widetilde{\mathbf{p}}_d$ is a function only of two spherical co-ordinates r and θ, equation (B.14) in this case may be reduced to the system of equations:

$$\rho_0\frac{\partial\widetilde{\mathbf{u}}_r}{\partial t} = -\frac{\partial\widetilde{\mathbf{p}}_d}{\partial r}, \tag{D.12}$$

$$\rho_0\frac{\partial\widetilde{\mathbf{u}}_\theta}{\partial t} = -\frac{1}{r}\frac{\partial\widetilde{\mathbf{p}}_d}{\partial\theta}, \tag{D.13}$$

where $u_r = \mathrm{Re}(\widetilde{\mathbf{u}}_r)$ and $u_\theta = \mathrm{Re}(\widetilde{\mathbf{u}}_\theta)$ are, respectively, the radial and tangential components of the particle velocity vector $\vec{u} = [u_r, 0, u_\theta] = u_r\vec{e}_r + u_\theta\vec{e}_\theta$ in the spherical (polar) co-ordinate system. Next, insertion of expression (D.11) for the sound pressure $\widetilde{\mathbf{p}}_d$ into equations (D.12) and (D.13) results in the expressions for $\widetilde{\mathbf{u}}_r$ and $\widetilde{\mathbf{u}}_\theta$, respectively. Thus

$$\widetilde{\mathbf{u}}_r = \frac{\mathbf{Q}_d}{2\pi r^3}(1 - k^2 r^2/2 + ikr)\cos\theta e^{i(\omega t-kr)} \tag{D.14}$$

or,

$$\tilde{\mathbf{u}}_r = \frac{\mathbf{Q}_d}{4\pi r^3}\sqrt{4+k^4r^4}\cos\theta e^{i(\omega t-kr+\varphi_r)}, \tag{D.15}$$

where

$$\cos\varphi_r = \frac{2-k^2r^2}{\sqrt{4+k^4r^4}} \quad \text{and} \quad \sin\varphi_r = \frac{2kr}{\sqrt{4+k^4r^4}}. \tag{D.16}$$

Similarly,

$$\tilde{\mathbf{u}}_\theta = \frac{\mathbf{Q}_d}{4\pi r^3}(1+ikr)\sin\theta e^{i(\omega t-kr)} \tag{D.17}$$

or,

$$\tilde{\mathbf{u}}_\theta = \frac{\mathbf{Q}_d}{4\pi r^3}\sqrt{1+k^2r^2}\sin\theta e^{i(\omega t-kr+\varphi_\theta)}, \tag{D.18}$$

where

$$\cos\varphi_\theta = \frac{1}{\sqrt{1+k^2r^2}} \quad \text{and} \quad \sin\varphi_\theta = \frac{kr}{\sqrt{1+k^2r^2}}. \tag{D.19}$$

From equations (D.15) and (D.18) we can deduce that the ratio of the amplitudes of the tangential and the radial components of the particle velocity is

$$\frac{u_{\theta 0}}{u_{r0}} = \tan\theta\frac{\sqrt{1+k^2r^2}}{\sqrt{4+k^4r^4}}. \tag{D.20}$$

Equations (D.15) and (D.18) may be presented in other forms, namely,

$$\tilde{\mathbf{u}}_r = \frac{2}{\rho_0\omega r(1+k^2r^2)}\sqrt{k^6r^6/4+(1+k^2r^2/2)^2}e^{i\vartheta_r}\tilde{\mathbf{p}}_d, \tag{D.21}$$

where

$$\cos\vartheta_r = \frac{k^3r^3}{2\sqrt{k^6r^6/4+(1+k^2r^2/2)^2}}$$

and

$$\sin\vartheta_r = \frac{-1-k^2r^2/2}{\sqrt{k^6r^6/4+(1+k^2r^2/2)^2}} \tag{D.22}$$

and

$$\tilde{\mathbf{u}}_\theta = \frac{1}{\rho_0 \omega r} \tan\theta e^{i3\pi/2} \tilde{\mathbf{p}}_d. \tag{D.23}$$

Equation (D.23) implies that the phasor $\tilde{\mathbf{p}}_d$ leads the phasor $\tilde{\mathbf{u}}_\theta$ by $\pi/2$ radians, hence the sound pressure and the tangential component of the particle velocity are out of phase in the whole field of the acoustic dipole.

Introducing the notation $u_r = \text{Re}(\tilde{\mathbf{u}}_r)$, $u_\theta = \text{Re}(\tilde{\mathbf{u}}_\theta)$ and remembering that the equation,

$$\frac{dr}{r} = \frac{u_r}{u_\theta} d\theta \tag{D.24}$$

is the differential equation of the streamlines, we have

$$2\frac{dr}{r} \sqrt{\frac{1 + k^2 r^2}{4 + k^4 r^4}} \frac{\cos(\omega t - kr + \varphi_\theta)}{\cos(\omega t - kr + \varphi_r)} = 2\frac{d(\sin\theta)}{\sin\theta}, \tag{D.25}$$

which is the equation of streamlines in the acoustic dipole field. Equation (D.25) can be presented in another form,

$$\frac{2[\cos(\omega t - kr) - kr\sin(\omega t - kr)]dr}{2[r\cos(\omega t - kr) - kr^2 \sin(\omega t - kr)] - k^2 r^3 \cos(\omega t - kr)} = 2\frac{d(\sin\theta)}{\sin\theta}. \tag{D.26}$$

In the near field of the acoustic dipole, where $kr \ll 1$, the differential equation for the streamlines is

$$\frac{dr}{r} = 2\frac{d(\sin\theta)}{\sin\theta}. \tag{D.27}$$

The solution of equation (D.27) is in the form,

$$r = C\sin^2\theta, \tag{D.28}$$

where C is a constant. Streamlines for the near field in the plane xz are shown in Figure D.2.

D3 Sound intensity of a dipole field

In the sound field of an acoustic dipole the active sound intensity vector is the radial sound intensity vector $\vec{I}_r = \text{Re}(\tilde{\mathbf{p}})\text{Re}(\tilde{\mathbf{u}}_r)\vec{e}_r$, see also equation (B.9). The vector of the time–averaged radial sound intensity is given by:

$$\overline{\vec{I}}_r = \overline{\text{Re}(\tilde{\mathbf{p}})\text{Re}(\tilde{\mathbf{u}}_r)}\vec{e}_r = \frac{1}{2}\text{Re}(\tilde{\mathbf{p}}\tilde{\mathbf{u}}_r^*)\vec{e}_r. \tag{D.29}$$

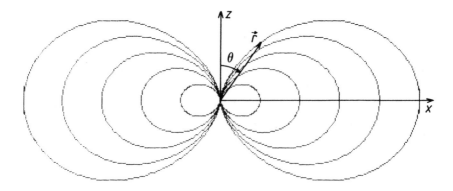

Figure D.2 Streamlines in the near field of an acoustic dipole.

Since the sound pressure phasor $\tilde{\mathbf{p}}_d$ may be presented in the form,

$$\tilde{\mathbf{p}}_d = \frac{\omega \rho_0 \mathbf{Q}_d \cos \theta}{4\pi r^2} \sqrt{1 + k^2 r^2}\, e^{i(\omega t - kr + \varphi_p)}, \tag{D.30}$$

where

$$\cos \varphi_p = -\frac{kr}{\sqrt{1 + k^2 r^2}} \quad \text{and} \quad \sin \varphi_p = \frac{1}{\sqrt{1 + k^2 r^2}} \tag{D.31}$$

and

$$\tilde{\mathbf{u}}_r^* = \frac{\mathbf{Q}_d^*}{4\pi r^3} \cos \theta \sqrt{1 + k^4 r^4}\, e^{-i(\omega t - kr + \varphi_r)}.$$

(See also equations (D.15) and (D.16)). Thus

$$\bar{I} = \bar{I}_r = \frac{Q_d^2 \omega \rho_0 \cos^2 \theta}{16\pi^2 r^5} \sqrt{(1 + k^2 r^2)(4 + k^4 r^4)} \cos(\varphi_p - \varphi_r) \tag{D.32}$$

or,

$$\bar{I} = \frac{Q_d^2 \omega \rho_0 k^3 \cos^2 \theta}{32\pi^2 r^2} \tag{D.33}$$

and finally

$$\bar{I} = \frac{Q_d^2 \omega^4 \rho_0 \cos^2 \theta}{32\pi^2 c_0^3 r^2}, \tag{D.34}$$

where $Q_d = |\mathbf{Q}_d|$.

The time–averaged sound intensity is maximal along the dipole axis, but there is no radiation in the direction transverse to the axis. The radiation pattern of the acoustic dipole is shown in Figure D.3.

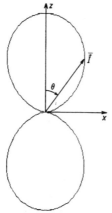

Figure D.3 Directivity pattern of acoustic dipole.

Comparison of equations (D.34) and (1.211) leads to the conclusion that the dipole source is relatively more efficient at high frequencies than the point source for equal source strengths, since the sound intensity of the dipole field is proportional to ω^4 instead of ω^2 for the point sound source.

Finally, the sound power W_d emitted by the acoustic dipole is

$$W_d = \int_0^{2\pi} \int_0^{\pi} \overline{I}_r r^2 \sin\theta d\theta d\psi \qquad (D.35)$$

or,

$$W_d = \frac{4\pi}{3} \frac{Q_d^2 \omega^4 \rho_0}{32\pi^2 c_0^3} \qquad (D.36)$$

or,

$$W_d = \frac{\rho_0 Q_d^2 \omega^4}{24\pi c_0^3}. \qquad (D.37)$$

D4 Far and near fields of the acoustic dipole

For the near field, where $kr \ll 1$, equation (D.11) for the sound pressure

$\tilde{\mathbf{p}}_d$ is reduced to:

$$\tilde{\mathbf{p}}_d = \frac{i\rho_0\omega}{4\pi r^2}\mathbf{Q}_d\cos\theta e^{i\omega t} \tag{D.38}$$

and the equations for the radial (D.14) and the tangential (D.17) components of the particle velocity vector are, respectively,

$$\tilde{\mathbf{u}}_r = \frac{\mathbf{Q}_d}{2\pi r^3}\cos\theta e^{i\omega t}, \tag{D.39}$$

$$\tilde{\mathbf{u}}_\theta = \frac{\mathbf{Q}_d}{4\pi r^3}\sin\theta e^{i\omega t}. \tag{D.40}$$

The expressions (D.38) to (D.40) indicate that in the near field the progressive waves disappear. The medium around the dipole source behaves like an incompressible fluid. The amplitude of the sound pressure varies in the near field in inverse proportion to r^2 but the amplitude of the particle velocity decreases as $1/r^3$, more quickly than the pressure amplitude.

In the near field the radial and tangential components of the particle velocity are in phase and the sound pressure phasor $\tilde{\mathbf{p}}_d$ leads both phasors $\tilde{\mathbf{u}}_r$ and $\tilde{\mathbf{u}}_\theta$ by $\pi/2$ radians, see Figure D.4.

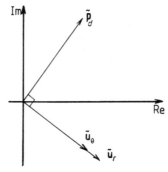

Figure D.4 Configuration of phasors in the complex plane for the acoustic dipole near field.

In the far field, where $kr \gg 1$, the sound pressure amplitude decays proportionally to $1/r$, since equation (D.11) is reduced to:

$$\tilde{\mathbf{p}}_d = \frac{\mathbf{Q}_d\rho_0\omega^2\cos\theta}{4\pi rc_0}e^{i(\omega t-kr+\pi)}. \tag{D.41}$$

For the far field the expressions for the radial and tangential components of the particle velocity are, respectively,

$$\tilde{\mathbf{u}}_r = \frac{\mathbf{Q}_d\omega^2}{4\pi rc_0^2}\cos\theta e^{i(\omega t-kr+\pi)} \tag{D.42}$$

and

$$\widetilde{\mathbf{u}}_\theta = \frac{\mathbf{Q}_d \omega}{4\pi r^2 c_0} \sin\theta\, e^{i(\omega t - kr + \pi/2)}. \tag{D.43}$$

The amplitude of the tangential component of the particle velocity diminishes proportionally to $1/r^2$, whereas the amplitude of the radial component decreases as $1/r$. Hence, in the far field the radial component dominates over the tangential and the particle velocity vector has a radial direction. From equations (D.41) and (D.42) we can conclude that in the far field of the acoustic dipole the relation,

$$\widetilde{\mathbf{u}}_r = \frac{\widetilde{\mathbf{P}}_d}{\rho_0 c_0} \tag{D.44}$$

holds, so the sound wave in the far distance from the dipole source behaves like a plane wave, see equations (1.104) and (1.105). Similarly, comparison of equations (D.41) and (D.34) leads to the conclusion that in the far field the relation between time–averaged intensity and sound pressure amplitude is the same as for plane, freely progressing waves, see equation (1.145), namely,

$$\overline{I} = \frac{p_0^2}{2\rho_0 c_0} = \frac{p_{\mathrm{rms}}^2}{\rho_0 c_0}, \tag{D.45}$$

where p_0^2 is the square of the amplitude of the harmonic sound wave and is given by

$$p_0^2 = \left(\frac{Q_d \rho_0 \omega^2 \cos\theta}{4\pi r c_0}\right)^2.$$

Finally, in the far field the sound pressure phasor and the phasor of the particle velocity radial component, which dominates over the tangential component, are in phase, see equations (D.41) and (D.42) and Figure D.5.

It should be noted that in contrast to the point source field, where the amplitude of the sound pressure decreases proportionally to $1/r$, the behaviour of the sound pressure in the acoustic dipole field has a dual character. Namely, in the near field the amplitude of the sound pressure varies as $1/r^2$, but in the far field as $1/r$. This dual structure of the sound field, as well as its dependence on $\cos\theta$, are the main features of the acoustic dipole field.

D5 Physical realisation of the acoustic dipole field

Two simple sound sources

Let us consider two simple sound sources of identical strengths, but opposite in phase, situated at a small distance d apart and emitting sound

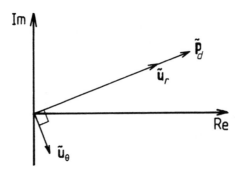

Figure D.5 Configuration of the $\tilde{\mathbf{u}}_r$, $\tilde{\mathbf{u}}_\theta$ and $\tilde{\mathbf{p}}_d$ phasors for the far field of an acoustic dipole.

with the same frequency ω, as was shown in Figure D.1 for the case of two point sources. The sound pressure of the sound field generated by these two simple sources is expressed by equation (D.1), where $\tilde{\mathbf{p}}_p$ denotes the sound pressure generated by the simple source. Since, from Taylor's theorem,

$$\tilde{\mathbf{p}}_p(\vec{r} + \vec{d}) = \tilde{\mathbf{p}}_p(\vec{r}) + \frac{\partial \tilde{\mathbf{p}}_p}{\partial \vec{e}_d} d + \tilde{\mathbf{R}}(d), \qquad (D.46)$$

where $d = |\vec{d}|$, $\vec{e}_d = \vec{d}/|\vec{d}|$ and $\tilde{\mathbf{R}}(d)$ is the remainder. From equation (1.210)

$$\tilde{\mathbf{p}}_p = \frac{i\rho_0 \omega Q_p}{4\pi r} e^{i(\omega t - kr)},$$

where Q_p (real) is the strength of the simple source. The equation (D.46) leads to

$$\tilde{\mathbf{p}}_p(\vec{r} + \vec{d}) = \tilde{\mathbf{p}}_p(\vec{r}) + \left(\vec{e}_d \cdot \nabla \tilde{\mathbf{p}}_p\right) d + \tilde{\mathbf{R}}(d), \qquad (D.47)$$

or,

$$\tilde{\mathbf{p}}_p(\vec{r} + \vec{d}) = \tilde{\mathbf{p}}_p(\vec{r}) + \vec{e}_d \cdot \vec{e}_r \frac{\partial \tilde{\mathbf{p}}_p}{\partial \vec{e}_r} d + \tilde{\mathbf{R}}(d), \qquad (D.48)$$

where $\vec{e}_r = \vec{r}/|\vec{r}|$. Finally,

$$\tilde{\mathbf{p}}_p(\vec{r} + \vec{d}) = \tilde{\mathbf{p}}_p(\vec{r}) + \cos\theta \frac{\partial \tilde{\mathbf{p}}_p}{\partial \vec{e}_r} d + \tilde{\mathbf{R}}(d), \qquad (D.49)$$

then, see equation (D.1),

$$\tilde{\mathbf{p}} = -\cos\theta \frac{\partial \tilde{\mathbf{P}}_p}{\partial r} d - \tilde{\mathbf{R}}(d), \tag{D.50}$$

where

$$\frac{\partial \tilde{\mathbf{P}}_p}{\partial r} = -\frac{i\rho_0\omega Q_p}{4\pi r^2}(1+ikr)e^{i(\omega t - kr)}. \tag{D.51}$$

Equation (D.50) may be presented in the forms,

$$\tilde{\mathbf{p}} = \frac{i\rho_0\omega Q_p d \cos\theta}{4\pi r^2}(1+ikr)e^{i(\omega t - kr)} - \tilde{\mathbf{R}}(d), \tag{D.52}$$

$$\tilde{\mathbf{p}} = \frac{i\rho_0\omega Q_d \cos\theta}{4\pi r^2}(1+ikr)e^{i(\omega t - kr)} - \tilde{\mathbf{R}}(d), \tag{D.53}$$

where $Q_d = |\vec{Q}_d| = |Q_p\vec{d}|$. Note that $\tilde{\mathbf{R}}(d) = o(d)$ for infinitely small d. The symbol $o(d)$, applied to the function $\tilde{\mathbf{R}}(d)$, $\tilde{\mathbf{R}}(d) = o(d)$, denotes that $\lim_{d \to 0} \tilde{\mathbf{R}}(d) = 0$ and $\lim_{d \to 0} \tilde{\mathbf{R}}(d)/d = 0$.

For d approaching zero equation (D.53) is reduced to the expression for the sound pressure for the acoustic dipole, see equation (D.11). However, in the case of simple sources, situated at a certain small but finite distance apart, the question arises about the conditions which should be fulfilled to treat this pair of sound sources as an acoustic dipole. From equation (D.53) we can deduce that if the absolute value of the second term of the right side of equation (D.53), $|\tilde{\mathbf{R}}(d)|$, is much smaller than the absolute value of the first term

$$\left| \frac{i\rho_0\omega Q_p d \cos\theta}{4\pi r^2}(1+ikr)e^{i(\omega t - kr)} \right|$$

then equation (D.53) is reduced to the formula for the sound pressure in the acoustic dipole field. Comparison of both terms leads to the conclusion that, if the source region is acoustically compact, i.e., $d \ll c_0/\omega = \lambda/(2\pi)$ or $kd \ll 1$, then at $r \gg d$ the sound field generated by two simple sources can be treated as an acoustic dipole. It should be noted that two conditions must be fulfilled, namely, $kd \ll 1$ and $r \gg d$.

Since for the acoustic dipole created by a pair of simple sources $Q_d = Q_p d$, the power emitted by this acoustic dipole, see equation (D.37), is

$$W_d = \frac{Q_p^2 d^2 \omega^4 \rho_0}{24\pi c_0^3}. \tag{D.54}$$

Comparing equation (D.54) with formula (1.212), we can deduce that the ratio of powers W_d/W_p, where W_p denotes the power emitted by a point source (simple source), is

$$\frac{W_d}{W_p} = \frac{Q_p^2 d^2 \omega^4 \rho_0}{24\pi c_0^3} \cdot \frac{8\pi c_0}{\rho_0 \omega^2 Q_p^2} = \frac{1}{3}\frac{\omega^2 d^2}{c_0^2} \tag{D.55}$$

or,

$$\frac{W_d}{W_p} = \frac{1}{3}k^2 d^2. \tag{D.56}$$

Since $kd \ll 1$, we can conclude that the power emitted by the dipole is much smaller than the sound power generated by the simple source. This effect is the result of the self–cancellation between fields from the two simple sources.

Oscillating sphere

Another physical realisation of the acoustic dipole field is the sound field created by an oscillating rigid sphere. A sphere of radius a which oscillates along a certain axis, say the z axis, as illustrated in Figure D.6, generates an acoustic dipole field [4,11,15,16]. The displacements of the centre of the rigid sphere are small and the velocity of the sphere centre (along the z axis) in the complex notation is assumed to be $U_0 \vec{e}_z e^{i\omega t}$. The dipole field generated by the oscillating sphere is expressed by the sound pressure phasor and the phasors of the particle velocity components as follows:

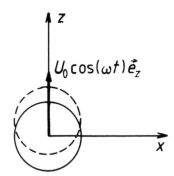

Figure D.6 Oscillating sphere.

$$\tilde{p} = \frac{i\rho_0 \omega U_0 \cos\theta a^3 (1 + ikr)}{r^2 (2 - k^2 a^2 + i2ka)} e^{i[\omega t - k(r-a)]} \tag{D.57}$$

or,

$$\widetilde{\mathbf{p}} = \frac{\rho_0 \omega U_0 \cos\theta a^3 \sqrt{1+k^2 r^2}}{r^2 \sqrt{4+k^4 a^4}} e^{i[\omega t - k(r-a) + \varphi_p - \varphi_a]} \qquad (D.58)$$

and

$$\widetilde{\mathbf{u}}_r = U_0 \left(\frac{a}{r}\right)^3 \cos\theta \frac{2 - k^2 r^2 + i2kr}{2 - k^2 a^2 + i2ka} e^{i[\omega t - k(r-a)]} \qquad (D.59)$$

or,

$$\widetilde{\mathbf{u}}_r = U_0 \left(\frac{a}{r}\right)^3 \cos\theta \frac{\sqrt{4+k^4 r^4}}{\sqrt{4+k^4 a^4}} e^{i[\omega t - k(r-a) + \varphi_r - \varphi_a]} \qquad (D.60)$$

and

$$\widetilde{\mathbf{u}}_\theta = U_0 \left(\frac{a}{r}\right)^3 \sin\theta \frac{1 + ikr}{2 - k^2 a^2 + i2ka} e^{i[\omega t - k(r-a)]} \qquad (D.61)$$

or,

$$\widetilde{\mathbf{u}}_\theta = -iU_0 \left(\frac{a}{r}\right)^3 \sin\theta \frac{\sqrt{1+k^2 r^2}}{\sqrt{4+k^4 a^4}} e^{i[\omega t - k(r-a) + \varphi_p - \varphi_a]}, \qquad (D.62)$$

where φ_p is defined by equation (D.31), φ_r is defined by equation (D.16) and $\varphi_a = \varphi_r(a)$ can be determined from equation (D.16) by putting $r = a$.

Taking into account formulae (D.58) and (D.59), we have for the time–averaged radial sound intensity the following expression:

$$\overline{\vec{I}_r} = \overline{I_r}\vec{e}_r = \frac{1}{2}\mathrm{Re}(\widetilde{\mathbf{p}}\widetilde{\mathbf{u}}_r^*)\vec{e}_r$$

$$= \frac{1}{2}\rho_0 c_0 U_0^2 \left(\frac{a}{r}\right)^2 \cos^2\theta \frac{k^4 a^4}{4+k^4 a^4}\vec{e}_r. \qquad (D.63)$$

Hence, the sound power W_{os} emitted by the oscillating rigid sphere may be obtained by integrating $\overline{I_r}$ over a sphere of radius r $(r \gg a)$, see equation (D.63):

$$W_{os} = 2\pi \int_0^\pi \overline{I_r} r^2 \sin\theta d\theta = \frac{2}{3}\pi\rho_0 c_0 U_0^2 a^2 \frac{k^4 a^4}{4+k^4 a^4}. \qquad (D.64)$$

From equation (D.63) it is seen that for the far field $(kr \gg 1)$ $\overline{I} = \overline{I_r} = p_0^2/(2\rho_0 c_0)$, see equation (1.145), where

$$p_0 = \frac{\rho_0 \omega U_0 \cos\theta a^3}{r^2} \frac{\sqrt{1+k^2 r^2}}{\sqrt{4+k^4 a^4}} \qquad (D.65)$$

is the amplitude of the sound pressure, equation (D.58). Namely, since

$$p_0^2 = \frac{\rho_0^2 \omega^2 U_0^2 \cos^2 \theta a^6}{r^4} \frac{1 + k^2 r^2}{4 + k^4 a^4}$$

$$= \rho_0^2 c_0^2 U_0^2 \cos^2 \theta \left(\frac{a}{r}\right)^2 \frac{1 + k^2 r^2}{k^2 r^2} \frac{k^4 a^4}{4 + k^4 a^4}$$

for $kr \gg 1$ p_0^2 is reduced to the expression:

$$p_0^2 = \rho_0^2 c_0^2 U_0^2 \cos^2 \theta \left(\frac{a}{r}\right)^2 \frac{k^4 a^4}{4 + k^4 a^4}. \tag{D.66}$$

Note that the relation $\overline{I} = \overline{I_r} = p_0^2/(2\rho_0 c_0)$ holds for the whole sound field of the pulsating sphere, see equations (C.9) and (C.15), whereas the same relation only holds in the far field of the oscillating rigid sphere.

References

[1] G. K. Batchelor (1970). *An Introduction to Fluid Mechanics*. Cambridge University Press, Cambridge.

[2] P. A. Thompson (1972). *Compressible Fluid Dynamics*. McGraw–Hill, New York.

[3] L. D. Landau and E. M. Lifshitz (1987). *Fluid Mechanics*, second edition. Pergamon Press, Oxford.

[4] S. Temkin (1981). *Elements of Acoustics*. John Wiley, New York.

[5] Ya. B. Zeldovich and Yu. P. Raizer (1967). *Physics of Shock Waves and High Temperature Hydrodynamic Phenomenon*. Academic Press, New York.

[6] G. Arfken (1985). *Mathematical Methods for Physicists*, second edition. Academic Press, Orlando.

[7] R. Aris (1962). *Vectors, Tensors and Basic Equations of Fluid Mechanics*. Dover Publications, New York.

[8] P. M. Morse and H. Feshbach (1953). *Methods of Theoretical Physics*. McGraw–Hill, New York.

[9] F. J. Fahy (1989). *Sound Intensity*. Elsevier Applied Science, London.

[10] J. A. Mann, J. Tichy and A. J. Romano (1987). Instantaneous and time–averaged energy transfer in acoustic fields, *Journal of the Acoustical Society of America*, **82** (1), pp 17–30.

[11] S. N. Rschevkin (1963). *The Theory of Sound*. Pergamon Press, Oxford.

[12] L. L. Beranek (1954). *Acoustics*. McGraw–Hill, New York.

[13] Y.Rocard (1960). *General Dynamics of Vibration*. Crosby Lockwood, London.

[14] R.B.Lindsay (1960). *Mechanical Radiation*. McGraw–Hill, New York.

[15] J.Lighthill (1978). *Waves in Fluids*. Cambridge University Press, Cambridge.

[16] P.E.Doak (1968). An introduction to sound radiation and its sources, in *Noise and Acoustic Fatigue in Aeronautics* (eds E. J. Richards and D. J. Mead), John Wiley, London.

Glossary of Symbols

Bold type, as $\mathbf{z}$, indicates a complex quantity (complex vector)

Bold type with tilde above, as $\widetilde{\mathbf{p}}$, indicates rotating complex vector (phasor)

Arrow above symbol, as in $\vec{u}$, indicates vector

Bar above symbol, as in $\overline{p^2}$ indicates in general averaging with respect to time, except in the case of $\overline{\alpha}$

Angular brackets, as in $\langle L \rangle$, indicate spatial averaging or ensemble averaging

Magnitude of a real vector or modulus of a complex vector is indicated by $|\ \ |$, as in e.g. $|\vec{u}|$ or $|\widetilde{\mathbf{p}}|$

Re, Im denote, respectively, real and imaginary parts of complex quantity

a radius of a pulsating sphere (p 60); radius of circular duct (p 284); radius of a bubble (p 389)

$\vec{a}$ unit vector with direction of wave vector (p 44)

a_j real part of Fourier component F_j (p 127)

A amplitude (real) of particle displacement for plane sound wave travelling in the direction of increasing values of x (p 33); amplitude of the harmonic oscillation in a standing wave (p 48); real constant used in expression for sound pressure of a harmonic spherical wave from a point source (p 54); amplitude of sound pressure of incident plane sound wave (p 191)

A_p amplitude of piston displacement in oscillatory motion (p 32)

A_1, A_2 constants (real) in the expression for the particle displacement for a plane sound wave travelling in the direction of increasing values of x (p 33)

$\mathbf{A}$ complex amplitude of sound pressure for plane sound wave travelling in the direction of increasing values of x (p 36); complex constant used in expression for sound pressure of harmonic spherical wave from a point source (p 54)

$\mathbf{A}_i$ complex amplitude of sound pressure of incident plane wave approaching the transient region in the duct with the side branch silencer (p 285)

$\mathbf{A}_r$ complex amplitude of sound pressure of plane wave travelling in the main duct after reflection from side branch silencer (p 285)

$\mathbf{A}_{\text{sb}}$ complex amplitude of sound pressure standing wave in side branch (p 286)

466

$\mathbf{A}_t$ complex amplitude of sound pressure of plane wave transmitted downstream of a duct with side branch silencer (p 285)

$\mathbf{A}_0$ complex constant (p 37)

$\mathbf{A}_1$ complex amplitude of sound pressure of incident plane wave, transmitted towards upstream change in section of expansion chamber (p 278)

$\mathbf{A}_2$ complex amplitude of sound pressure of plane wave transmitted downstream of expansion chamber (p 278)

$\widetilde{\mathbf{A}}$ rotating vector (phasor) (p 38)

APL   assumed protection level

b_j imaginary part of Fourier component F_j (p 127)

B amplitude of particle displacement for plane sound wave travelling in the direction of decreasing values of x (reflected wave) (p 34); bandwidth of bandpass filter

$\mathbf{B}$ complex amplitude of plane sound pressure wave travelling in the direction of decreasing values of x (p 36, p 191)

$\mathbf{B}_1$ complex amplitude of sound pressure of plane wave reflected from upstream change in section of expansion chamber (p 278)

BPL   band (sound) pressure level (p 114)

c local velocity of sound (pp 25-26); damping constant (effective mechanical)

c_a averaged velocity (over time) of sound wave front (p 22)

c_b phase velocity of bending waves (p 213)

c_L longitudinal wave velocity (p 213)

c_p specific heat of gas at constant pressure

$c_{p,ph}$ phase velocity for sound pressure wave (p 33, p 68)

c_s velocity of sound wave front (p 24)

$c_{u,ph}$ phase velocity for particle velocity sound wave (p 68)

c_v specific heat of gas at constant volume

c_0 velocity of sound in undisturbed medium (velocity of sound wave front)

C constant (family of characteristics) (p 30); correction used in subtraction of decibels (p 77); capacitance of lowpass filter (p 94); tonal correction (p 164)

$C_1, C_2 \ldots$ constants for family of sound wave fronts

C_+ characteristics of sound waves travelling in the direction of increasing values of x

C_- characteristics of sound waves travelling in the direction of decreasing values of x

$C(f)$ convolution of Fourier transforms $W(f)$ and $P(f)$ (p 140)

$C(x)$ sound pressure amplitude for standing wave (p 195)

$C_1(x)$ particle velocity amplitude (multiplied by $\rho_0 c_0$) for standing sound wave (p 196)

$\mathbf{C}$ complex amplitude of plane transmitted sound wave (p 208)

d distance between measurement points A and B (p 146); mean free path; distance from noise source; distance apart of simple sources in a dipole

$\vec{d}$ particle displacement vector in physical space (p 40)

$\tilde{\mathbf{d}}$ particle displacement phasor (p 56)

$\tilde{\mathbf{d}}_a$ phasor of displacement of the surface of a pulsating sphere (p 447)

D duration correction (p 164); duct diameter; maximum dimension of a source (p 251, p 375); diameter of circular nozzle (p 381)

DI   directivity index (p 258)

$\mathrm{DI}_{\mathrm{axis}}$ directivity index on a designated axis (p 258)

e distance of mass centre of rotating parts from the axis of rotation (p 413)

$\vec{e}_d$ unit vector with direction of vector $\vec{d}$ (p 146, p 452)

$\vec{e}_r$ unit vector with direction of vector $\vec{r}$ (p 146)

$\vec{e}_u$ unit vector with direction of particle velocity vector (p 42)

$\vec{e}_x, \vec{e}_y, \vec{e}_z$ unit vectors in the x, y and z directions of the rectangular cartesian co–ordinate system (p 40)

E internal energy of thermodynamic system; Young's modulus; averaged (in space) sound energy density

E_a sound (acoustic) energy density (p 42)

E_0 averaged (in space) initial sound energy density (p 236)

f frequency (Hz); fractional exposure (p 339)

f_a, f_b lower/upper band limits of one third octave filters (p 108)

f_{ac} cutoff frequency for anechoic chamber (p 248)

f_{alias} aliased frequency (p 133)

f_c critical frequency of wave coincidence (p 214)

f_j frequency referring to the jth Fourier component

f_{li}, f_{ui} lower/upper band limits of the ith filter in the sequence of the $(1/q)$th octave filters (p 109)

f_{ln}, f_{un} lower/upper band limits of the $(1/q)$th octave filters (p 110)

f_m centre frequency of band pass filter (including octave and one third octave filters)

f_{mi} centre frequency of ith filter in the sequence of $(1/q)$th octave filters (p 109)

f_{mn} centre frequency of the filter in the sequence of $(1/q)$th octave filters (p 109)

$f_{m,\,\text{oct}}$ centre frequency of the octave filter (p 109)

f_n undamped natural frequency (p 414)

f_p blade passing frequency (p 378)

f_r resonant frequency of Helmholtz resonator (p 297)

f_{rms} root mean–squared value of a time–dependent quantity (p 46)

f_{sig} signal frequency (p 133)

$f_{\text{sig},c}$ frequency of a component in a signal (p 134)

f_1, f_2 lower and upper band limits of bandpass filters

F constant (0.3 for octave and 0.15 for one third octave analysis) applied in loudness calculation (p 159) or in perceived noise calculation (p 162)

F_j jth Fourier component

F_{max} maximum frequency which can be detected in time series; also called Nyquist or folding frequency

F_s sampling rate $(F_s = 1/\Delta t)$ (p 133)

F_0 Fourier component; amplitude of applied harmonic force (p 447)

F_1, F_2 forces on fluid element (p 12); arbitrary functions in the solution of the wave equation (p 29); Fourier components

$\tilde{\mathbf{F}}_a$ radiation reaction phasor (force that a pulsating sphere exerts on the surrounding medium) (p 447)

$\mathcal{F}_j$ jth Fourier component (p 122)

$\{\mathcal{F}_j\}$ discrete Fourier transform (DFT) of the time series $\{p_n\}$ (p 122)

$G(f, B)$ the area under the curve of power spectrum $\mathcal{G}(f)$ between the frequencies $f - B/2$ and $f + B/2$, $G(f,B) = \mathcal{G}(f) \cdot B$ (p 116)

$\mathcal{G}(f)$ one–sided power spectral density function, PSD (p 115)

h thickness of a wall (p 207); half the distance of separation between slabs of sound–absorbent material in a rectangular silencer (p 305); height of chimney stack above burner (p 393); specific enthalpy (p 433)

h_0 specific enthalpy of undisturbed (by sound) medium (p 434)

i integer (p 73)

i $\sqrt{-1}$

$\vec{I}$ instantaneous sound intensity vector (sound energy flux density vector) (p 42)

$I = |\vec{I}|$ instantaneous sound intensity (magnitude)

$\overline{I}$ time–averaged sound intensity

$\overline{I_A}$ A–weighted time–averaged sound intensity

$\vec{I}_{ac}$ active sound intensity vector

I_{a0} amplitude of the active sound intensity magnitude (p 445)

$\vec{\mathbf{I}}_c$ complex sound intensity vector (p 441)

$\overline{I}_i$ time–averaged sound intensity of plane incident wave (p 194)

$\overline{I}_r$ time–averaged sound intensity of plane reflected wave (p 194)

$\vec{I}_{rc}$ reactive sound intensity vector

I_{ref} reference sound intensity (p 72)

$\vec{I}_{r0}$ imaginary part of complex intensity vector (p 441)

$\overline{I}_1, \overline{I}_2$ time–averaged sound intensities at entrance and exit of a silencer (p 304)

$\overline{I}_1, \overline{I}_2, \overline{I}_3$ time–averaged sound intensities from a point source located on the floor of a room, at the intersection of a wall and the floor and at the corner of the room, respectively

IL   intensity level (p 72)

j integer

k wave number (p 33); equivalent stiffness (p 413)

$\vec{k}$ wave vector (or propagation vector) (p 33)

k_x, k_y, k_z direction cosines of wave vector in the rectangular cartesian co–ordinate system (p 41)

k_f constant for constant bandwidth band pass filters (p 104)

K constant (p 115, p 382)

$\mathcal{K}$ isentropic (adiabatic) bulk modulus of elasticity; effective stiffness of a vibrator (p 447)

$\mathcal{K}_0$ isentropic bulk modulus (of elasticity) at $\rho = \rho_0$ (p 21)

l number of segments of stationary random signal (p 131); length of duct

l_{eff} effective length of Helmholtz resonator neck (p 293)

l_p length of perimeter of cross–sectional area which is adjacent to sound–absorbent material in a silencer (p 305)

l_s length of silencer (p 304)

l_w length of anechoic wedge (p 248)

l_1, l_2 lengths of projecting tubes in expansion chamber silencer (p 301)

L length of probe traverse (p 206); length of duct; sound pressure level

L_1, L_2 sound pressure levels

L_A A–weighted sound level

$L_{Aeq,T}$ A–weighted equivalent continuous sound level (with integration time T)

$L_{Aeq,15s}$ A–weighted equivalent continuous sound level (with integration time 15 seconds) (p 166)

$L_{Aeq,300s}$ A–weighted equivalent continuous sound level (with integration time 300 seconds) (p 167)

$L_{Aeq,24h}$ A–weighted equivalent continuous sound level (with integration time 24 hours) (p 165)

$L_{AN,T}$ A–weighted percentile level measured over time T (p 168)

$L_{Ar,T}$ rating level (p 182)

L_{atten} sound attenuation (p 305)

L_{axis} sound pressure level on a designated axis (p 257)

L_d daytime equivalent continuous sound level (p 165)

L_{dn} day–night equivalent sound level (p 165)

$L_{EP,d}$ daily personal noise exposure (p 328)

$L_{EP,w}$ weekly average of daily personal noise exposures (p 328)

L_{EPN} effective perceived noise level (p 164)

L_{eq} equivalent continuous sound level

$L_{eq,T}$ equivalent continuous sound level (with integration time T)

L_{IL} insertion loss (p 272)

L_{mi} maximum perceived noise level of ith aircraft movement (p 171)

L_n sampled sound pressure level (p 100)

L_n night–time equivalent continuous sound level (p 165)

L_N percentile levels, where N = 10, 30, 90, etc. (p 168)

L_{NP} noise pollution level (p 169)

L_{PN} perceived noise level (p 161)

$L_{PN\,max}$ maximum perceived noise level during an aircraft flyover (p 170)

L_{TL} transmission loss (p 270)

$L_{TL\,max}$ maximum value of transmission loss

L_{TPN} tone–corrected perceived noise level (p 164)

$L_{TPN\,max}$ maximum value of tone–corrected perceived noise level (p 164)

L_W sound power level

L_{WA} A–weighted sound power level

L_{10} sound level exceeded for 10% of the measurement time

L_{90} sound level exceeded for 90% of the measurement time

m natural number; sound attenuation coefficient (p 244, p 303); total mass of a machine (p 413)

m_a attached mass (due to inertia of medium surrounding a pulsating sphere) (p 448)

m_0 mass of fluid displaced by pulsating sphere (p 449)

m_1 mass of rotating parts (p 413)

M natural number

M_c Mach number of convected turbulent eddies (p 384)

$\mathcal{M}$ mass of vibrator (p 447)

n integer; number of sound sources (p 73); sample number; number of noise events

n_T number of reflections occurring during reverberation time T (p 240)

$\vec{n}$ unit vector normal to a surface; unit vector along the direction of wave propagation

N natural number; number of samples in signal; perceived noise index (p 162); number of aircraft movements (p 170); rotational speed (p 378)

N_m maximum perceived noise index (p 162)

N_t total perceived noise (p 162)

NC   noise criterion (p 172)

NNI   noise and number index (170)

NPL   noise pollution level (p 169)

NR   noise rating (p 178)

NRR   noise reduction rating (p 358)

p sound pressure

p_A, p_B sound pressure at positions A and B in a sound field (p 146)

$p(t)$ sound pressure as a function of time t

$p_A(t)$ A–weighted sound pressure as a function of time t

$p_{f,B}$ sound pressure of a signal component with bandwidth B and centre frequency f

$p_{f_m,B}$ sound pressure of signal filtered by band pass filter of centre frequency f_m and bandwidth B

p_i sound pressure generated by ith source (p 73)

p_n nth sample of the time series

$\{p_n\}$ time series representing sound pressure signal

p_{rms} root mean square of the sound pressure (p 46, p 71)

p_{rms_i} root mean square of the sound pressure for the ith source (p 74)

p_{ref} reference sound pressure (p 71)

$p_T(t)$ signal (sound pressure as a function of time) segmented by rectangular time window of length T (p 139)

p_0 amplitude of a plane, harmonic sound pressure wave travelling in the direction of increasing x

p_+ sound pressure for plane waves travelling in the direction of increasing values of x

p_- sound pressure for plane waves travelling in the direction of decreasing values of x

$\overline{p_d^2}$ mean–squared pressure of direct sound (p 230); mean–squared pressure of sound at distance d (p 393)

$\overline{p_{d,\,axis}^2}$ mean–squared sound pressure on a designated axis (p 232)

$\overline{p_r^2}$ mean–squared pressure of reflected sound (p 228)

$\mathbf{p}$ complex amplitude of sound pressure phasor (p 37)

$\widetilde{\mathbf{p}}$ phasor of sound pressure (p 36)

$\widetilde{\mathbf{p}}_d$ phasor of sound pressure for a dipole field (p 452)

$\widetilde{\mathbf{p}}_i$ phasor of sound pressure of incident plane sound wave

$\widetilde{\mathbf{p}}_p$ phasor of sound pressure in a sound field generated by a point source (p 451)

$\widetilde{\mathbf{p}}_r$ sound pressure phasor of reflected plane sound wave

$\widetilde{\mathbf{p}}_{sb}$ sound pressure phasor of the standing wave in the side branch duct (p 286)

$\widetilde{\mathbf{p}}_t$ sound pressure phasor of transmitted plane sound wave

$\mathbf{P}_0$ complex amplitude of sound pressure at position x in duct (p 276)

$\widetilde{\mathbf{p}}_1$ phasor of sound pressure in front of transition region in duct (p 272); phasor of sound pressure due to point source located at point P_1 (p 451)

$\widetilde{\mathbf{p}}_2$ phasor of sound pressure behind transition region in duct (p 272); phasor of sound pressure due to point source located at point P_2 (p 451)

P fluid pressure; sound pressure amplitude (p 244); loudness level in phons (p 157)

P_0 pressure of fluid undisturbed by sound; static pressure surrounding a bubble (p 389)

$P(f)$ Fourier transform of signal $p(t)$ (p 140)

$P_T(f)$ Fourier transform of the segment of record $p_T(t)$ (p 140)

P_W absorbed sound power (p 238)

q heat supplied per unit mass (p 15); a natural number ($(1/q)$th octave filters) (p 108)

Q heat supplied to the thermodynamic system

Q_{irrev} heat supplied to the thermodynamic system in irreversible process (p 15)

Q_{rev} heat supplied to the thermodynamic system in reversible process (p 15)

Q sound source strength of a pulsating sphere (p 62)

Q_p modulus of sound source strength of point sound source (p 62)

Q_s sound source strength of a simple source (p 62)

Q_H sound source strength of a pulsating sound source of hemispherical shape (p 63)

$\vec{\mathbf{Q}}_d$ strength of dipole source (dipole strength) (p 452)

$\mathbf{Q}_d$ magnitude of dipole strength

$\mathbf{Q}$ complex sound source strength (p 61)

$\mathbf{Q}_p$ complex sound source strength of a point source (p 62)

$\mathbf{Q}_{pH}$ complex sound source strength of a point hemispherical source (p 63)

$\mathcal{Q}$ directivity factor (p 231)

$\mathcal{Q}_{axis}$ directivity factor for a particular axis (p 232)

$\vec{r}$ position vector (p 3)

r spherical co–ordinate (distance from the origin of the spherical co–ordinate system) (p 51); modulus of complex reflection ratio $\mathbf{r}$ (p 193)

r_n resistive (real) part of specific acoustic impedance of a wall (p 192)

r_z real (resistive) part of the specific acoustic impedance of spherical waves from a point sound source (p 65)

$\mathbf{r}$ complex reflection ratio (p 193)

R specific gas constant; resistance of low pass filter (p 94); resistive part of input acoustic impedance of Helmholtz resonator (p 295)

R_a radiation resistance for a pulsating sphere (increase in effective damping of a pulsating sphere due to radiation of sound) (p 448)

R_c room constant (p 228)

$R_{c,\,after}$ room constant after treatment (p 423)

$R_{c,\,before}$ room constant before treatment (p 423)

R_{field} field incidence sound reduction index (p 211)

R_{mean} arithmetic mean of the sound reduction indices in dB in one third octave bands from 100 Hz to 3150 Hz (p 212)

R_0 sound reduction index at normal incidence (p 188); real part of input acoustic impedance of Helmholtz resonator (p 292)

R_1 piston resistance function (p 283)

$R_{[0,\,\Theta]}$ random incidence sound reduction index for angles of incidence from 0 to Θ (p 211)

$R_{[0°,\,90°]}$ random incidence sound reduction index for angles of incidence from 0° to 90° (p 211)

$\mathcal{R}$ effective mechanical damping constant of a vibrator (p 447)

$\mathbf{R}$ remainder (p 460)

s specific entropy (p 9); variable used in sound transmission theory (p 211)

s_1 integer used in locating the position of anti–nodes in standing wave (p 199)

s_2 integer used in locating the position of nodes in standing wave (p 199)

S cross–sectional area of tube (p 10, p 22); surface enclosing a fluid volume (p 43); surface area of interior of room (p 229); cross–sectional area of interior of duct (p 305); loudness of pure tones in sones (p 157); loudness index for broad band sound (p 158)

$S(f)$ two–sided power spectral density function (p 116)

S_m maximum loudness index of broad band sound (p 158)

S_r scan rate for swept narrow band filters (p 121)

S_{sb} cross–sectional area of side branch (p 285)

S_t total loudness of broad band sound (p 158); Strouhal number (p 384)

S_0 cross–sectional area of tube (p 290)

$S_1,\,S_2$ cross–sectional areas in front of and behind transition region in a duct (p 272); cross–sectional areas of inlet and outlet ducts of expansion chamber silencer (p 277)

$S_1,\,S_2\ldots$ surface areas in a room (p 229)

$\mathcal{S}$ entropy of the thermodynamic system (p 15)

SEL single event sound exposure level (p 165)

SIL speech interference level (p 171)

SL   spectrum level (p 113)

SPL   sound pressure level (p 71)

SPL_{bg} sound pressure level of background noise (p 77)

SPL_i sound pressure level of sound generated by ith source (p 74)

SPL_{mach} sound pressure level of noise from a machine (p 77)

SPL_t sound pressure level of noise from machine and background (p 77)

SWR   standing wave ratio (p 202)

t time

t_r average time between reflections (p 240)

t_s specific instant in time (p 130); period associated with measurement (p 339)

t_1 time (p 236)

T period; averaging time; time of record (digital signal analysis); reverberation time

T_A averaging time

T_{ab} notional reverberation time of room containing sound–absorbent material (p 424)

T_{after} reverberation time after treatment (p 424)

T_{before} reverberation time before treatment (p 424)

T_e daily duration of worker's exposure to noise (p 328)

T_{ref} reference time (p 166)

T_0 length of working day of 8 hours (p 328)

T_1 reverberation time after introduction of sound–absorbent material (p 261)

T   temperature

TNI   traffic noise index (p 170)

TR   transmissibility (p 414)

u particle velocity (magnitude), piston velocity (p 22)

$\vec{u}$ particle velocity vector

$\vec{u}_{ac}$ active part of particle velocity vector (p 438)

u_p piston velocity (p 32)

u_r radial component of the particle velocity (particle velocity in the sound wave from a point source) (p 54)

$\vec{u}_{rc}$ reactive part of particle velocity vector (p 438)

u_{r0} amplitude of the radial component u_r of particle velocity in the sound field of dipole source (p 454)

u_x, u_y, u_z x, y and z components of the particle velocity vector in the rectangular cartesian co–ordinate system (p 40)

$u_{\theta 0}$ amplitude of the tangential component u_θ of particle velocity in the sound field of dipole source (p 454)

u_0 particle velocity amplitude (p 57)

$\tilde{u}$ particle velocity phasor

$\tilde{\vec{u}}$ particle velocity phasor–vector

$\tilde{u}_a$ phasor of particle velocity on surface of pulsating sphere (p 447)

$\tilde{u}_i$ particle velocity phasor of incident plane sound wave

$\tilde{u}_n$ phasor of particle velocity component normal to a surface (p 191)

$\tilde{u}_r$ phasor of radial component of particle velocity (phasor of particle velocity in the sound wave from a point source) (p 55); particle velocity phasor of reflected plane sound wave

u_{r0} complex amplitude of phasor of particle velocity in sound field of a point sound source (p 62)

$\tilde{u}_t$ particle velocity phasor of transmitted plane sound wave

$\tilde{u}_w$ phasor of velocity of a wall (p 207)

$\tilde{u}_\theta$ phasor of tangential component of particle velocity (p 453)

$\tilde{u}_1$ phasor of particle velocity in front of transition region in duct (p 272)

$\tilde{u}_2$ phasor of particle velocity behind transition region in duct (p 272)

U_0 amplitude of the particle velocity of a plane, harmonic wave travelling in the direction of increasing x (p 44); normal component of the velocity vector amplitude on the surface of a pulsating sphere (p 62); velocity of jet at nozzle exit (p 381)

$\tilde{U}$ volume velocity phasor

$\tilde{U}_i$ volume velocity phasor of incident sound wave (p 275)

$\tilde{U}_r$ volume velocity phasor of reflected sound wave (p 275)

$\tilde{U}_{sb}$ volume velocity phasor of the standing wave in the side branch duct

$\tilde{U}_t$ volume velocity phasor of transmitted sound wave (p 286)

U_0 complex amplitude of volume velocity in duct (p 276)

v mean flow velocity (magnitude) in airway within silencer (p 311)

V volume of fluid element disturbed by sound wave; fluid volume fixed in space ; volume of room; volume of chamber of Helmholtz resonator (p 291)

V_0 volume of fluid element undisturbed by sound

V_1 input voltage

V_2 output voltage

w width of silencer (p 305)

$w(t)$ rectangular time window (p 139)

W sound power (acoustic power); work done on a thermodynamic system

W_A A–weighted sound power

W_burner sound power from a burner (p 393)

W_chimney sound power from a chimney (p 393)

W_d sound power of acoustic dipole (p 457)

W_i sound power of incident plane wave (p 194)

W_p sound power of a point source (p 63)

W_{pH} sound power of a hemispherical point sound source (p 64)

W_{os} sound power of oscillating rigid sphere (p 463)

W_{ps} sound power of pulsating sphere (p 446)

W_r sound power of reflected plane wave (p 194)

W_rad sound power radiated by a machine component (p 373)

W_ref reference sound power (p 73)

W_t sound power of transmitted plane wave (p 270)

$W(f)$ Fourier transform of window function $w(t)$ (p 140)

$W_H(t)$ Hanning function (p 143)

x cartesian co–ordinate; distance travelled by sound wave

x_p piston displacement (p 32)

x_max position of anti–node in standing sound wave (p 197)

x_min position of node in standing sound wave (p 197)

x_n reactive (imaginary) part of specific acoustic impedance of a wall (p 192)

x_z imaginary (reactive) part of the specific acoustic impedance of spherical waves from a point sound source (p 65)

X_a radiation reactance for the pulsating sphere (p 449)

X_0, X_0' imaginary part of input acoustic impedance of Helmholtz resonator (p 292)

X_1 piston reactance function (p 283)

y cartesian co–ordinate

z cartesian co–ordinate, cylindrical co–ordinate

z_1 real part of complex vector $\mathbf{z}$ (p 38)

z_2 imaginary part of complex vector $\mathbf{z}$ (p 38)

$\mathbf{z}$ complex vector (p 38); specific acoustic impedance (p 64)

z_{in} specific acoustic impedance of inner wall surface (p 210)

z_n specific acoustic impedance of the wall (p 191)

z_{out} specific acoustic impedance of outer wall surface (p 210)

Z acoustic impedance (p 64)

Z_l acoustic impedance (of plane waves) in a duct at $x = l$ (p 276)

Z_m mechanical impedance (p 65)

Z_{ms} mechanical impedance of a source (p 66)

Z_r radiation impedance (p 66)

Z_{ra} radiation impedance for the pulsating sphere (p 449)

Z_{sb} acoustic impedance at entrance to side branch duct (p 287)

Z_x acoustic impedance (of plane waves) at cross–section $S(x)$ of a duct (p 275)

Z_0 acoustic impedance (of plane waves) in a duct at $x = 0$ (p 276); input acoustic impedance for a duct closed by a rigid cap (p 281);input acoustic impedance of a duct terminated by an infinite flange (p 283)

Z_0, Z_0' input acoustic impedance of Helmholtz resonator (p 292)

Z_1 acoustic impedance in front of transition region in duct (p 273)

Z_2 acoustic impedance behind transition region in duct (p 273)

Z^+ acoustic impedance of tube defined in terms of incident wave (sound wave travelling in direction of increasing values of x) (p 273)

α sound absorption coefficient

α_A sound absorption coefficient at normal incidence

α_d ratio of absorbed sound power to incident sound power (p 189)

α_r sound (power) reflection coefficient

α_t sound (power) transmission coefficient

$\alpha_{t,\theta}$ sound transmission coefficient for sound waves with an angle of incidence θ

$\alpha_{t[0,\Theta]}$ sound transmission coefficient (random incidence; angles of incidence from 0 to Θ) (p 210)

$\alpha_1, \alpha_2 \ldots$ sound absorption coefficients

$\overline{\alpha}$ average sound absorption coefficient (p 228)

β phase angle (p 54)

γ ratio of specific heats (c_p/c_v)

δ static deflection (p 414)

Δ attenuation of octave and one third octave filters

Δl end correction for Helmholtz resonator neck (p 293)

Δl_{inn} inner end correction for Helmholtz resonator neck (p 293)

Δl_{out} outer end correction for Helmholtz resonator neck (p 293)

ΔL difference in levels used in addition of decibels (p 75); noise reduction

ΔP static pressure loss in a silencer (p 311)

$\Delta \rho$ small change in density

Δf spectrum resolution

Δt sampling time

ϵ specific internal energy; normalised standard error

ϵ_0 specific internal energy of undisturbed medium (p 434)

$\mathcal{E}$ sound exposure

$\mathcal{E}_0$ reference sound exposure

ζ damping ratio (p 414)

θ variable $t - x/c_0$ (p 30); spherical co–ordinate (the polar angle); angle of incidence of plane sound waves on microphone diaphragm (p 88) and on wall (p 210); argument of complex reflection ratio $\mathbf{r}$ (p 193)

Θ upper limit of angle of incidence in expression for sound transmission coefficient (p 210)

λ wavelength

ν Poisson's ratio

ξ_+ particle displacement for plane waves travelling in the direction of increasing x (p 31)

ξ_- particle displacement for plane waves travelling in the direction of decreasing x (p 31)

ξ, η, ζ x, y and z components of the particle displacement vector in the rectangular cartesian co–ordinate system (p 40)

ρ density

ρ_0 density of undisturbed (by sound) fluid

σ constant (p 37); cylindrical co–ordinate (perpendicular distance from the z axis) (p 51); common ratio in geometric series (p 108); sample standard deviation for A–weighted sound level (p 170); surface density (p 207)

σ_{rad} radiation efficiency (p 373)

τ variable $t + x/c_0$ (p 30); averaging time (p 43); equivalent time (p 164); time constant (p 236)

τ_{ref} reference time (p 164)

ϕ_a phase angle for particle velocity at the surface of pulsating sphere (p 443)

ϕ_p phase angle of sound pressure wave

ϕ the difference in phase between the particle velocity and sound pressure phasors (sound field from point source)

ϕ_A phase angle for plane sound pressure wave travelling in the direction of increasing values of x (p 37)

ϕ_B phase angle for plane sound pressure wave travelling in the direction of decreasing values of x (p 37)

ϕ_0 phase angle (p 33)

Φ phase (p 33)

Φ_p phase of sound pressure wave (p 33, p 68)

Φ_u phase of particle velocity wave (p 68)

Φ_0 initial phase (p 33, p 195)

Φ_{os} initial phase (p 61)

ψ cylindrical and spherical co–ordinate (the azimuth angle) (p 51)

ω circular (angular) frequency (rad/s)

ω_r undamped circular natural frequency of Helmholtz resonator (p 291)

φ_p phase angle associated with sound pressure phasor (dipole sound source) (p 456)

φ_r phase angle associated with phasor of radial component of particle velocity (dipole sound source) (p 454)

φ_θ phase angle associated with phasor of tangential component of particle velocity (dipole sound source) (p 454)

ϑ difference in phase between particle displacement and sound pressure phasors (sound field from point source) (p 56)

ϑ_r difference in phase between phasor of radial component of particle velocity and phasor of sound presure (dipole sound source) (p 454)

Index